SIXTH EDITION

Fundamentals of Electricity

Basics of Electricity, Electronics, Controls and Computers

Robert G. Seippel

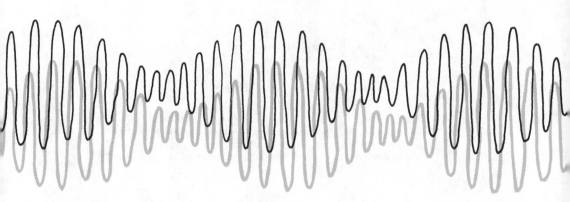

AMERICAN TECHNICAL PUBLISHERS, INC.
HOMEWOOD, ILLINOIS 60430

Preface to the Sixth Edition

The new Sixth Edition of *Fundamentals of Electricity* has been written to provide the electrician with an up-to-date look at the electrical field and the present state of the art. In addition, the scope of the book has been enlarged to reflect the increasing interrelationship between electricity and electronics.

In recent years it has become increasingly difficult to pinpoint the dividing line between *electricity* (concerned primarily with low-frequency power applications) and *electronics* (high-frequency circuits and devices for communications, controls, computers, and the like). It is not unusual for an electrician to be called upon to install and maintain electronic equipment, and he must therefore have a basic knowledge of both fields.

With these thoughts in mind, this revised Sixth Edition of *Fundamentals of Electricity* was developed to provide the student-electrician with an up-to-date look at the fundamental circuits used in electronic applications as well as power circuits for motors, lighting, and ordinary appliances.

Chapter 4, *Electrical Circuits*, has been rewritten to add advanced circuit types, including the wye to delta circuits and others. Full explanation of phase angles was added to Chapter 8, *Alternating Current Principles*. Since the oscilloscope and the signal generator are essential in trouble-shooting modern-day electrical/electronic systems, they have been included in Chapter 11, *Electrical Measuring Instruments*.

Chapter 14, *Solid State Devices*, is expanded to include special diodes, transistors, and integrated devices. A new Chapter 15, *Solid State Circuits*, is added as a parallel to *Electron Tube Circuits* of Chapter 13.

Previous Chapter 15, entitled *Automation Processes and Computers*, has been divided into two separate chapters. The first part is now Chapter 16, *Automatic Control Systems*, and the second part, now Chapter 17, covers the basics of *Computer Technology*. A small final Chapter 18, *Electrical/Electronic Safety*, points out the main hazards to avoid and emergency procedures. The Occupational Safety and Health Act of 1970, which makes this subject everyone's responsibility, points out the need for some instruction in this subject at all levels.

Five appendices have been added. These are: Appendix A, *Letter Symbols, Abbreviations, and Prefixes*; Appendix B, *Electrical/Electronic Symbols*; Appendix C, *Electrical/Electronic Formulas*, Appendix D, *Solving Electrical Problems with Trigonometry Tables or the Slide Rule* (both text and decimal-trig tables); and Appendix E, *Squares, Cubes, Roots, and Reciprocals*.

The Sixth Edition of *Fundamentals of Electricity* is the culmination of many outstanding teachers in the science of electricity and electronics. These include the late Kennard C. Graham, author of the previous edition, and earlier collaborators Wynne L. McDougal, Richard R. Ranson, and Carl H. Dunlap. The author of this revised edition gratefully acknowledges the contributions of these men.

The Publishers

Contents

Chapter **1**

The Atom

The intelligent study of electrical theory with its modern applications requires a knowledge of the atom since it forms the basic foundation for all electrical principles. The atom represents a field of study in itself, but the electrician and technician must have a knowledge of its basic structure if they are to understand and become proficient in their fields. The details essential to the study of electrical theory will be presented here.

Matter and Molecules

All substances, whether gas, liquid, or solid, are composed of *matter*. If we analyze the composition of matter we find that it is made up of molecules, which may be defined as the smallest division of matter that can be made and have the substance retain its chemical identity. Molecules are in turn made up of atoms. The molecular composition of matter has three classifi-

cations; elements, compounds, and mixtures, depending on the kinds of atoms that make up the molecule. Figure 1 illustrates these classifications.

Elements are composed of molecules made up of only one kind of atom.

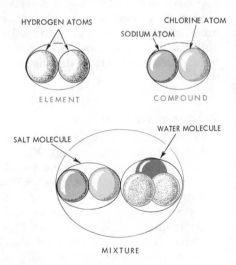

Fig. 1. **The three classifications of the composition of matter.**

Copper is classed as an element because if the molecule is divided it is found to contain only copper atoms. When the term "element" is used, it is in reference to the atom rather than the molecule of the element.

Compounds are composed of molecules made up of different kinds of atoms, or elements. A molecule of salt is a compound because it is made up of two different elements: an atom of sodium, and an atom of chlorine, Fig. 1. The composition of any particular compound never varies. A molecule of salt anywhere in the universe will always be made up of one sodium atom and one chlorine atom.

Mixtures are composed of different kinds of molecules, which may be either compounds or elements. An example would be a mixture of water and salt, both of which are compounds. An easy way to make a distinction between mixtures and compounds is to remember that a mixture can be chemically separated into the different substances that went into it. The salt and water mixture mentioned above can be separated into pure water and pure salt, but neither the water nor the salt can be separated without destroying the molecules.

Structure of the Atom

In its structure the atom is often compared to the solar system, Fig. 2, in which planets revolve around a central body, or sun, following paths known as orbits. The central body of the atom is made up of protons and neutrons and is called the *nucleus*. The planetary bodies of the atom are electrical particles called electrons. Fig. 3 illustrates the structure of the hydrogen atom which has one proton in the

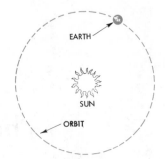

Fig. 2. The earth revolving in orbit around the sun.

nucleus and one orbiting electron. Hydrogen is one of the simplest atoms of all the elements; more complex atoms having a large number of electrons in orbit as well as a greater number of protons and neutrons in the nucleus.

The analogy of the atom to the solar system was used only to illustrate the basic structure of the atom and beyond this comparison no points of similarity will be found. The essential point to remember now is that the electrons rotate in fixed orbits around the nucleus. We will return to the matter of orbits after taking a closer look at the particles that make up the atom.

The Electron

The electron has been established by research as being a particle of electricity having a negative (—) charge. It is an extremely small particle in comparison with the proton, and due to its orbital speed around the nucleus carries a considerable amount of energy. Since all electrons are negative in charge, they will repel other electrons, and have a force of attraction for the positively charged protons.

The Proton

The proton is a part of the nucleus. It has a mass approximately 1836 times that of the electron, and has a positive (+) electrical charge opposite and equal to that of the electron. Protons, all being positive, repel each other but have a force of attraction for the negative electron.

The Neutron

The neutron is a particle found in the nucleus of most atoms. (The hydrogen atom, Fig. 3, does not have a neutron). It has the same approximate mass as the proton but is neutral in its electrical charge.

The Nucleus

Basically, the nucleus is made up of a compact mass of protons and neutrons. There are also many other particles in the nucleus, but they contribute little to the mass or effect of the atom and are of interest only to the atomic scientist.

Since it was stated earlier that protons having the same charge would repel each other, the fact that they group tightly together to form the nucleus seems an odd contradiction. It is, and one for which scientists at the moment have no definite answers, but theorize that the neutron could be made up of a proton and an electron whose charges have been neutralized and may represent the bonding power between protons that holds the nucleus together. The number of protons and neutrons in the nucleus varies for the different elements depending on the particular atom involved.

Atomic Number

In the normal atom, the number of protons in the nucleus is exactly equal to the number of electrons in orbit. The

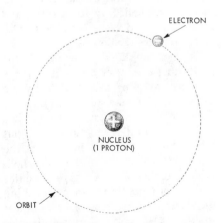

Fig. 3. An illustration of the Hydrogen atom showing similarity to the solar system.

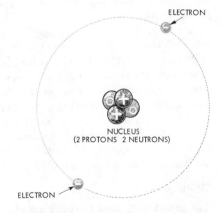

Fig. 4. Drawing of the helium atom showing the nucleus and orbiting electrons.

number of protons in the nucleus never varies for any particular atom and is called the *atomic number* of the atom. The atomic number has been established by scientists for each element and these are listed in what is usually called a "Table of the Elements." The atomic number as well as the atomic symbol and the atomic weight are given as shown in Table I. For example, the helium atom, Fig. 4, has two protons in the nucleus and an atomic number of 2.

Atomic Weight

The atomic weight of an atom is equal to the sum of the protons and neutrons in the nucleus. It should be noted that the number of protons *does not* always equal the number of neutrons in the nucleus, and can only be determined by reference to the table showing the atomic weights, Table I. For example, helium has an atomic weight of 4 which should equal two protons and two neutrons in the nucleus, which is the case. However, copper has 29 protons and an atomic weight of 63.54. Subtracting the number of protons from the atomic weight gives the number of neutrons as approximately 34.

Isotopes. When a particular atom has different forms and weights it is called an *isotope.* An isotope is formed when a neutron is added or taken away from the nucleus of an element. The normal hydrogen atom has only one proton in the nucleus, and an atomic weight of *1*, but deuterium, an isotope of hydrogen, has a neutron in the nucleus and its atomic weight is *2*.

Atomic Orbits

It would seem that the orbiting electron, having a negative charge would fall into the positively charged nucleus. It does not do so because the extreme speed of its orbital motion tends to throw it out of orbit around the nucleus, similar to a space capsule escaping the earth's gravitational pull. The attraction of the positively charged nucleus is just strong enough to hold the electron in orbit, balancing the forces of motion and attraction.

The number of electrons in orbit around the nucleus varies of course, depending upon the particular type of atom. However, it has been established by research that no matter how many electrons are involved, the number of electrons in any orbit (or shell), follows a specific pattern. For example, the first orbital shell never contains more than 2 electrons regardless of the total number of electrons in the atom. The second can vary depending again on the type of atom, but has a maximum number of 8 electrons. The third shell can contain from 1 to 18 electrons. The fourth shell contains the remainder of the electrons in the atom after the first three shells have been filled. The copper atom, Fig. 5, illustrates the shell arrangement in having 2 electrons in the first shell, 8 in the second, and 18 in the third. The remaining 29th electron forms the outer shell.

In the simplified drawing of the orbits in the copper atom, it appears that electrons in each shell follow an orderly rotational sequence in their arrangement. Actually, each electron has a different orbit from the others, but the distance from the nucleus is the same for all electrons in the same shell.

The number of electrons in the outer shell is of importance in establishing the electrical properties of a substance; whether it is a conductor or an insulator. Copper which is a good conductor

TABLE 1   THE ELEMENTS

Atomic Number	Name of Element	Symbol of Element	Atomic Weight	Atomic Number	Name of Element	Symbol of Element	Atomic Weight
1	Hydrogen	H	1	53	Iodine	I	127
2	Helium	He	4	54	Xenon	Xe	131
3	Lithium	Li	7	55	Cesium	Cs	133
4	Beryllium	Be	9	56	Barium	Ba	137
5	Boron	B	11	57	Lanthanum	La	139
6	Carbon	C	12	58	Cerium	Ce	140
7	Nitrogen	N	14	59	Praseodymium	Pr	141
8	Oxygen	O	16	60	Neodymium	Nd	144
9	Fluorine	F	19	61	Promethium	Pm	147
10	Neon	Ne	20	62	Samarium	Sm	150
11	Sodium	Na	22	63	Europium	Eu	152
12	Magnesium	Mg	24	64	Gadolinium	Gd	157
13	Aluminum	Al	27	65	Terbium	Tb	159
14	Silicon	Si	28	66	Dysprosium	Dy	162
15	Phosphorus	P	31	67	Hilmium	Ho	165
16	Sulfur	S	32	68	Erbium	Er	167
17	Chlorine	Cl	35	69	Thulium	Tm	169
18	Argon	A	39	70	Ytterbium	Yb	173
19	Potassium	K	39	71	Lutecium	Lu	175
20	Calcium	Ca	40	72	Hafnium	Hf	179
21	Scandium	Sc	45	73	Tantalum	Ta	181
22	Titanium	Ti	48	74	Tungsten	W	184
23	Vanadium	V	51	75	Rhenium	Re	186
24	Chromium	Cr	52	76	Osmium	Os	190
25	Manganese	Mn	55	77	Iridium	Ir	193
26	Iron	Fe	56	78	Platinum	Pt	195
27	Cobalt	Co	59	79	Gold	Au	197
28	Nickel	Ni	59	80	Mercury	Hg	201
29	Copper	Cu	64	81	Thallium	Tl	204
30	Zinc	Zn	65	82	Lead	Pb	207
31	Gallium	Ga	70	83	Bismuth	Bi	209
32	Germanium	Ge	73	84	Polonium	Po	210
33	Arsenic	As	75	85	Astatine	At	211
34	Selenium	Se	79	86	Radon	Rn	222
35	Bromine	Br	80	87	Francium	Fr	223
36	Krypton	Kr	84	88	Radium	Ra	226
37	Rubidium	Rb	85	89	Actinium	Ac	227
38	Strontium	Sr	88	90	Thorium	Th	232
39	Yttrium	Y	89	91	Protactinium	Pa	231
40	Zirconium	Zr	91	92	Uranium	U	238
41	Columbium	Cb	93	93	Neptunium	Np	239
42	Molybdenum	Mo	96	94	Plutonium	Pu	239
43	Technetium	Tc	99	95	Americium	Am	241
44	Ruthenium	Ru	102	96	Curium	Cm	242
45	Rhodium	Rh	103	97	Berkelium	Bk	245
46	Palladium	Pd	107	98	Californium	Cf	246
47	Silver	Ag	108	99	Einsteinium	E	253
48	Cadmium	Cd	112	100	Fermium	Fm	256
49	Indium	In	115	101	Mendelevium	Mv	256
50	Tin	Sn	119	102	Nobelium	No	254
51	Antimony	Sb	122	103	Lawrencium	Lw	257
52	Tellurium	Te	128	104 & 105	Under Study		

NOTE: Elements 1 through 92 occur normally in nature. Elements 93 and above are those discovered by man as a result of transmutation.

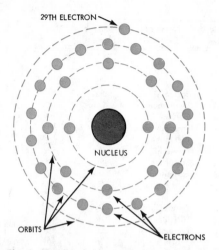

29TH ELECTRON

NUCLEUS

ORBITS

ELECTRONS

Fig. 5. Simplified drawing of the copper atom illustrating the numeric arrangement of electrons in the various shells. The colored lines represent orbital distance of the electrons from the nucleus and is the same for each electron in the shell.

of orbit. Substances made up of these kinds of atoms are called insulators, since they will not conduct electrical current at all, or offer great resistance to its flow.

It is possible to remove one or more electrons from such an outer shell by attraction of adjacent atoms, by action of a magnetic field, or by other means, such as mechanical friction. Loss of an electron destroys the neutral state of the atom, giving it an excess positive charge of one unit.

STATIC ELECTRICITY

Static Charge

When two objects are rubbed together, electrons are removed from surface atoms, relatively more from one material than from the other. The object whose atoms acquire an excess of electrons is said to have a negative static charge, the other object gaining a positive static charge. The term *static* means fixed, or at rest.

If a rubber rod is scoured with a piece of wool, electrons tend to collect on the surface of the rubber rod, imparting a negative static charge, while positively-charged atoms at the surface of the wool give it a positive static charge. When a glass rod is rubbed with a piece of silk, the rod becomes positively charged, the silk negative. Results in these and similar cases depend upon the particular physical and electrical nature of the materials involved.

has only one electron in the outer shell. This electron being farthest from the nucleus where the attractive force is greatly decreased, is comparatively easy to pull out of its orbit by an electrical pressure. An electron pulled out of orbit is called a *free* electron. An atom which has lost an electron from its outer shell is said to be positively charged, and will attempt to pull other free electrons into its orbit to complete the outer shell. This movement of free electrons is what constitutes a flow of electric current, more often referred to as *electron flow*.

In atoms which have a large number of electrons in the outer orbit, the combined force of attraction for the nucleus is much stronger, thus it is more difficult to force an electron out

The Law of Static Attraction and Repulsion

Fig. 6 represents a small ball of very light material such as plant pulp

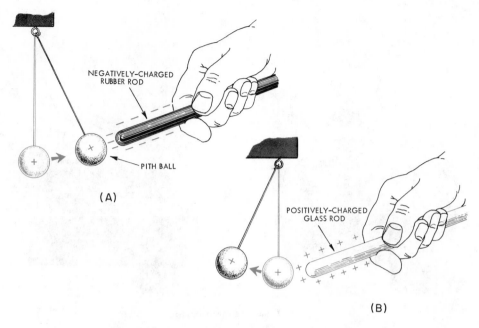

Fig. 6. Unlike charges attract; like charges repel.

(known as pith), suspended by a string. If the ball is given a positive charge, and a charged rubber rod is brought near as in Fig. 6(A), the positively-charged ball will be attracted, swinging toward the negatively-charged rod, and thus showing that unlike charges attract. If the charged glass rod is brought near, as in Fig. 6(B), the ball will be repelled, swinging toward the left, showing that two positively charged objects avoid one another. When the experiment is repeated, substituting a negatively-charged ball, the positively-charged glass rod attracts the ball, again illustrating the rule that unlike charges attract, and the rubber rod repels it. The rule for interaction between positive and negative electrical charges may be stated as follows: *unlike charges attract; like charges repel.*

Induced Charges

When small bits of paper are scattered upon a flat surface and a positively-charged glass rod is brought near them as in Fig. 7(A), they are drawn toward it. If, instead of a glass rod, a negatively-charged rubber rod is used, Fig. 7(B), the bits of paper will act in the same manner. In the first case, they are attracted by a positive charge; in the second, by a negative one. If the glass rod, alone, had attracted them, it might have been assumed that the bits were negatively charged, but

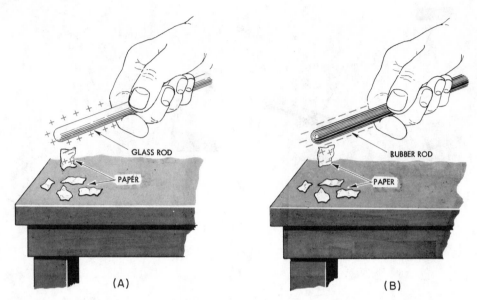

Fig. 7. Induced static charges.

since they were not repelled by the rubber, they could not have been negatively charged.

The reason for identical action in both cases may be seen by referring again to the figures. In Fig. 7(A), the positively-charged rod draws electrons to the upper surfaces of the paper bit, dislodging them from atoms, and repelling the now positively-charged atoms from which they came. In Fig. 7(B), the negatively-charged rod repels electrons from their atoms, and attracts the then positively-charged atoms, the bits consequently moving toward the rods in both instances. When the rods are taken away, it is found that the papers do not have any elec-

trical charge, the atoms having reverted to a *neutral state*. The rods are said to have induced a temporary state of charge in the bits of paper, the process being known as static induction.

Common Examples of Static Charge

Displacement of electrons, with resulting static charge, occurs every time one object moves with respect to another, although effects are not always noticeable. The act of walking across the floor removes electrons from the carpet and transfers them to the soles of the shoes, combing one's hair transfers electrons from the hair to the comb, wind friction draws electrons from walls of houses, sloshing of gasoline in a speeding gasoline truck creates a charge in the metal tank. One sometimes feels a distinct electrical shock when stepping from his automobile onto the pavement, and everybody is familiar with the crackle of a nylon garment as a woolen jacket is removed from contact with it. Most of these effects are slight and unimportant, but a few significant ones are noted in the following paragraphs.

Static Charge on Moving Belts

Static electricity is sometimes generated unintentionally and its presence can be highly undesirable since it may cause severe electrical damage or even prove hazardous to human life. A common source of trouble is from static created by friction between belts and pulleys. Fig. 8(A) shows a motor driving its load by means of a belt and two pulleys. Friction between the surface of the belt and the surfaces of the pulleys, friction between the particles of the material of which the belt is composed, and even friction between the belt and the surrounding air, builds up a charge upon the surface of the belt as indicated in the figure.

When the static charge attains a high value, it will find some means of neutralizing itself, perhaps by jumping through the air in the form of a spark. In doing so, it selects the easiest available path. If this path happens to include the motor windings, the spark may rupture the insulation, causing an expensive breakdown.

Such trouble is often avoided by arranging a metallic brush to remove the charge from the belt as it is formed. Fig. 8(B) shows how this is accomplished. The brush is placed near the belt, and it carries off excess electrons to the earth. The brush may be in the form of a comb with sharp teeth, which project close to the surface of the belt, or it may have short, flexible springs, the springs making direct contact with the belt. Connection to the earth is called a *ground connection* or, more simply, a *ground*. The belt is said to be *grounded*.

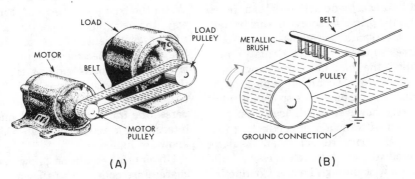

(A)   (B)

Fig. 8. Method of removing static charges from a moving belt.

Static Hazards. Static electricity often causes fires or explosions in grain elevators. Friction between grain and metal chutes builds up a negative charge upon the grain and a positive charge upon the metal. When the charge becomes sufficiently great, a spark from a mass of grain to a piece of metal may cause fire or explosion. The remedy here, of course, is to make sure that metal objects are well grounded.

In paper mills and in printing establishments, static charges on the surface of the paper may cause sheets to stick together. Sometimes this condition results in great expense. It is often counteracted by air-conditioning the plant to keep the air moist. When moist, air is a better conductor than when dry. Static charge can neutralize itself by passing through moist air directly to the earth. Where this method is impractical or ineffective, the paper is caused to pass over combs or brushes which are connected to a negative or a positive source of supply. Discharge from the brushes neutralizes any charge upon the paper.

Serious, and often fatal, explosions have occurred in hospital operating rooms because of static discharge. Static built up by friction of rubber-soled shoes with the floor, or by friction of the anaesthetic with nozzles or hoses, was the principal source of this trouble. A static spark near explosive gas sometimes injured the attendants or even killed the patient. Grounding, here, did not provide complete protection because the problem was so complicated. Today, however, the difficulty has been solved with the aid of conducting rubber. Ordinary rubber is an insulator which allows static charges to build up on it or through its use. Conducting rubber is a form of rubber which is

a sufficiently good conductor that static charges pass to ground before they can become sufficiently concentrated to cause a spark.

Fires in gasoline filling stations sometimes result from static sparks. Rubber automobile tires, speeding over asphalt pavements, acquire a strong negative charge, which is transferred to the body of the car. A spark between the body of the car and the metal nozzle of the gasoline hose may occur if the charge on the car is not removed before the hose is inserted into the gas tank. The charge is usually removed, however, by touching the nozzle to the metal of the car, the electrons being carried off by a grounding wire embedded in the fabric of the hose.

Lightning. One of the most frequent static hazards is not man-made. Lightning, in some areas, is decidedly hazardous to property and life. Fig. 9 will help explain its nature. Fig. 9(A) shows a storm cloud passing above the earth. This cloud has a negative charge on the lower surface and a positive charge on the upper surface. Different theories explain how these charges accumulate, but they are beyond the scope of this book. Therefore, we will say that under certain conditions the charges are collected in this manner. When the charges are arranged in this way, the accumulation continues until there is sufficient voltage to cause a heavy flow of electrons from the negative to the positive side of the cloud through the moist interior. This flow of current may amount to a million amperes for a fraction of a second, and the intense spark makes the flash which is known as lightning.

Under other conditions, however, the charges are collected as shown in Fig. 9(B). Investigations have indicated

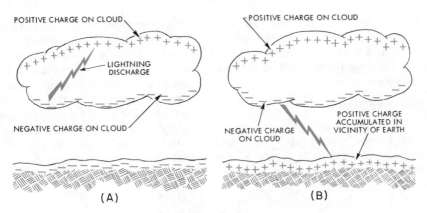

Fig. 9. Lightning discharges.

that when the charges have this arrangement the lightning is more apt to travel between the earth and the cloud than between the two oppositely charged surfaces of the cloud. This very heavy flow through the air develops intense heat and heats the air through which it travels. The heated air expands rapidly and moves away. Cold air rushes in to take its place, and they meet with the resounding crash known as thunder.

When a lightning discharge takes place between a cloud and earth, or something on it, considerable damage can result from the intense heat. The discharge generally strikes the highest object in the vicinity in which the cloud and the earth have built up pressure.

Structures are sometimes protected by means of *lightning rods*. A lightning rod is a sharp-pointed metal rod which projects above the highest point of a structure. Fig. 10 shows a rod fastened to the peak of a gable roof. The rod

is connected to a heavy wire which follows along the roof and down the side of the house to the ground. The wire should make a connection between the rod and ground as nearly vertical as possible. The rod provides an easy path

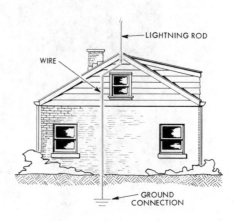

Fig. 10. The lightning rod.

for the high momentary current which results when lightning strikes near by. If the rod were not present, the lightning would strike the gable and damage the house.

Dust Precipitator

On occasion static charges may be used to great advantage. Many large manufacturing plants create a great deal of dust as a waste product. The dust sometimes accumulates upon the surface of the country for miles around, ruining farm lands and damaging the appearance of the landscape. Costly lawsuits sometimes result from this matter, and communities have passed laws to combat the nuisance. Cement plants, and others of a similar nature, have managed to cure the ailment with the aid of *dust precipitators*.

A precipitator, such as in Fig. 11, is

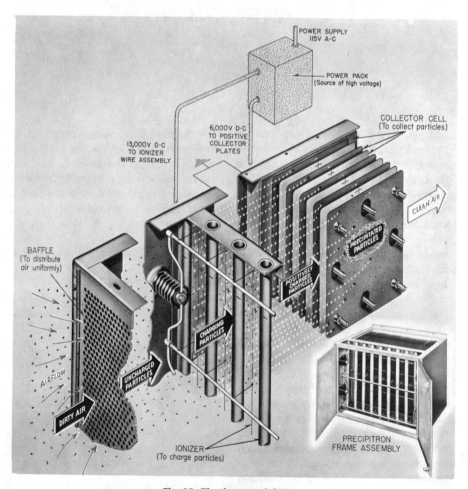

Fig. 11. The dust precipitator.

inserted in a vent which carries off fumes from the plant. The dirty air enters the unit through a baffle and goes through a grill which is attached to a source of negative charges by means of an insulated wire. The dust particles become negatively charged through contact with, or nearness to, the grill. The positive connection from a power supply is attached to the collector cell. The negatively charged particles of dust are attracted to the positively charged walls and cling there while the clean gas or air passes out through the vent pipe. When the particles of dust

have built up a thick layer upon the walls, the weight of the mass becomes greater than the attractive force, and the thick lumps drop through the chute for disposal.

Van de Graaff Static Generator

A static machine widely used today is the Van de Graaff generator, Fig. 12. A hollow insulating column held in position by a supporting base and covered with a metal dome. Within the base is a high-voltage device which supplies a continuous stream of electrons to a metallic comb. It is placed a short dis-

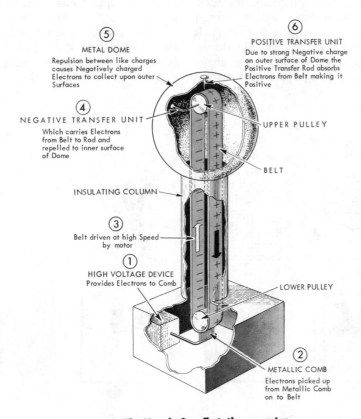

Fig. 12. The Van de Graaff static generator.

tance from the belt which moves, in the direction of the arrows, around two pulleys. Within the dome is the negative transfer unit, which carries electrons from the moving belt to the inner surface of the dome. There are air spaces separating the unit from both dome and the belt. Another transfer unit is fastened to the dome directly above the upper pulley, its lower end approaching quite close to the belt.

In operation, the lower pulley is driven at high speed by an electric motor. The device in the base supplies electrons to the comb. These electrons are repelled continuously from the pointed tips of the comb onto the surface of the belt, which carries them upward toward the negative transfer rod. Here, electrons pass from the belt onto the rod, and are repelled onto the inner surface of the dome. Repulsion between like charges causes the nega-

tively charged electrons to collect upon the outer surface of the dome and to build up a heavy charge there. Since a strong negative charge is built up on the outer surface of the dome by electrostatic repulsion, a positive charge tends to build up on the inner surface. The positive transfer rod fastened to this inner surface, becomes charged. Since it is very close to the belt, it absorbs electrons from it so that the belt will be positively charged as it passes. When the belt reaches the bottom pulley, the positive charge on its surface is neutralized, and an excess of electrons is supplied by the comb.

Extremely high voltages are built up by this machine for use in special X-ray machines and for experimental work in atomic research operations. It should be noted that the dome may be made to take a positive charge rather than a negative one, if so desired.

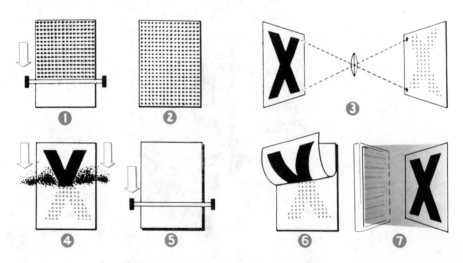

Fig. 13. The operation of the Xerox duplicator.

Xerox Duplicator

A recent addition to the field of electrostatic devices is the *Xerox duplicator* employed in business offices throughout the United States and Europe. The principle of operation is illustrated in Fig. 13. (1) Shows a selenium-coated metal plate being charged by passing under a network of negatively-charged wires. (2) This gives the plate a positive charge. (3) Copy X is projected onto the charged plate through a camera lens. The small plus marks show the projected image made up of positive charges. These charges disappear in areas exposed to light. (4) A negatively charged powder is sprinkled over the positively charged plate which adheres to the image. (5) Paper is passed over the plate. (6) The positively charged paper attracts the powder which forms a positive image. (7) The paper is heated to fuse the powder and form a permanent print.

ATOMIC ENERGY

Importance of the Nucleus

In earlier discussion of the atom, the nucleus was described as being positively charged and composed of protons or protons and neutrons. Yet, an atom may be deprived of one or more electrons without losing its chemical identity, whereas any slight alteration of the nucleus changes the nature of the atom drastically so that it becomes an atom of some other substance than the original material. The alchemist's dream of the Middle Ages, transmutation of base metal into gold, can be done theoretically and experimentally, although it is not economically possible of achievement. Scientists in the past twenty years have made a concentrated study of the nucleus, with important results. Early attempts to disrupt it met with failure, and it soon became apparent that components adhered together with unimaginable force. The advent of more powerful laboratory tools and apparatus finally accomplished the desired result, giving rise to investigations which have added steadily to the understanding of nucleonic structure.

The Neutron

Atoms which acquire an electric charge because of the loss of an electron are termed *ions*. The cyclotron and allied apparatus employ ions to bombard target materials for the purpose of disrupting nuclei. Experimenters discovered that secondary particles ejected in this way seemed to have greater power than did the original high-speed ions. These secondary particles proved to be neutrons, their greater penetrating quality resulting from lack of electrical charge. They were not repelled as were the ions by the normally positive charge on the nucleus. It was discovered, too, that certain heavy and unstable metals which broke down spontaneously and gradually into other substances, discharged neutrons in the process.

Chain Reaction

Investigation of neutron-producing materials led to the development of the atomic bomb, whose tremendous destructive force is based upon chain reaction. Fig. 14 illustrates the meaning of the term *chain reaction*, only the nuclei of atoms being suggested in the

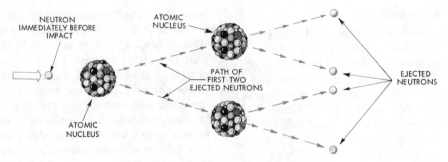

Fig. 14. A chain reaction of atomic energy.

figure. When a single neutron strikes an atomic nucleus, at the left, the atom ejects two neutrons which strike other atoms, each of which emits a pair of neutrons that can bombard four atoms. This process, called *fission,* multiplies with incredible speed despite the fact that every neutron does not succeed in dislodging one or more others, releasing an infinite volume of energy in the form of heat and radiation. A chain reaction cannot be maintained, however, unless a certain minimum quantity of fissionable material is present.

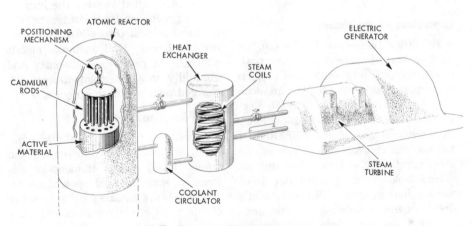

Fig. 15. An atomic power plant.

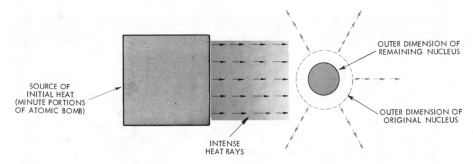

Fig. 16. The process of fusion in the atomic nucleus.

Useful Atomic Power

Methods have been developed for utilizing the immense energies of chain reaction to produce useful power. Fig. 15 shows an atomic power plant consisting mainly of a reactor, a heat exchanger, a steam turbine, and an electrical generator. A coil of water pipes in the exchanger is exposed to liquid or gaseous "coolant" pumped from the reactor. Coolant heat turns the water into steam, thus enabling the turbine to drive the generator.

The principal difficulty met by the scientific experimenters was to control the speed of chain reaction so a destructive cycle would not suddenly occur, vaporizing the reactor itself. The danger was overcome by the use of *cadmium* rods which may be raised or lowered in proximity to the tubes which contain active material, cadmium having the property of rapidly absorbing neutrons. When the rods are lowered all the way, absorption of neutrons is so complete that practically no fission takes place. When they are raised to a certain higher position, fission carries on at the maximum desirable rate. Intermediate positions provide graded rates of heat production, the positioning mechanism being automatically controlled according to requirements of the load.

Thermonuclear Fusion

The term *fusion* refers to partial destruction of an atomic nucleus, a portion of it being transformed into pure energy. Fig. 16 illustrates what is meant by the term, the spherical volume of a nucleus being reduced from the original dimension to a smaller dimension the remainder having been turned into raw energy. Intense heat is required to initiate the process which, once it is started, carries on at the same incredible speed as a chain reaction.

The hydrogen bomb functions on this principle, a small atomic bomb being used to start the activity. Although researchers are working on the problem, no safe method of harnessing fusion to generate power has yet been discovered.

REVIEW QUESTIONS

1. What is an element?

2. Distinguish between atoms and molecules.

3. What is a proton?

4. What, generally, is the shape of an atomic orbit?

5. What is an atomic shell?

6. Is the normal copper atom positively- or negatively-charged?

7. How many electrons are there in the outer shell of the copper atom?

8. What is the polarity of a piece of silk after rubbing a glass rod?

9. State the rule for static attraction and repulsion.

10. Explain static induction and give some of its applications.

11. How may static charge be dissipated from a moving belt?

12. How is the danger of static charge combatted in a hospital operating room?

13. Tell how a dust precipitator works.

14. Explain the operation of a Van de Graaff Static Generator.

15. Describe briefly the Xerox process.

16. What is an ion?

17. What portion of the nucleus is active in a chain reaction?

18. What is meant by the term chain reaction?

19. Of what material are reactor control rods made?

20. Tell the difference between atomic fusion and fission.

Chapter 2

Dynamic Electricity

ELECTRONS IN MOTION

In this chapter we shall study the factors that control the flow of electrons. Electron flow is analogous to molecular flow in liquids. We may suspect that many of the principles that control fluid flow should be applicable to the behavior of electrons in motion.

The amount and force of the flow of fluids in a pipe (conductor) are regulated by several mechanical properties of the pipe system. For example, the inner size of a hose influences both the speed and pressure of water flow. Other mechanical factors which affect fluid flow are the density and thickness of screens in ducts or pipes, pressure relief valves, number of branches, pumping pressure, etc.

There are also non-mechanical factors that affect the flow of molecules in fluids. For example, the gravitational field of the earth affects the flow of water in rivers and in plumbing systems. Temperature affects molecular motion and the mechanical properties of the conduction system.

The term *dynamic* means in motion, dynamic electricity referring to electrons in motion. The preceding chapter dealt with the *static* charge set up by electrons at rest. This chapter gives attention to moving electrons which constitute an electric current. The electrons are identical in both instances, but those which form an electric current are termed "free," being so called because they have escaped from their atoms and although still moving, no longer circulate about a central nucleus.

The manner in which electrons are removed from atoms will be discussed shortly. Fig. 1, represents the path of a *free electron* in a length of copper wire. The highly irregular path results from the necessity for the electron to thread its way amid the millions of atoms that compose the material. Only a single electron is shown in this illustration, but the wire would be seething with activity of countless electrons set in motion by the generating force. As they pass along the wire, electrons collide with one another or with atoms,

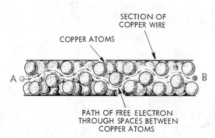

SECTION OF
COPPER WIRE

COPPER ATOMS

A

B

PATH OF FREE ELECTRON
THROUGH SPACES BETWEEN
COPPER ATOMS

Fig. 1. The movement of a free electron in a copper wire.

constantly detouring, some of them being captured by atoms which have lost one or more electrons. This opposition to passage of electrons is called the *resistance* of the conductor, a subject to be discussed later on.

The *apparent* speed of an electron is the speed of light (186,000 miles per second). Suppose the copper wire shown in Fig. 1 were 186,000 miles long. If an electrical pressure was applied at *A,* electron flow would take place over the entire wire (even at *B*) in one second. Since one electron looks just like another, it would be difficult to say where it started. It is believed that actual electron movement is rather slow. A guess may be two or three inches per minute.

Electrical Force

An electrical force which frees electrons from their atoms and causes them to move along a wire, may be generated by chemical action, by heat, by action of magnetic lines, or in a number of ways that will be considered in due course. In all cases, the wire must form an endless path or *circuit* from one side or terminal of the generator to the other, Fig. 2.

The generating force, of whatever nature, causes a mass of electrons to be set free at one terminal where atoms then take on a positive charge, and propels them toward the opposite terminal which thus takes on a negative charge. The *generating force* prevents free electrons at the negative terminal from slipping back through the generator itself to the positive terminal. Meanwhile, positively-charged atoms at the other terminal draw electrons from atoms in the conducting wire, trying to become neutral again.

This effect carries all along the wire, atoms which have lost electrons drawing those from adjacent atoms and perhaps losing them again; this activity progressing with the speed of a chain reaction. Also, many free electrons appearing at the negative terminal are drawn all the way around the circuit by the strong force of attraction massed at the positive terminal of the generator. Each positively-charged atom tries to acquire and keep an electron to satisfy its need, however, the generating force will likely take the electron away again. This operation repeats itself so long as the device is in operation. Summarizing the whole process, the generating force initiates a flow of electrons which complete a circuit of the wire despite resistance offered to their passage.

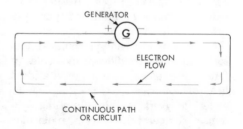

GENERATOR

+ G −

ELECTRON
FLOW

CONTINUOUS PATH
OR CIRCUIT

Fig. 2. When the circuit is complete, electrical force is present at all points along the circuit.

The Ampere

The gain and loss of free electrons from the main stream along the wire averages out to the extent that the same volume of electrons will be found in any section chosen at random around the whole circuit. The unit quantity of electrons is the *coulomb,* a term seldom heard in electrical discussions, but which represents a total of 6.28×10^{18} electrons. One *ampere* of current is said to flow when one coulomb of electricity passes a given point in the conductor during a period of one second.

If only one-half this number of electrons pass in one second, the current is said to be one-half ampere. If ten times the unit quantity flow in one second, the current is said to be ten amperes. The density or strength of electron flow, then, is measured in amperes, the actual number of electrons being of no practical concern because amperes are indicated directly by an electrical instrument called an ammeter.

The Ohm and the Volt

The term representing opposition to flow of electrons in a wire is the *ohm.* This is an arbitrary unit, once based upon the resistance of a certain standard column of mercury. Today, the ohm is defined as the resistance of a wire in which an electrical pressure of one volt causes an electric current of one ampere to flow.

The *volt* is the unit of electrical force, and is defined as that force which causes an electron flow of one ampere through a resistance of one ohm. Thus, the terms volt and ohm are complementary, both dependent upon the one quantity capable of precise physical definition: the ampere.

The voltage of a generating device is usually a set value which does not change while the device is operating under normal conditions, and altogether independent of the wire or circuit to which it is connected. Resistance offered to flow of electrons, however, varies according to the physical dimensions and atomic nature of the material through which the electric current passes.

Conductors and Insulators

Materials in which electron flow is readily established are known as *conductors.* In copper, for instance, the single electron in the fourth shell is loosely held, and easily removed to become a free electron. Copper therefore, is termed a good conductor. Metals, generally, fall into this class, silver being superior to copper in this respect, aluminum and iron inferior.

Substances which strongly oppose flow of electrons through them are termed *insulators.* Paper is considered a good insulator because electrons may be drawn from atomic orbits only with extreme difficulty. Insulators, too, vary in quality. Thus, mica and bakelite are superior to paper.

Factors That Determine Conductor Resistance

The major factors influencing resistance of a conductor are: length, cross-sectional area, temperature, and, as mentioned above, the substance of which the conductor is made.

Length. When an electromotive force of 1v is applied to the ends of the conductor in Fig. 3(A), a current of 1 amp will flow through the conductor as long as the resistance remains at a value of

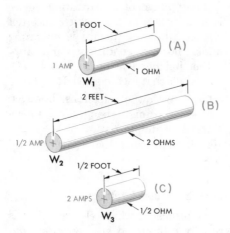

Fig. 3. The resistance of a conductor varies as its length.

1 ohm. The manner in which this force is applied and the nature of its source have been omitted from the illustration for the sake of clarity.

Suppose the length of W_1 doubled as in Fig. 3(B). Here, the conductor is designated as W_2, and the two conductors are the same except for length. The laws of nature are such that a free electron traveling through a certain length of conducting material is subject to the same forces of resistance as in any other similar length of the material. Where the length of the path is doubled, each free electron meets twice the resistance. Under this condition, the pressure of 1 volt is able to maintain a continuous stream of free electrons equal to only ½ ampere.

Since it is twice as difficult for any one electron to travel the length of the conductor, it is said that the resistance of the conductor has been doubled. And, since the resistance of W_1 is 1

ohm, the resistance of W_2 must be 2 ohms.

If the length of the conductor is increased to four times that of W_1, an electromotive force (emf) of 1v will be able to maintain a flow of electrons equal to only ¼ amp. The resistance of this conductor, therefore, is four times that of conductor W_1, or 4 ohms.

If the length of the conductor is reduced to one-half, Fig. 3(C), the resistance will be reduced by one-half. Therefore, 1v of pressure is able to maintain a flow of twice as many free electrons or a number equal to a current flow of 2 amps. Thus, we can state this as a rule: *The resistance of an electrical conductor varies directly as its length.*

Cross-Sectional Area of a Conductor. The *cross section* of a conductor is its end surface. When the conductor, Fig. 4(A), is cut squarely across its diameter, the end surface, or cross section, is

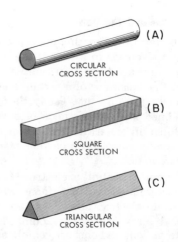

Fig. 4. Electrical conductors having three types of cross section.

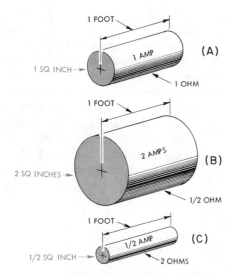

Fig. 5. Resistance of conductor varies inversely as cross sectional area.

circular. The *cross-sectional area* is the area of this squarely cut, circular end. If the conductor is square, Fig. 4(B), its cross-sectional area is that of the square formed by cutting across the conductor. If the conductor is triangular, Fig. 4(C), its cross-sectional area is that of a triangle. The cross section of the conductor may have any shape, but it is usually circular.

Suppose the cross-sectional area of the conductor of Fig. 5(A) is doubled, as in Fig. 5(B). Here, any certain number of free electrons traveling through the conductor will have more room to spread out than in the conductor of Fig. 5(A). The chances of collision and capture on the part of these electrons are much less than if they were crowded into the smaller cross-sectional area of Fig. 5(A). The exactness of nature's laws is such that when the cross-sectional area is twice as great, there are just one-half as many forces of resistance acting on the electron. In other

words, when the area is doubled, the resistance of the conductor is halved. If the area is made four times as great, the resistance of the conductor is only one-fourth.

If the cross-sectional area of the conductor is decreased to one-half its original value, the stream of electrons will be crowded into one-half the space, Fig. 5(C). There will be twice as many chances for capture of individual electrons, and the resistance of the conductor will be twice as great.

The resistance of the conductor, then, varies in the opposite direction from the change in cross-sectional area, resistance decreasing when the cross-sectional area is increased and the resistance increasing when the cross-sectional area is decreased. When two quantities vary in opposite directions with respect to each other, they are said to vary *inversely*. In view of these facts, it is possible to state the rule of area as follows: *Resistance of an electrical conductor varies inversely as its cross-sectional area.*

Designating Conductor Size. Copper wire, which has a circular cross section, is the most common conductor. As an aid to understanding how wire sizes are designated, refer to Table I, which lists copper wire sizes from No. 16 to No. 40, inclusive. The first column gives sizes based upon the American Wire Gage (AWG) system of designations. The second column lists the diameters of the wires in *mils*. (The mil is $\frac{1}{1000}$ part of an inch.) The third column is headed *Area in Circular Mils,* and the last column gives the resistance of the wire in ohms per 1,000 feet.

Circular-Mil Area. Number 36 wire has a diameter of 5.0 mils, and a circular mil area of 25. Number 31 wire has a diameter of 8.9 mils and a circular-mil area of 80. A like relationship exists with respect to each size of conductor, so it becomes obvious that *circular mils equal diameter in mils squared,* that is, multiplied by itself. Thus, 5 squared means 5 × 5 and is usually written 5². However, diameter in mils squared is a cumbersome expression, and the term circular-mil area

was devised to replace it.

One other point should be noted about the term circular-mil area. It is not a true expression of area, for area is expressed in square units: square inches, square mils, square feet, and the like. Here is a square, Fig. 6, with the sides drawn to represent 5 mils on each side, and the circle inside of the square represents the cross section of a wire which has a diameter of 5

TABLE 1 COPPER WIRE TABLE

AWG*	Diameter in Mils	Area Circular Mils	Ohms per 1000 Feet (20°C)
16	50.8	2583	4.00
17	45.3	2048	5.00
18	40.3	1624	6.39
19	35.9	1288	8.05
20	32.0	1021	10.15
21	28.5	810	12.80
22	25.4	643	16.14
23	22.6	510	20.36
24	20.1	405	25.67
25	17.9	321	32.37
26	15.9	254	40.80
27	14.2	202	51.47
28	12.6	160	64.90
29	11.3	127	81.84
30	10.0	101	103.20
31	8.9	80	130.10
32	8.0	63	164.10
33	7.1	50	206.90
34	6.3	40	260.90
35	5.6	32	329.00
36	5.0	25	414.80
37	4.5	20	523.00
38	4.0	16	660.00
39	3.5	12	832.00
40	3.1	10	1049.00

*American Wire Gage

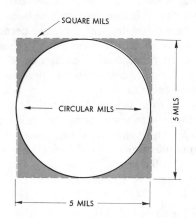

Fig. 6. Showing relative areas of square mils and circular mils.

mils. The area of the square is found by squaring the length of the side: $5 \times 5 = 25$ square mils. To find the area of the circle, we apply the rule:

$$A = \frac{\pi d^2}{4}$$

The Greek letter π (pi) is equal to 3.1416, and d is the diameter of the circle. Therefore,

$$A = \frac{3.1416 \times 5^2}{4}$$

Dividing 3.1416 by 4 gives 0.7854, so that

$$A = 0.7854 \times 5^2$$

Squaring the diameter, we have

$$A = 0.7854 \times 25$$

and

$$A = 19.6 \text{ square mils}$$

From this example it is evident that area of a wire in square mils can be found by multiplying diameter-squared by 0.7854. This number is a *conversion factor,* for changing circular mils to square mils. Square mils divided by 0.7854 gives area in circular mils.

Conductor Material. Certain materials conduct electrical current more easily than others. Copper is a very good conductor. Lead, which conducts current much less easily, is a poor conductor. Although silver is a better conductor than copper, the higher cost excludes it from general use.

The less the resistance of a material, the better it serves as a conductor. The property of a material which describes its resistance to the flow of electricity is known as its *resistivity*.

One accepted method of expressing the resistivity of a material is to state the resistance of a material which has a length of 1 foot and a cross-sectional area of 1 circular mil. For annealed copper, the resistance of a 1-foot length having a cross-sectional area of one mil is 10.37 ohms. Therefore, we say that its resistivity is 10.37 *ohms per mil-foot.* The resistivity of silver is 9.9 ohms per mil-foot. It is of interest to note the resistivity of several other conducting materials: aluminum, 17; brass, 42; constantan, 295; gold, 14; cast iron, 54; lead, 132; manganin, 265; mercury, 577; nichrome, 602; and tungsten, 33.

Calculating the Resistance of a Copper Wire. The practical relationship between these factors may be expressed in a simple formula:

$$R = \frac{KL}{A}$$

R = resistance of conductor in ohms.

L = length of conductor in feet.

A = area of conductor in circular mils. This is equal to d^2.

K = resistivity of conductor in ohms per mil-foot.

For example, suppose that a piece of copper wire is 200 feet long and 0.032 of an inch in diameter. The diameter of this wire is 32 mils (1,000 × 0.032 = 32), for there are 1,000 mils per inch. The cross-sectional area, in circular mils, is equal to 32^2, or 1,024 circular mils (32 × 32 = 1,024). Copper has a resistivity of 10.37 ohms per mil-foot. Substituting these values in the formula.

$$R = \frac{KL}{A}$$

$$R = \frac{10.37 \times 200}{1,024}$$

Multiplying 10.37 by 200,

$$R = \frac{2,074}{1,024}$$

and dividing 2,074 by 1,024,
$$R = 2.025 \text{ ohms}$$

The formula shows that 200 feet of a 32-mil diameter wire has a resistance of 2.025 ohms. We can compare the resistance of this wire to the resistance of a No. 20 wire, as given in Table I, for the No. 20 wire has a diameter of 32 mils. The fourth column gives the resistance of a No. 20 wire as 10.15 ohms per 1,000 feet. Since 200 feet is 1/5 of 1,000 feet ($^{200}/_{1008} = ^2/_{10} = ^1/_5$), we merely divide the resistance per 1,000 feet by 5. This gives us 2.03 ohms (10.15 ÷ 5 = 2.03), which checks closely with the calculated value.

Effect of Temperature on Conductor Resistance

The atoms of a copper wire are in a constant state of vibration, the extent of which depends upon the temperature of the conductor, increasing as temperature rises. The more strongly the atoms vibrate, the harder it becomes for electrons to pass through, because relative spacings are continually changing. In other words, resistance of the conductor increases with increase of temperature.

For each degree Centigrade rise in temperature above zero, the resistance of a copper conductor increases by an amount equal to .00427 times the resistance at 0 degrees. At boiling point, the resistance of the conductor would be: 100 × .00427 or .427 higher than at the freezing point, and a conductor whose resistance was 10 ohms at 0° would be equal to: 10(1 + .427) or 14.27 ohms. Aluminum has a somewhat lower coefficient, .0039, even though the resistance of aluminum is basically much higher than that of copper, as already noted.

It is worthwhile observing that a few substances have a negative *temperature coefficient of resistance*, resistance decreasing with rise in temperature. Carbon is such a material. The reason, though not clearly understood, is likely a more rhythmic oscillation of atoms which tends to assist rather than impede movement of free electrons.

The resistances of copper conductors given in Table 1 were taken at a temperature of 20° Centigrade, or 68° Fahrenheit. These values are used in making ordinary electrical calculations.

Temperature Scales

Fig. 7 shows the relationship between the *Centigrade scale* commonly used to measure temperature in scientific experiments, and the *Fahrenheit scale* employed in the ordinary household

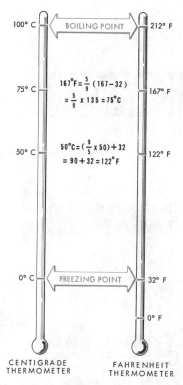

Fig. 7. Centigrade and Fahrenheit scales.

Fahrenheit reading *before* ~~r~~
by 5/9, when converting fr~~om~~
heit to Centigrade. In maki~~ng~~
sion from Centigrade to ~~F~~
the multiplier 9/5 must firsт ɓe used,
and 32 degrees added *after* the calculation has been made.

Thus, when the Fahrenheit reading
is 158 degrees, the conversion is made
as follows: 158 −32 = 126. And 5/9
× 126 = 70 degrees Centigrade. If,
on the other hand, the Centigrade reading is 60 degrees, the conversion is
made in this way: 9/5 × 60 = 108.
And 108 + 32 = 140 degrees Fahrenheit.

Increase of Heat with Flow of Current

The loss of electrons and their re-
capture by atoms when current flows
under the force of an electrical voltage,
requires expenditure of energy; and expenditure of energy is always accompanied by the creation of heat. The
heavier the flow of electrons, the greater
the atomic activity, and consequently
the greater the temperature rise. This
increased heating is usually undesirable since it hastens breakdown of the
plastic or rubber insulation with which
indoor conductors are usually covered.

There are instances, however, where
creation of heat by flow of electricity is
highly desirable, as in the familiar devices shown in Fig. 8. Heat generated
in conductor material which offers very
high resistance to flow of electrons, such
as tungsten, makes possible the flood of
light from a modern incandescent lamp.
Conductors of *nickel-iron alloy,* not so
highly resistant to flow of electrons as
tungsten, but far more resistant than
copper, are wound into coils which provide heat for electrical appliances.

thermometer. The temperature of melt-
ing ice is marked at 0° on the Centigrade scale, 32° on the Fahrenheit,
while the boiling point of water is at
100° Centigrade, and 212° Fahrenheit.

Between freezing point and boiling
point, there are 100 Centigrade degrees
or 180 Fahrenheit degrees, the spacing
between adjacent Centigrade degrees
being 9/5 that between Fahrenheit
degrees. Expressed another way, the
width of Fahrenheit degrees is 5/9 that
of Centigrade degrees.

Reference to Fig. 7 shows that 32
degrees must be subtracted from a

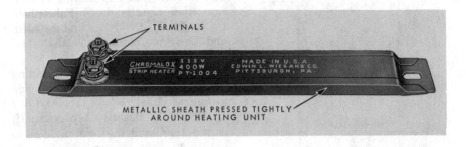

TERMINALS

115V
CHROMALOX 400W   MADE IN U.S.A.
STRIP HEATER PT-1004   EDWIN L. WIEGAND CO.
                        PITTSBURGH, PA.

METALLIC SHEATH PRESSED TIGHTLY
AROUND HEATING UNIT

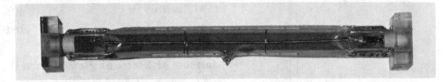

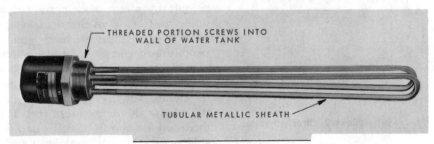

THREADED PORTION SCREWS INTO
WALL OF WATER TANK

TUBULAR METALLIC SHEATH

Fig. 8. Types of heating elements.

Fuses

Because conductor insulation is subject to damage from heat, it is necessary to provide means for interrupting the flow of current when its value exceeds the amount which the wire may be allowed to carry. The earliest form of such protection, and one still widely used today, is to install a weak point in the circuit. This weak spot is in the form of an easily replaceable fuse whose conducting member is usually made of zinc alloy which has much higher resistance than copper, and which melts at a comparatively low temperature, Fig. 9, shows the location of the *fuse* in the circuit between a generator and the electrical device, or load, which it supplies. Fig. 10 shows the inner construction and outer appearance of a plug type fuse.

Superconductivity

Conductivity is a term used to denote the opposite of resistance, so that a conductor which has high resistance

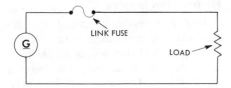

LINK FUSE

G

LOAD

Fig. 9. Fuse connected in a circuit.

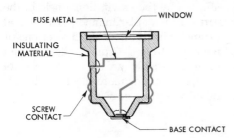

FUSE METAL — WINDOW

INSULATING MATERIAL

SCREW CONTACT

BASE CONTACT

Fig. 10. Cross section of a plug fuse.
Bussman Manufacturing Co.

is said to have low conductivity, or one having low resistance is said to have high conductivity. In experiments with

extremely low temperatures, under the scientific heading of *cryogenics,* investigators have tested electrical resistance

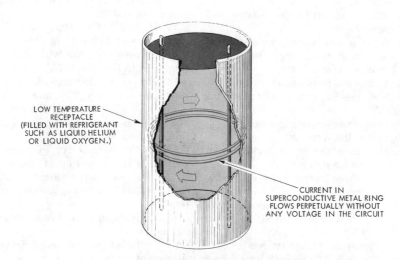

LOW TEMPERATURE RECEPTACLE (FILLED WITH REFRIGERANT SUCH AS LIQUID HELIUM OR LIQUID OXYGEN.)

CURRENT IN SUPERCONDUCTIVE METAL RING FLOWS PERPETUALLY WITHOUT ANY VOLTAGE IN THE CIRCUIT

Fig. 11. The basic arrangement for producing superconductivity in a conductor.

near the point of absolute zero, which is —273.1° Centigrade.

At temperatures approaching this value, obtained through the evaporation of liquid helium, atomic vibration in a conductor practically ceases, and resistance to flow of electrons borders on zero. It is believed that, under conditions of absolute zero, a current induced in a conductor, Fig. 11, will continue to circulate forever even though the generating force has been removed, the conductor having acquired the state of *superconductivity*.

Electron Flow in Space

Just like fluids, electrical currents may exist in open space. For example, a flow of electrons occurs in electrical discharges of lightnings and between electrodes in TV tubes and fluorescent lamps.

The flow of fluids in open spaces is affected by the gravitational field of the earth. In a similar manner, electric and magnetic fields affect the flow of one or more electrons in space. Electric and magnetic fields are used to control beams of one or more electrons in TV tubes and atom smashers.

REVIEW QUESTIONS

1. Distinguish between planetary and free electrons.

2. What is meant by the resistance of a conductor?

3. What is the underlying nature of the force which causes free electrons to pass through a circuit?

4. Define the term ampere.

5. How does the ampere differ from the coulomb?

6. How is resistance expressed?

7. Define the term volt.

8. How do atoms of conductors differ from those of insulators?

9. State three factors which determine conductor resistance.

10. If 200 ft of a certain conductor have a resistance of 1 ohm, what is the resistance of 500 ft?

11. Suppose the conductor has a cross-sectional area only one-half its original value. What is the resistance of 300 ft?

12. If the wire had twice its original cross-sectional area, what would be the resistance of 600 ft?

13. If 1200 ft of a certain conductor has a resistance of 2 ohms, how does its cross-sectional area compare with that of the conductor in Question 10?

14. How many circular mils in a square conductor measuring 1-in. on a side?

15. What is the Fahrenheit temperature of absolute zero? (the Centigrade temperature is approximately —273°)

16. Is the resistivity of aluminum greater or less than that of copper?

17. What causes the generation of heat in a conductor carrying an electric current?

18. Define temperature coefficient of resistance.

19. Is the temperature coefficient of resistance of copper positive or negative?

20. Define the term conductivity.

Ohm's Law in Direct Current Applications

OHM'S LAW

Formulation of Ohm's Law

Definition of the term volt brought out the fact that an electrical *pressure* of 1 volt caused an electron flow of 1 ampere through a resistance of 1 ohm. When a pressure of 2 volts is applied to the 1-ohm circuit, current flow doubles. When a pressure of ½ volt is used, the current flow is cut in half. In other words, current through a given resistance changes the same way as the voltage, increasing or decreasing in direct proportion to the voltage. This observation can be expressed in the form of a rule: Current flow in a circuit varies directly as the applied voltage.

The study of resistance shows that when voltage is held constant, the value of current changes when resistance is altered, current decreasing as resistance increases, and increasing as resistance decreases. When two related quantities change in this manner, one is said to vary inversely as the other. Here, the

rule may be stated: Current flow varies inversely as the resistance of the circuit.

George Simon Ohm, an early German scientist, combined the two rules into a single one that bears his name: Current in an electrical circuit varies directly as the applied voltage, and inversely as the resistance. *Ohm's Law* is the foundation upon which are based all electrical calculations, however simple or however complicated.

Basic Symbols. Three basic symbols are used in dealing with electrical quantities. The first of these indicates electrical force, which is known as electromotive force and is measured in volts. Therefore, the expression

$$E = 120 \text{ v}$$

means that the electromotive force (emf) active in a given circuit is equal to 120 volts. The unit of force (the volt) is represented by the letter E.

Electrical resistance is designated by the letter R and is measured in ohms. Thus,

$$R = 5 \text{ ohms}$$

states that the resistance present in a particular circuit amounts to 5 ohms. The unit of resistance may also be represented by the Greek letter Ω (omega):

$$R = 5 \ \Omega$$

Electron flow in an electrical circuit is indicated by the letter I, representing *intensity* of current flow. The letter C, which might be expected to be the abbreviation for current, is used to designate a different quantity. The unit of current flow, the ampere, is abbreviated *amp*. Thus:

$$I = 24 \text{ amps}$$

means that the current flowing in the circuit amounts to 24 amps.

APPLICATIONS OF OHM'S LAW

Ohm's Law can be used in all applications of direct current as shown, defined and illustrated in this chapter. It also may be used for alternating current applications that are purely resistive in nature. If the circuits being considered have inductive or capacitive components, however, the applications require more stringent analysis. These applications are covered in detail in Chapter 8.

Finding Resistance

The practical value of Ohm's law to the technician is that if he knows any two values in an electrical circuit, the third may be easily determined because the value of one quantity has a definite relationship to the other two.

For example, a technician has measured the conditions in several circuits and has listed the results as shown in Table I. However, in a few of the circuits he has been unable to measure the resistance. Of course, in such a situation he might disassemble the circuits and measure the resistance using an ohmmeter, but this is time consuming and unnecessary if he knows Ohm's law.

Refer to Table I·and study the examples where all the values are given and see if you can determine the value of the resistance where these are not given.

If we refer to the first line of the table we find that $E = 12$ volts, $I = 2$ amps, and $R = 6$ ohms. It is obvious then that if we didn't know the value of R, that according to Ohm's law, there must be a relationship between E and I that will give a·value of 6 ohms for the circuit. Study Table I and see if you can determine this relationship.

An examination of the table will show, that for the first three lines, if we divide the voltage (E), by the current (I), we arrive at the figures shown for resistance in the last column of the table. To put it into a simple formula, the relationship is:

$$R = \frac{E}{I}$$

Thus, in order to find resistance if we know current and voltage, it is necessary only to divide the voltage (E), by the current (I).

Now, to determine the unknown values of resistance in Table I using Ohm's law, the first step is to write the formula for resistance.

$$R = \frac{E}{I}$$

TABLE I

E Voltage	I Current	R Resistance
12 volts	2 amps	6 ohms
6 volts	3 amps	2 ohms
100 volts	25 amps	4 ohms
10 volts	2 amps	? ohms
8 volts	4 amps	? ohms
20 volts	10 amps	? ohms

[handwritten left: $R = \dfrac{20 V.}{10 \text{ Amps}}$ $R = 2$ ohms]

[handwritten right: $R = \dfrac{8 \text{ volts}}{4 \text{ Amps}}$ $R = 2$ ohms]

The next step is to replace the symbols with known quantities. In the fourth line of the table, $E = 10$ volts, $I = 2$ amps.

$$R = \frac{10 \text{ volts}}{2 \text{ amps}}$$

[handwritten: devide 2/10 ohms]

$$R = 5 \text{ ohms}$$

The other resistances in the table may be calculated in the same manner as the previous examples.

Finding Current

It has been shown in the previous paragraphs that if we know two circuit quantities we can easily determine the other. If we want to know the current flow in a circuit, we must first know the voltage, E, and the resistance R. Let us use the same method that showed us how to calculate resistance and see how current flow can be determined. Table II gives values for voltage, resistance, and current that might be found in a typical circuit. Study the table carefully and see if you can discover the relationship be-

tween the values of voltage and resistance in the first two columns and how they must be arranged to equal the current in the last column.

Let us examine one of the examples of Table II and see what the relationship is. Referring to the values for the first line, we find $E = 10$ volts, $R = 5$ ohms, and $I = 2$ amps. Since the combination of 10 volts and 5 ohms must equal 2 amps, it is obvious that the only way we can obtain the correct answer is to divide 10 volts by 5 ohms, which equals 2 amps. Putting it in a simple formula we have:

$$I = \frac{E}{R}$$

Substituting our known values,

$$I = \frac{10 \text{ volts}}{5 \text{ ohms}}$$

$$I = 2 \text{ amps}$$

For a further example of how the formula works, lets check the second line of Table II and solve for current. Our first step is to refer back to our formula

[handwritten bottom: Current flow = Amps = I]

33

TABLE II

E Voltage	R Resistance	I Current
10 volts	5 ohms	2 amps
20 volts	4 ohms	5 amps
16 volts	2 ohms	8 amps
12 volts	3 ohms	? amps
30 volts	6 ohms	? amps
100 volts	10 ohms	? amps

[handwritten: $I = \frac{12V}{3\,ohm}$]
[handwritten: $I = 4\,amp$]
[handwritten: $\frac{R}{PS} = $ ... 5 AMPS]

and then substitute for the formula using our known values of voltage and resistance. From the second line of Table II, we find $E = 20$ volts, and $R = 4$ ohms, so we write it out according to Ohm's law:

$$I = \frac{20 \text{ volts}}{4 \text{ ohms}}$$

$$I = 5 \text{ amps}$$

Now, using what you have learned, see if you can solve for the unknown values of current in Table II. Just follow the examples already given and you should encounter no difficulty.

Finding Voltage

Having discovered how to find current and resistance using Ohm's law, we come to the last and easiest quantity to calculate—voltage. Table III gives example values of current, resistance, and voltage from typical circuits. Since we

[handwritten: to find voltage, "multiply."]

TABLE III

I Current	R Resistance	E Voltage
2 amps	3 ohms	6 volts
3 amps	3 ohms	9 volts
4 amps	4 ohms	16 volts
2 amps	6 ohms	? volts *[12]*
6 amps	5 ohms	? volts *[30]*
3 amps	4 ohms	? volts *[12]*

know that to find voltage we must know current and resistance, a careful study of Table III will show how I and R must be arranged to give E.

Beginning with the first line in the table, we have a current, I, of 2 amps, and a resistance, R, of 3 ohms, which results in a value of 6 volts in the last column. How this is obtained is quite clear: 2 amps *times* 3 ohms = 6 volts. Stating this according to Ohm's law we have:

$$E = I \times R$$

or, as it is usually written:

$$E = IR$$

Following down through Table III using the formula for the first two columns easily verifies the values found in the third column. Thus, the values not given in the table may be easily determined.

The generator in Fig. 1 develops an electromotive force of 50 V and feeds a small floodlight. Assume that resistance of the wires connecting the generator to the floodlight is so small that it need not be considered. How much current will the 50-V generator force through the light which has a resistance of 10 ohms? Figure 1(A) shows a pictorial view of the generator and floodlight. Figure 1(B) presents a schematic diagram, the light indicated by a wavy line, which is the symbol for a resistor in an electrical drawing.

The first step is to write the formula for finding current. The next step is to replace the symbols with known quantities wherever possible. Since the value of E is given as 50 V, and that of R, 10 ohms, the formula is written:

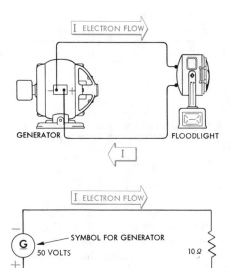

Fig. 1. A floodlight supplied by an electrical generator.

$$I = \frac{E}{R}$$

$$I = \frac{50}{10}$$

$$I = 5 \text{ amps}$$

Thus 5 amps will flow through the 10-ohm light when 50 V is applied.

If the light has a resistance of 5 ohms, while the electromotive force remains at 50 V, Fig. 2, then:

$$I = \frac{50}{5}$$

$$I = 10 \text{ amps}$$

This example shows that reducing the resistance in a circuit while holding emf constant causes an increase in current.

35

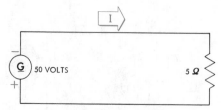

Fig. 2. Schematic of circuit containing 50-volt generator and 5-ohm resistor.

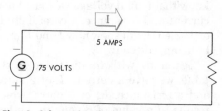

Fig. 4. Schematic of circuit in which resistance is unknown.

Now increase the resistance to 10 ohms, and the emf to 100 V, Fig. 3. Here the resistance and applied voltage have been doubled. Increasing the resistance alone would bring a reduction in current; while increasing only the emf increases current flow, but:

$$I = \frac{100}{10}$$

$$I = 10 \text{ amps}$$

Doubling the resistance would have reduced current to one-half, or 5 amps, at 50 volts. Doubling the emf however, raises current from 5 amps to 10 amps, the same value as in the preceding example.

Further Applications of Ohm's Law

It is obvious that the value of current flow may be determined in every case, provided that the voltage and the resistance of the circuit are known. Sometimes voltage and current are given, and it is desired to find the resistance.

In Fig. 4 the applied voltage is 75 V, the current is 5 amps, and the resistance of the circuit is required. A rearrangement of Ohm's Law states that:

$$R = \frac{E}{I}$$

$$R = \frac{75}{5}$$

$$R = 15 \text{ ohms}$$

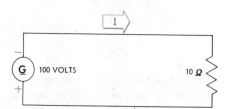

Fig. 3. Schematic of circuit with 100-volt generator and 10-ohm resistor.

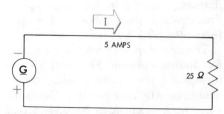

Fig. 5. Schematic of circuit in which voltage is unknown.

In the third type of problem, current and resistance are known, and voltage unknown. These problems can be solved with the third arrangement of Ohm's law. In Fig. 5 the current is 5 amps and the resistor has a value of 25 ohms.

$$E = IR$$

$$E = 5 \times 25$$

$$E = 125 \, V$$

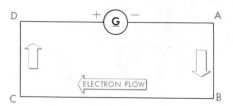

Fig. 6. Voltage loss around a circuit.

The voltage required to send a current of 5 amps through a resistance of 25 ohms is found to be 125 V.

Voltage Drop

Electrical pressure or voltage was described earlier as an attractive force established by positively-charged atoms whose outer electrons have been torn away by action of the generating device. Pressure really has two aspects: repulsion at the negative generator terminal between the mass of liberated electrons and any one free to move into the circuit wire; and attraction at the positive generator terminal, seeking to draw electrons out of the circuit wire. These forces of repulsion and attraction are present not only in the immediate vicinity of the generator, but throughout the whole circuit. Nature's laws are so exact that force is equally distributed along the conductor between negative and positive terminals.

Starting at negative *G*, Fig. 6, and proceeding in the direction of electron flow, some of the total force is used up in sending electrons to point *A*, still more of it from *A* to *B*, *B* to *C*, *C* to *D*, and finally *D* to positive *G* at which point the last of the generated pressure will have been consumed. This loss of voltage or pressure in various sections of the conductor is termed *voltage drop*.

Fig. 7 shows a resistor connected to a generator by means of conducting wires, which are often called *leads*. The resistor in this illustration is a long piece of high-resistance wire wound spirally on an insulating form. Resistance of the leads is negligible. The generator develops an *emf* of 200 volts and the resistor has a value of 50 ohms. According to Ohm's Law,

$$I = \frac{200}{50}$$

$$I = 4 \text{ amps}$$

The value of the current, 4 amps, is marked beside the arrows on the conductors, indicating the direction of current flow.

An *emf* of 200 volts appears at the negative terminal of the generator. This voltage is also present at point *A* on the resistor, because there is no resistance in the lead, and accordingly, no loss in pressure. When the current flows through the resistor, however, there will be a loss. The resistor shown is divided into four sections. Between point *A* and point *B*, there is 5 ohms of resistance. If the current is 4 amps, there

37

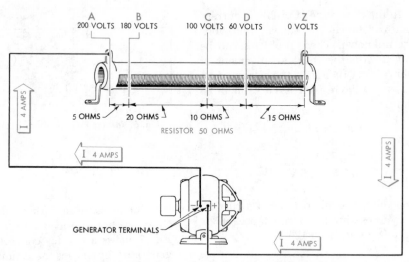

Fig. 7. Illustrating how voltage drop occurs along a resistor.

will be a voltage drop between points A and B. Employing Ohm's Law,

$$E = IR$$
$$E = 4 \times 5$$
$$E = 20 \text{ volts}$$

Therefore, the voltage remaining at point B is 180 volts (200 − 20 = 180) instead of the original 200 volts.

As current passes through the resistor from point B to point C, there will be additional voltage drop. With 20 ohms of resistance between these points, voltage drop determined by Ohm's Law is:

$$E = 4 \times 20$$
$$E = 80 \text{ V}$$

The voltage drop at point C is 100 volts

(180 − 80 = 100), and the voltage drop between C and D will be 40 volts:

$$E = 4 \times 10$$
$$E = 40 \text{ V}$$

the voltage at point D being 60 volts. Finally, voltage drop between points D and Z is 60 volts:

$$E = 4 \times 15$$
$$E = 60 \text{ V}$$

so the voltage at point Z is 0 volts (60 − 60 = 0).

The different points along the resistor in Fig. 7 were chosen at random. Additional ones could be selected if desired to show voltage drop at closer intervals.

Practical Example of Voltage Drop. Fig. 8 shows an electrical circuit consisting of a generator, an electric lamp, and the connecting wires. The voltage of the generator is 120 V, and the current required by the lamp is 10 amps. That portion of the conductor between the negative terminal of the generator and point *A,* at one end of the lamp, has a resistance of 1 ohm. The other conductor from the positive terminal to the *B* end of the lamp also has a resistance of 1 ohm. Since brightness of the lamp depends upon the voltage applied, it is desired to know how much voltage is present between points *A* and *B.*

Here, 10 amps flow through the lamp. Since the current is the same throughout every portion of the circuit, the generator must send 10 amps of current from its negative (−) terminal and receive 10 amps at its positive (+) terminal. To find the voltage drop in each of these conductors, it is necessary to apply Ohm's law as follows:

$$E = IR$$
$$E = 10 \times 1$$
$$E = 10 \, V$$

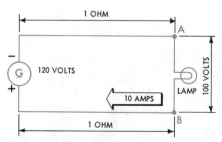

Fig. 8. Showing voltage drop along 1-ohm conductors which are supplying an electric lamp.

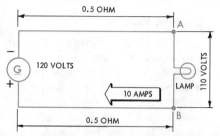

Fig. 9. Showing voltage drop along ½-ohm conductors which are supplying an electric lamp.

From the formula, it is seen that the voltage drop from the negative terminal to point *A* is 10 V. The voltage drop from *B* to the positive terminal is also 10 V. Thus, 20 V are consumed in the conductors, and only 100 V will be present across *A* and *B.* If this remaining voltage is insufficient for proper lighting output, it may be necessary to replace the conductors with ones of larger cross section which have less resistance.

Suppose that larger conductors are substituted, Fig. 9, having a resistance of 0.5 ohm each instead of 1 ohm. What is the effect upon the voltage applied to the lamps? Referring again to the formula,

$$E = IR$$
$$E = 10 \times 0.5$$
$$E = 5 \, V$$

the voltage drop in each conductor is 5 V; in the two conductors, twice this amount, or 10 V. In this case, the voltage applied to ends *A* and *B* is 110 (120 − 10 = 110) which should prove more satisfactory than the original value of 100 volts.

Electrical Power

An electrical current represents a flow of energy, and energy can never be created—it is only transformed. The familiar rotating electromagnet generator requires some form of mechanical driving unit in order to produce voltage and the resulting electric current. If the mechanical driver is a steam turbine, more steam will have to be admitted to the turbine blades when the generator is required to send out a heavy current of electrons, than would be needed for a smaller value. That is, the turbine has to provide greater turning effort to take care of a higher energy demand. The physical effort exerted by the driving unit is covered by the term *work,* and the rate at which it performs this work is covered by the term *power.*

The term power is applied not only to mechanical work, but to electrical work as well, the unit of electrical power being the *watt.* The voltage causes electrons to flow through the resistance offered by a circuit, the heavier the stream of electrons the greater the energy moving in the conductor. When a pressure of 1 volt sends a current of 1 ampere through a resistance, the power in the circuit is said to be 1 watt. This value is arrived at by multiplying 1 volt by 1 amp, so that: 1 volt $\times$ 1 amp $=$ 1 watt. If voltage is doubled, the current rising to 2 amps, the power in the circuit is equal to: 2 volts $\times$ 2 amps $=$ 4 watts.

Power Formulas

The relationship between power, voltage, and current is expressed in the formula: $P = E \times I$, or $P = EI$, where P stands for power, E for voltage, and I for amperes. This form may be changed as desired, to:

$$E = \frac{P}{I} \text{ or to } I = \frac{P}{E}.$$

Another useful power formula is:

$$P = I^2R$$

This formula is derived from the original one,

$$P = EI,$$
$$\text{or, } P = E \times I$$

by noting that from Ohm's law

$$E = IR$$

Then, $\qquad P = E \times I$

becomes: $\quad P = IR \times I$

or, $\qquad\quad P = I \times I \times R$

$$P = I^2R$$

Examples of Power Calculations

Fig. 10 shows a generator connected to a load resistor which takes 5 amps,

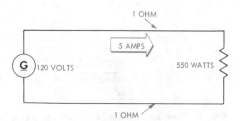

Fig. 10. Power and power loss in electric circuits.

40

each of the connecting wires having a resistance of 1 ohm. Power output of the generator may be found by using the formula: $P = 120$ volts $\times$ 5 amps $= 600$ watts. The power delivered to the load will not be quite so large. Although the current through the load is the same as that throughout the whole circuit, the voltage across its terminals is less than that of the generator by the amount lost in the connecting wires.

Voltage drop in the upper wire is equal to: 5 amps $\times$ 1 ohm $= 5$ volts, that in the lower wire equal to the same amount. Voltage across the terminals of the load resistor, therefore, equals the generator voltage of 120 minus 10 volts, or 110 volts. Since $P = EI$, the power delivered to the load is equal to: 110 volts $\times$ 5 amps $= 550$ watts. The difference between 600 watts at the generator and 550 watts at the load, or 50 watts, must have been dissipated in the supply conductors. Applying the second formula, the power lost in one wire is: $P = I^2R = 5^2 \times 1 = 25$ watts, and that of the two wires twice this amount, or 50 watts, as noted above.

The generator in Fig. 11 delivers a current of 20 amps at a pressure of 240 volts, the resistance of each supply conductor being .5 ohm. Proceeding as before, the power output of the generator amounts to: 240 volts $\times$ 20 amps, or 4800 watts. Voltage drop in the two supply wires equals: 2×20 amps $\times$ .5 ohm $= 20$ volts. The voltage across the load terminals is: 240 volts minus 20 volts, or 220 volts. Power consumed in the load is: 220 volts $\times$ 20 amps $= 4400$ watts. The difference, 400 watts, must have been consumed as heat in

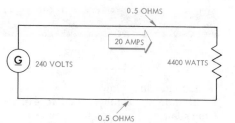

Fig. 11. Another example of power and power loss in electric circuits.

the supply wires. A check shows $I^2R = 20^2 \times 1 = 400$ watts, proving the fact.

It is worth noting that *power loss* in the supply wires might have been determined from the original formula. Voltage drop in the two supply wires was found, above, to be 20 volts. Since power is volts $\times$ amps, the power loss is: 20 volts $\times$ 20 amps, or 400 watts.

The Kilowatt

Power output of the generator in Fig. 11 was expressed as 4800 watts. In practice, the term *kilowatt* (abbreviated kw) is customarily employed for large values of power, a kilowatt being equal to 1000 watts. The output of the generator would be expressed as 4.8 kilowatts, the input to the load, 4.4 kilowatts.

Nameplates on household electric ranges are always marked in kilowatts, viz.: 14 kw, 16 kw, etc. Electric dryers are marked 4.8 kw or 5 kw as the case may be. Electric lamps, on the other hand, are labeled in watts, such as 100 watts, 150 watts, and so on.

Fundamentals of Electricity

1. State Ohm's law.

2. State the symbols used to express Ohm's law.

3. Write the formula for Ohm's law in three different ways.

4. If the current in a circuit is ½ amp and the resistance 10 ohms, what is the voltage?

5. If 2 amps flow under a pressure of 15 volts, what is the resistance of the circuit?

6. If the circuit voltage is 75 volts and the resistance 60 ohms, what current will flow in the circuit?

7. If 24 volts pressure causes a flow of 5 amps, what voltage will send 4½ amps through the circuit?

8. When a certain resistor is inserted in circuit A, 6 amps flow through it. When installed in circuit B, the current is only 3 amps. How does the voltage of circuit B compare with that of circuit A?

9. What forces cause electrons to flow in a conductor?

10. An electric heater draws 10 amps and the voltage across its terminals is 120 volts. If the resistance of the supply conductors totals 2 ohms, what is the voltage drop in the supply wires?

11. What is the generator voltage?

12. What is meant by the term *power*?

13. Define the electrical unit of power.

14. How much power is consumed by the heater of Question 10?

15. Write the power formula in two ways.

16. How much power is dissipated as heat in the conductors of Question 10?

17. How much power is produced by the generator in Question 10?

18. Define the term *kilowatt*.

19. A 4.8 kw electric dryer is connected to supply wires whose resistance totals ½ ohm. If the power loss in these wires is 200 watts, what is the current flowing to the dryer?

20. What is the voltage of the circuit in Question 19.

Electrical Circuits

Series or End-to-End Arrangement

The three basic electrical circuit arrangements are: *series, parallel,* and *series-parallel.* In Fig. 1, a lead wire from lamp No. *1* is connected to a supply conductor, the other lead to lamp No. *2.* The second lead from lamp No. *2* is attached to lamp No. *3,* and the remaining lead from this lamp to the other supply wire. When circuit devices such as resistors or lamps are connected end-to-end in this way, they are said to be in series.

A distinguishing feature of the series circuit is that a break in continuity at any point interrupts current flow in the whole circuit. If the filament in one of the three lamps of Fig. 1 should break, all three would go out. A second characteristic of the circuit is that current through each of the end-to-end devices is exactly the same. Lamp No. *2,* for example, cannot have a greater or lesser flow of electrons than lamp No. *1,* because any current received by this lamp must pass through No. *1* on the way from negative generator terminal, and all of it must be passed on to lamp No. *3.* It is obvious, too, that current flow in the supply wires is the same as that

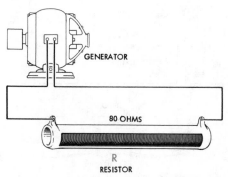

Fig. 2. Total resistance of a series circuit represented by one resistor.

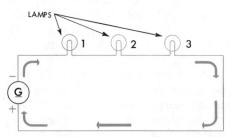

Fig. 1. Series circuit with three lamps.

through any device connected in the circuit, illustrating the rule: *Current is precisely the same at any and every point in a series circuit.*

Resistance in a Series Circuit

In Fig. 2, sufficient high-resistance wire is wound upon an insulating cylinder to provide a total resistance of 80 ohms. Neglecting resistance of supply wires, this value represents the complete resistance in circuit. In Fig. 3, the resistor has been separated into four sections of 20 ohms each, by cutting through the cylindrical support, but leaving the wire intact. The four sections are equivalent to four distinct units, R_1, R_2, R_3, and R_4, which have been connected together, and it is evident that the value of resistance present in the circuit is still exactly 80 ohms.

This conclusion, written in formula style:

$$R_T = R_1 + R_2 + R_3 + R_4,$$

illustrates the rule that total resistance in a series circuit is equal to the sum of individual resistances.

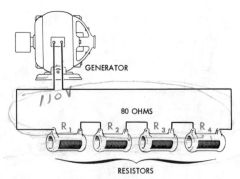

Fig. 3. Total resistance of a series circuit with four resistors.

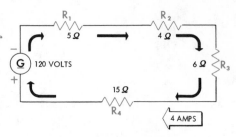

Fig. 4. Series resistors.

Voltage and Power in Series Circuit

The generator in Fig. 4 is rated at 120 volts, and the values of series resistors are: $R_1 = 5$ ohms, $R_2 = 4$ ohms, $R_3 = 6$ ohms, and $R_4 = 15$ ohms. The resistance of supply conductors in this and following examples, will be assumed negligible unless otherwise stated. Circuit resistance, as shown above, is the sum of individual resistances, or 30 ohms.

Since the voltage is 120, Ohm's law shows the current to be:

$$\frac{120 \text{ volts}}{30 \text{ ohms}} = 4 \text{ amps.}$$

Voltage across the terminals of resistor R_1 may be found through the relationship:

$$4 \text{ amps} \times 5 \text{ ohms} = 20 \text{ volts.}$$

$$E = IR$$

Voltage across the terminals of resistor R_2 is found in the same way, to be 16 volts, R_3, 24 volts, and R_4, 60 volts. Adding: 20 volts + 16 volts + 24 volts + 60 volts = 120 volts, the voltage of the circuit.

Power consumed in resistor R_1 may be determined in one of two ways: multiplying voltage across its terminals by the current, or, if the voltage were not yet known or calculated, multiplying current-squared by resistance. Using the first method:

$$P = 20 \text{ volts} \times 4 \text{ amps} = 80 \text{ watts.}$$

By the second method:
$$P = (4 \text{ amps})^2 \times 5 \text{ ohms.}$$
$$16 \times 5 = 80 \text{ watts.}$$

Power used by each of the resistors may be determined in a like manner, showing that R_2 consumes 64 watts, R_3 96 watts, and R_4 240 watts, the sum of all four amounting to 480 watts. The amount of power delivered to the circuit by the generator equals: $P = EI$ $= 120 \text{ volts} \times 4 \text{ amps} = 480 \text{ watts}$, proving the result noted above. Summing up, voltage, resistance, and power in a series circuit equal the sum of respective values for the individual units connected in series.

PARALLEL CIRCUITS

Arrangement of Parallel Circuit

The parallel circuit is employed in the home; lights, television, refrigerator, and kitchen appliances being connected in this manner. A parallel circuit has as many paths as there are devices. In Fig. 5, three resistors R_1, R_2, and R_3 connected across the line wires from the generator, offer three separate paths to the flow of current.

If the generator is rated at 24 volts, and the resistance of each resistor is 12 ohms, the current flowing through each path is equal to the voltage at its terminals divided by its resistance. Since the voltage here is 24 volts and the resistance is 12 ohms, the current

in $R_1 = \dfrac{24 \text{ volts}}{12 \text{ ohms}}$, or 2 amps. The

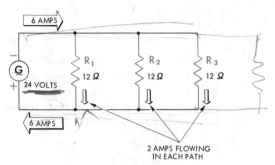

Fig. 5. Simple parallel circuit.

current flowing through R_2 and R_3 is likewise 2 amps in each case, the total current being: 2 amps + 2 amps + 2 amps, which is 6 amps.

Resistance of Parallel Circuits

With a voltage of 24 volts and a total current of 6 amps, Ohm's law reveals that the combined resistance of the three resistors must be equal to:

$\dfrac{24 \text{ volts}}{6 \text{ amps}}$, or 4 ohms. This value is

termed the equivalent resistance. If the voltage across the terminals of each resistor is given, it is unnecessary to calculate the equivalent resistance. But, in more complicated circuits where such voltages are unknown, it is impossible to determine the value of current without first obtaining the equivalent resistance by a method that will now be explained.

This method is sometimes called "taking the reciprocal of the sum of the reciprocals." The reciprocal of a number is equal to 1 divided by that number. Thus, the reciprocal of 2 is ½, that of 3 is ⅓, that of 4 is ¼, that of R_1 is $1/R_1$, that of R_2 is $1/R_2$. If two resistors, R_1 and R_2 are connected in par-

45

allel, the sum of the reciprocals is:

$$\frac{1}{R_1} + \frac{1}{R_2} \text{ which equals } \frac{R_1 + R_2}{R_1 \times R_2}$$

The reciprocal of the sum of the reciprocals is equal to the number 1 divided by this fraction, which is the same as turning the fraction upside down, so that it becomes:

$$\frac{R_1 \times R_2}{R_1 + R_2}$$

This formula is usually written:

$$R_T = \frac{R_1 R_2}{R_1 + R_2}$$

R_T being the equivalent resistance of the combination of resistors.

Use of the formula is illustrated in connection with Fig. 6, where two 4-ohm resistors are connected in parallel on a 24-volt circuit. If one unit is designated as R_1 and the other as R_2, the equivalent resistance R_T is equal to:

$$\frac{4 \times 4}{4 + 4}, \text{ or } \frac{16}{8} \text{ which is 2 ohms.}$$

Applying Ohm's law, the current flowing through the supply wires is:

$$\frac{24 \text{ volts}}{2 \text{ ohms}}, \text{ or 12 amps.}$$

This result may be checked in this simple example by determining the current in each resistor and then adding the two values. The individual currents are:

$$\frac{24 \text{ volts}}{4 \text{ ohms}}, \text{ or 6 amps.}$$

Adding them, 6 amps + 6 amps = 12 amps, the same result as obtained by using the formula.

In Fig. 7, a 24-volt generator supplies current to three parallel resistors, R_1, of 2 ohms; R_2, of 3 ohms, and R_3 of 6 ohms. The above formula may be extended to cover three resistors in this manner:

$$R_T = \frac{R_1 R_2 R_3}{R_1 R_2 + R_2 R_3 + R_1 R_3}$$

Applying the formula here:

$$R_T = \frac{2 \times 3 \times 6}{(2 \times 3) + (3 \times 6) + (2 \times 6)}$$

or $\frac{36}{36}$, which equals 1 ohm. According to Ohm's law, the current flowing in the circuit is: $\frac{24 \text{ volts}}{1 \text{ ohm}}$, or 24 amps.

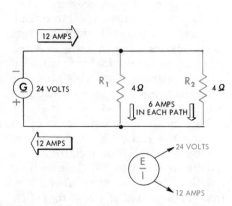

Fig. 6. Equivalent resistance in a parallel circuit.

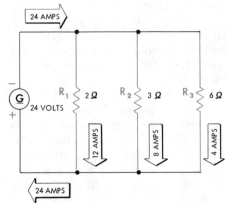

Fig. 7. Equivalent resistance—three resistors in parallel.

The method explained here can be extended to include any number of resistors. Thus, if four units are connected in parallel, there will be four multipliers in the numerator (*top*) of the fraction, while the denominator (*bottom*) will have four groups in each of which there are three multipliers.

Comparison With Series Connection

The parallel arrangement differs from the series in some important respects. If the path through one of a group of parallel units is interrupted, all others will continue to function. The voltage across each parallel device is that of the circuit, while the current in various portions may differ widely. Resistance in the series circuit is the sum of individual resistances, while in the parallel circuit it is the reciprocal of the sum of the individual reciprocals. Power calculation, however, is basically the same.

Power in the Parallel Circuit

Referring again to Fig. 7, the current through R_1 is 12 amps, its resistance 2 ohms, and the pressure across its terminals 24 volts. Using either of the two alternate formulas, the power consumed is found to be 288 watts. The power in R_2 is 192 watts, and that of

R_3 96 watts. Total power is the sum of:
$$288 + 192 + 96 = 576 \text{ watts}$$

Power output of the generator is found to be:
$$24 \text{ volts} \times 24 \text{ amps} = 576 \text{ watts},$$

confirming the above figure. It is evident that power in this circuit, as in the series type, is equal to the sum of power consumed by individual devices.

SERIES-PARALLEL CIRCUITS

A Common Type of Series-Parallel Circuit

The advantage of the formula method becomes apparent in dealing with series-parallel circuits. Fig. 8(A) represents a series-parallel arrangement in which two groups of paralleled resistors are connected in series. The equivalent resistance of each group must be determined first, and then the two equivalent resistances must be added to find the circuit resistance.

Consider first the group of two resistors. According to the formula, the equivalent resistance equals:

$$\frac{2 \times 4}{2 + 4}, \text{ or } \frac{4}{3} \text{ ohm.}$$

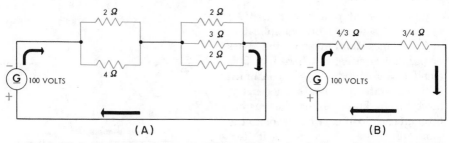

Fig. 8. (A) Series-parallel circuit; (B) Circuit which is electrically equivalent to (A).

The equivalent resistance of the group of three resistors equals:

$$\frac{2 \times 3 \times 2}{(2 \times 3) + (3 \times 2) + (2 \times 2)},$$

$$\text{or } \frac{3}{4} \text{ ohm.}$$

Adding the two equivalent series resistors, as illustrated in Fig. 8(B), the total resistance of the circuit equals:

$$\frac{4}{3} \text{ ohm} + \frac{3}{4} \text{ ohm, or } \frac{25}{12} \text{ohm.}$$

The amount of current flowing in the circuit is equal to:

$$\frac{100 \text{ volts}}{\frac{25}{12} \text{ ohm}}, \text{ or } 48 \text{ amps.}$$

This current passes from the negative terminal of the generator, through each of the two resistance groups in series, to the positive terminal of the generator. The voltage drop across the group of two resistors is equal to the equivalent resistance of the group multiplied by the current, which is:

$$48 \text{ amps} \times \frac{4}{3} \text{ ohm, or } 64 \text{ volts.}$$

The voltage drop across the group of three resistors is equal to:

$$48 \text{ amps} \times \frac{3}{4} \text{ ohm, or } 36 \text{ volts.}$$

Power Calculation

Calculation of power in Fig. 8(A) appears somewhat more complicated, at first glance, than in either the series or parallel circuits, but here too, it is equal to the sum of power consumed in the separate resistors. It is necessary, however, to know either the current through each load or, what leads to the same thing, the voltage across its terminals.

Since voltages for the two groups have already been found to be 64 volts and 36 volts respectively, currents in various units are readily determined. Current through the 2-ohm resistor in the two-parallel group is:

$$\frac{E}{R} = \frac{64}{2}$$
$$= 32 \text{ amps,}$$

and the power expended: 64 volts × 32 amps = 2048 watts. Current in the companion 4-ohm resistor is 16 amps, the power 1024 watts. Passing to the second group, current in each of the 2-ohm resistors is:

$$\frac{36}{2} = 18 \text{ amps,}$$

the power 648 watts. Current in the 3-ohm unit is 12 amps, and power 432 watts.

Circuit power is the sum of these amounts: 2048 + 1024 + 648 + 648 + 432 = 4800 watts. Generated power is: 100 volts × 48 amps = 4800 watts, or 4.8 kw, proving the method correct.

Another Form of Series-Parallel Circuit

Fig. 9 illustrates a type of circuit which comes under the general heading of series-parallel combinations, and which is sometimes called a parallel-series grouping. Two or more resistors

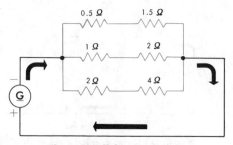

Fig. 9. Parallel-series circuit.

are connected in series, in each leg, and the legs are connected in parallel. A moment's reflection will show that the total resistance of each leg should first be obtained, and the problem solved as a simple combination of parallel resistors.

Adding the resistors in the upper leg of Fig. 9,

0.5 ohm + 1.5 ohm = 2 ohms.
The middle leg is equal to

1 ohm + 2 ohms = 3 ohms.
And the bottom leg is equal to

2 ohms + 4 ohms = 6 ohms.
Thus, three resistors of 2 ohms, 3 ohms, and 6 ohms, respectively, are connected in parallel, the identical problem that was presented and solved in connection with Fig. 7.

MORE ADVANCED CIRCUIT TYPES

The Three-Wire Circuit

Electric supply circuits to homes and other buildings are usually the three-wire type illustrated in Fig. 10. Although these circuits are obtained ordi-

narily by way of transformers, which will be covered later on under the head of alternating current theory, the two generators shown in the figure will suffice. The negative terminal of generator G_1 is connected to the upper supply wire, the positive terminal connected to the negative terminal of generator G_2 and a middle supply wire, the positive terminal of G_2 to the lower supply wire.

This method of connecting generating devices will be discussed in a subsequent chapter. For the present, it is merely necessary to observe that G_1 furnishes a pressure of 120 volts between upper and middle conductors, G_2 the same voltage between middle and lower conductors, while they combine to provide 240 volts between the outer conductors.

The consuming devices shown as resistors in the illustration represent lighting circuits, heaters, or other appliances.

The purpose of the three-wire circuit is shown by the illustration, lamps and other 120-volt loads spanning upper and middle wires, or lower and middle wires, while large heating devices or

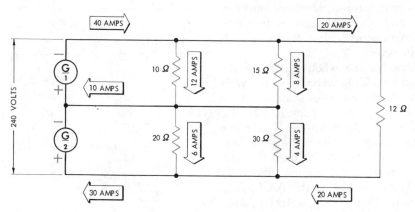

Fig. 10. Three-wire circuit.

motors rated at 240 volts, connect to the two outer wires. Current through the 10-ohm resistor equals 120 volts divided by 10 ohms, which is 12 amps. Current in the 15-ohm resistor is 8 amps, in the lower 20-ohm unit 6 amps, and in the 30-ohm unit 4 amps.

An interesting characteristic of the three-wire arrangement is that current from one two-wire portion flows also through the other two-wire portion, while the middle, or *neutral* conductor carries the difference between them, (if the currents are unequal) to or from the common generator terminal. In the present instance, current in the upper circuit is 40 amps, while that in the lower is 30 amps, the neutral conductor taking the difference of 10 amps from the junction point back to the generators.

If the loads were interchanged so the lower portion required 40 amps and the upper 30 amps, the neutral wire would again carry 10 amps, but in the opposite direction. If the upper and lower currents were equal, say both 40 amps, the neutral would carry no current. Turning to the range load, (represented by the 12 ohm resistor across the outer conductors) which is 20 amps at 240 volts, both outer supply wires must furnish 20 amps in addition to their 120-volt requirements, but the neutral is not involved.

Power in the whole circuit is obtained by adding wattages of the various units, regardless of whether 120 or 240 volts. The power consumed by the electric range is: 240 volts × 20 amps = 4800 watts. Power in the upper section is: 120 volts × 20 amps = 2400 watts; in the lower section: 120 volts × 10 amps = 1200 watts. Total power consumption is: 4800 watts + 2400 watts + 1200 watts = 8400 watts, or

8.4 kw. The same result is obtained, of course, by calculating generated power, the output of G_1 amounting to: 120 volts × 40 amps = 4800 watts, and that of G_2: 120 volts × 30 amps = 3600 watts. Adding them, the output is found to be 8.4 kw.

Kirchhoff's Laws

In 1847, a German physicist named Kirchoff stated two laws which have become invaluable in electrical circuit analysis. These laws are as follows:

1. *Voltage Law.* The algebraic sum of voltages in a closed loop is zero.
2. *Current Law.* The algebraic sum of currents meeting at a junction point is zero.

The voltage law can be illustrated as shown in Fig. 11. The reader will note that polarity has been assigned to each resistance in relation to its position with the power source. Kirchoff's voltage law states, that *the algebraic sum of voltages in a closed loop equals zero.* Therefore, let us begin at point *A* and look at the circuit in the direction of current

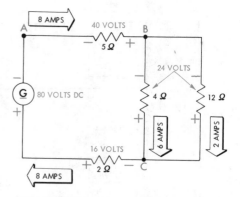

Fig. 11. Illustration of Kirchhoff's Laws.

flow ($-$ to $+$). There are two paths of current flow. These are: (1) through the 5 ohm, the 4 ohm and the 2 ohm resistors; and (2) through the 5 ohm, the 12 ohm, and the 2 ohm resistors.

The second or current law can also be illustrated in Fig. 11. Kirchoff's current law states that *the algebraic sum of currents meeting at a junction point is zero.*

On one side of point B, we have $+8$ amperes of current. On the other side of point B we have -6 amperes and -2 amperes of current, observing polarity signs. The algebraic sum of these currents are thus:

$+8$ amperes -6 amperes -2 amperes $= 0$ amperes

Another way to look at this is that current moves into point B and splits into two paths. The amount of current in each path is dependent on the resistance of the path.

Analyzing the current law at point C, we have $+6$ amperes and $+2$ amperes flowing into point C. On the other side of point C we have -8 amperes of current, observing polarity signs. The algebraic sum of these currents are as follows:

$+6$ amperes $+2$ amperes -8 amperes $= 0$ amperes

Networks

The term *network* includes any combination of resistors or other loads in which arrangement of components does not fall within the regular classifications: series, parallel, or series-parallel. Networks are commonly employed in radio, television, or hi-fi circuits, as well as in electronic motor control and servomechanisms. Solution of a network problem usually calls for determining the *equivalent resistance*. No mathematical preparation is needed if methods are carefully observed, and are used as a pattern in attacking similar problems which arise.

Mesh Circuit Solution

One of the more basic uses of Kirchoff's voltage law is in the solution of *mesh circuits*. As mentioned previously, Kirchoff's voltage law states that the algebraic sum of voltages in a closed loop is zero. Some circuits have two power supplies. Circuits of this type are best solved by use of *mesh currents* and simultaneous equations.

A mesh is the simplest loop possible in a circuit. Each mesh current is assumed to flow freely within its closed loop. In the Fig. 12, there are two mesh current loops. The first mesh loop is *FEBAF*. The second mesh loop is *DEBCD*. Normally current is assumed to flow from the negative side of the power supply to the positive. In following the loops, one must observe the polarity of each voltage drop, then algebraically add the voltage drops around the loops. In the first loop starting and ending at point F, we write our first equation (A) thusly:

$$-E_{R3} - E_{R1} + E_1 = 0$$

In the second loop starting and ending at point D, we write our second equation (B) thusly:

$$-E_{R3} - E_{R2} + E_2 = 0$$

We then substitute the values shown in Fig. 12, and solve the equations by the simultaneous method as follows:

Equation A:
$$-2(I_1 + I_2) - 4I_1 + 7 = 0$$
Equation B:
$$-2(I_1 + I_2) - 1I_2 + 7 = 0$$

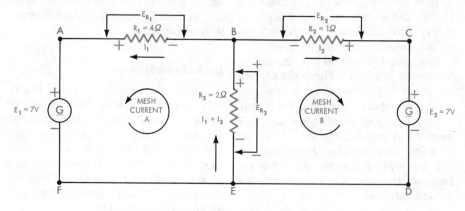

EQUATION A: $-E_{R_3} - E_{R_1} + E_1 = 0$ OR $-2(I_1 + I_2) - 4I_1 + 7 = 0$

EQUATION B: $-E_{R_3} - E_{R_2} + E_2 = 0$ OR $-2(I_1 + I_2) - 1I_2 + 7 = 0$

Fig. 12. Mesh current circuit.

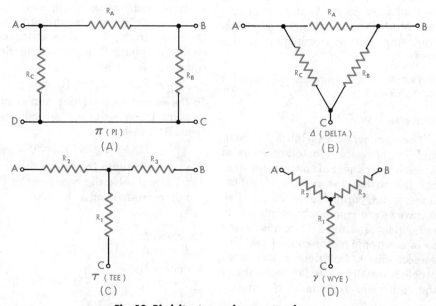

Fig. 13. Pi, delta, tee, and wye networks.

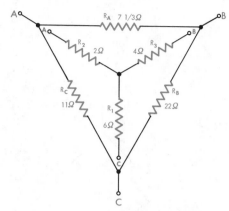

Then:

$$-2I_1 - 2I_2 - 4I_1 = -7$$
$$-2I_1 - 2I_2 - 1I_2 = -7$$

And:

$$-6I_1 - 2I_2 = -7$$
$$-2I_1 - 3I_2 = -7$$

Multiply the second equation terms by -3:

$$-6I_1 - 2I_2 = -7$$
$$+6I_1 + 9I_2 = +21$$

Algebraically add the terms:

$$7I_2 = 14$$
$$I_2 = 2 \text{ amps}$$

Substituting 2 amps for I_2 in the second equation and solve for I_1:

$$-2I_1 - 3\,(2) = -7$$
$$-2I_1 - 6 \quad\;\; = -7$$
$$-2I_1 = -1$$
$$I_1 = 0.5 \text{ amps}$$

Mesh circuits can be used for three current loops also, but this demands the use of three unknowns, therefore equations are more complex.

Fig. 14. Wye to delta and/or delta to wye conversion.

Wye and Delta Resistance Networks

Fig. 13 illustrates four configurations of three resistors each. These configurations are often used by the electrical and electronic industries in one form or another. The *pi* and *delta* configurations have a different form but are electrically identical. Likewise, the form of the tee and wye configurations are different in form but electrically identical. The names are derived from their shapes: *pi* for the Greek letter π, *delta* for the Greek letter Δ, T and Y for their English counterparts.

Wye to Delta Resistance Conversion

Analysis of a network may prove very difficult unless the circuit in study can be converted from wye to delta or vice versa. In Fig. 14, a wye network is converted to a delta network. It is accomplished quite simply with the use of the following formulas:

$$R_A = \frac{R_1R_2 + R_2R_3 + R_1R_3}{R_1}$$

$$R_B = \frac{R_1R_2 + R_2R_3 + R_1R_3}{R_2}$$

$$R_C = \frac{R_1R_2 + R_2R_3 + R_1R_3}{R_3}$$

Substituting for values shown in the circuits of Fig. 14, we have the following calculations:

$$R_A = \frac{(6 \times 2) + (2 \times 4) + (6 \times 4)}{6}$$

$$R_B = \frac{(6 \times 2) + (2 \times 4) + (6 \times 4)}{2}$$

$$R_C = \frac{(6 \times 2) + (2 \times 4) + (6 \times 4)}{4}$$

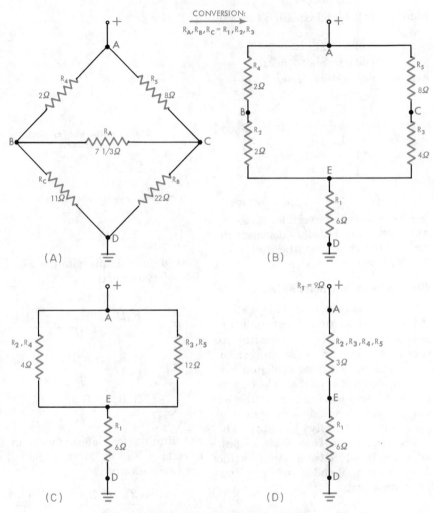

Fig. 15. Bridge network resistance solution.

$$R_A = \frac{12 + 8 + 24}{6} = \frac{44}{6} = 7\text{-}1/3 \text{ ohms}$$

$$R_B = \frac{12 + 8 + 24}{2} = \frac{44}{2} = 22 \text{ ohms}$$

$$R_C = \frac{12 + 8 + 24}{4} = \frac{44}{4} = 11 \text{ ohms}$$

Delta to Wye Resistance Conversion

In the Fig. 14, the values of the delta network are converted to the wye network with the use of the following formulas:

$$R_1 = \frac{R_B R_C}{R_A + R_B + R_C}$$

$$R_2 = \frac{R_C R_A}{R_A + R_B + R_C}$$

$$R_3 = \frac{R_A R_B}{R_A + R_B + R_C}$$

Substituting for values shown in the Fig. 14, we have the following calculations:

$$R_1 = \frac{11 \times 22}{7\text{-}1/3 + 22 + 11}$$

$$R_2 = \frac{11 \times 7\text{-}1/3}{7\text{-}1/3 + 22 + 11}$$

$$R_3 = \frac{7\text{-}1/3 \times 22}{7\text{-}1/3 + 22 + 11}$$

$$R_1 = \frac{242}{40\text{-}1/3} = 6 \text{ ohms}$$

$$R_2 = \frac{80\text{-}2/3}{40\text{-}1/3} = 2 \text{ ohms}$$

$$R_3 = \frac{161\text{-}1/3}{40\text{-}1/3} = 4 \text{ ohms}$$

Bridge Network Resistance Solution

The bridge network is probably the best use made of the delta to wye network conversion. It is accomplished in several steps. These steps are shown in Fig. 15. The reader should note carefully the positions of terminals A through E in each view. In view (A), a bridge network is shown. In view (B) one delta network has been converted to a wye network which changes the circuit to a series-parallel circuit. The terminal E represents the center of the wye network. The same values are used here as in Fig. 15, for simplicity sake. View (C) simplifies the series-parallel circuit and finally, in view (D), the series-parallel circuit is further simplified to a simple series circuit.

Other Networks

There are many solutions to networks. Each has its advantages. Each also must be selected by the user according to his ability to perform the mathematics necessary for their solution. It may prove interesting for the reader to look further into the following network solution methods: node voltage analysis, superposition theorem, Thevanin's theorem, Norton's theorem and Millman's theorem.

1. Name the three basic circuit arrangements.

2. Current in a 120-volt circuit is 5 amps. One of three series resistors is 4 ohms, another 15 ohm. What is the third?

3. How much power is consumed by this third resistor?

4. What is the voltage drop across the other two resistors?

5. If the third resistor in Q 2 is replaced by one which reduces circuit current to one-half the original amount, what is its value?

6. How much power will be dissipated by this new resistor?

7. How much power will be dissipated by the 15-ohm resistor?

8. The voltage of a series circuit is unknown, but in one part of it a 6-ohm resistor consumes 24 watts. How much current flows?

9. Two 100-watt, 120-volt lamps are connected in series on a 240-volt circuit. If a 150-watt, 120-volt, lamp is inadvertently substituted for one of the 100-watt lamps, which one is most likely to burn out? Why?

10. What is the voltage across the terminals of the 150-watt lamp?

11. What is the equivalent resistance of a ½-ohm, a 2-ohm, and a 3-ohm resistor connected in parallel?

12. If the circuit voltage is 12, how much power is consumed by the ½-ohm resistor?

13. Two 2-ohm resistors in parallel, are in series with three 3-ohm resistors in parallel. What current flows through each 2-ohm resistor under a circuit pressure of 10 volts?

14. A 120-240-volt three-wire circuit is carrying 30 amps in the upper portion, 20 amps in the lower. What changes will take place if the neutral wire breaks?

15. What is meant by the term *network?*

16. State Kirchhoff's laws.

17. State the rule of signs which applies when voltage drop is taken in the direction of electron flow.

18. What sign is given to a current which flows toward a junction point?

19. When a series circuit is substituted for a parallel circuit in a network calculation, which combination of resistors will have the larger set of values?

20. A current of 16 amps flows through a 3-ohm and a 5-ohm resistor connected in parallel. How much current flows through the 5-ohm unit?

21. What is the need for delta to wye network conversion?

22. What is a mesh circuit? What basic law is used in its solution?

Chapter 5

Magnetism and
Electromagnetism

MAGNETISM

Why Study Magnetism?

Ancients believed the invisible force of magnetism purely a magical quality, and hence of little practical interest. With Man's steadily increasing knowledge of things in general during the passing centuries, however, it assumed a larger and larger role. Magnetism is essential to the generation of electricity by rotating apparatus and the creation of mechanical power by the electric motor. It makes possible the operation of most electrical measuring instruments, and has innumerable industrial and commercial applications of its own. A thorough grasp of the subject is basic for the student of electricity, while even the layman is hampered in this age of electronics and space travel unless he is acquainted with at least fundamental concepts.

Magnetism and Atomic Quality

Scientists have found magnetism to be a property of the atom. More specifically, it is the result of normal elec-

tronic activity. Discussion in Chapter 1 explained that static attraction or repulsion depended upon either a concentration of electrons removed from atomic orbits, or a concentration of atoms whose orbits were in need of electrons. Material there also called attention to the fact that electrons are believed to *spin* on their axes in the same way the Earth turns on its axis, as they orbit around the nucleus. Both spinning and orbiting activities of electrons are concerned in the production of magnetic effects.

Effects of Electronic Motion

Fig. 1, shows an orbiting electron, representation of the nucleus and other electrons being omitted for sake of clarity. A moving electron is accompanied by a *line of magnetic force* whose exact nature is unknown, but which possesses two proven characteristics: it acts in a certain definite direction, and it appears endless, forming a complete circle or ellipse.

Direction of this force is parallel to the *axis of spin,* as indicated in the illustration, which shows an electron spinning in a counter-clockwise direc-

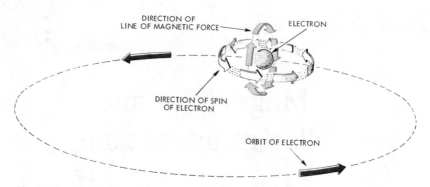

DIRECTION OF
LINE OF MAGNETIC FORCE

ELECTRON

DIRECTION OF SPIN
OF ELECTRON

ORBIT OF ELECTRON

Fig. 1. The colored line indicates the direction of the magnetic circle of force which accompanies electron motion.

tion when viewed from above, its magnetic force clockwise when observed from the right. This direction of spin may be termed positive. If the electron is caused to turn over, it will appear to spin in the opposite or negative direction when observed from above, its magnetic force now counter-clockwise from the right.

When an atomic shell is filled, or complete, the number of *positively-spinning electrons* is exactly equal to the number of *negatively-spinning electrons,* so that their magnetic effects cancel. For example, a shell with eight electrons will have four positive-spins and four negative-spins. One more fact to note here is that observed direction of spin may be reversed by causing the electron to turn over. That is, direction of spin may be reversed without necessity for coming to rest and starting again in the opposite rotation.

Classes of Materials Found in Nature

Some materials show no magnetic effects, and are commonly termed *non-magnetic.* Others show various degrees of magnetic tendency, some even re-

pelling magnetic force in a manner comparable to repulsion between like static charges. Of the scientific classifications for magnetic materials, the two most generally important are called: *paramagnetic,* and *ferromagnetic.* Paramagnetic substances possess a low order of magnetic qualities, ferromagnetic, a high order. The latter type includes mostly compounds of iron, cobalt, or nickel.

Magnetic lines of force pass more readily through ferromagnetic substances than through paramagnetic and other materials. Resistance to passage of magnetic lines is known as *reluctance,* but the term *permeability,* denoting the opposite magnetic quality or conductivity, is more often used in practice.

Explanation in Chapter 1 brought out the fact that copper owed its high electrical conductivity to the single electron in its fourth atomic shell, the other three shells being filled. In ferromagnetic substances, two of the four shells are complete, but the third and fourth shells are incomplete, and there is in the atom an excess of positive over negative spins, or vice versa. Atoms in these materials, therefore, have a resulting magnetic "charge."

Magnetic Polarities

Ferromagnetic substances which have a strong magnetic charge are said to be magnetized. When a thin strip of magnetized iron is suspended at its middle, one particular end always points to the approximate location of the Earth's north pole. Mariners, using such magnetized strips as compasses, termed the one end north-seeking, or *north*, polarity, the other end the *south*. These terms became shortened to *north pole* and *south pole*, ultimately to *N* pole and *S* pole. (Usually designated as "an *S* pole", or "an *N* pole.")

Pattern of Iron Filings

If a layer of iron filings is sprinkled over a piece of cardboard, and a magnetized strip or bar is laid upon it, a pattern like that of Fig. 2 will take shape when the cardboard is gently tapped. The filings arrange themselves in paths which are highly concentrated at the two ends, or poles, and extend from one end of the strip to the other. The form assumed by these paths is dictated by that of the endless mag-

netic lines of force established by spinning electrons in the material of which the strip is composed.

Powder Pattern

When an almost microscopically fine magnetic powder is spread over a slightly magnetized piece of ferromagnetic material, patterns of very fine lines are observed on the surface of the object as in Fig. 3(A), distinct groupings visible in many random directions, a few in end-to-end formation. Increased magnetization of the iron results in a changed pattern, Fig. 3(B), more of the groupings assuming an end-to-end state, the higher the degree of magnetization, the greater the number of such groups. Finally, when the iron has been magnetized as strongly as possible, substantially the whole surface will show a continuous structure of this kind. Under this condition, the piece of iron is said to be magnetically saturated.

Domains

Scientific investigations show these surface groupings to be characteristic of the atomic state inside the

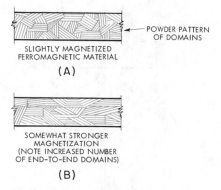

SLIGHTLY MAGNETIZED
FERROMAGNETIC MATERIAL

POWDER PATTERN
OF DOMAINS

(A)

SOMEWHAT STRONGER
MAGNETIZATION
(NOTE INCREASED NUMBER
OF END-TO-END DOMAINS)

(B)

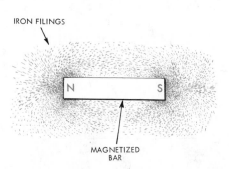

IRON FILINGS

N   S

MAGNETIZED
BAR

Fig. 2. Patterns assumed by iron filings around a bar magnet.

Fig. 3. Powder patterns formed by magnetic domains.

material. These groupings are termed *domains,* and are found in all ferromagnetic substances. A domain is formed by a drawing together of atoms whose magnetic forces are all in one direction, that is, either negative or positive. Where a given sample is not highly magnetized, the domains are arranged generally so that the magnetic effect of one group offsets that of another. As a higher degree of magnetization is imposed, magnetic polarities of the atoms are rotated to conform with that of the magnetizing force. In effect, the original domains are swung around, or oriented, to the desired state.

When the force is discontinued, some of the domains revert to the original condition, but others remain changed for a longer or shorter time, depending upon the atomic structure of the particular material. It should be noted that tests of non-ferromagnetic substances show that domains are not present in them.

Attraction and Repulsion

In Fig. 4(A), the N pole of a compass is attracted by that end of a bar magnet which is marked S. At the same time, the S pole of the compass is attracted by the N pole of the bar magnet. If another bar magnet is substituted for the compass, Fig. 4(B), the S end of the first magnet is found to attract the N pole of the second, and to repel the S pole, as in Fig. 4(C).

Continuing the experiment at the opposite end of the first unit, the N pole repels the N pole of the second, but attracts its S pole. Hence, a rule may be stated: *Unlike magnetic poles attract; like magnetic poles repel.* Note the similarity between this rule and the one for static attraction and repulsion. It is well to observe, here, that attraction

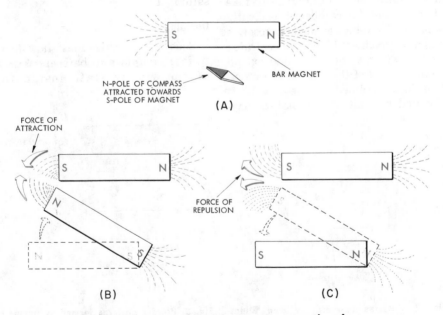

Fig. 4. Attraction and repulsion between magnetic poles.

of the N compass needle toward the North geographical pole shows that an S magnetic pole exists in this region.

Permanent and Temporary Magnets

If a bar of soft iron is magnetized, it tends to remain strongly magnetic only so long as the magnetizing force is present, quickly reverting to its original state. Such a piece is called a temporary magnet. If a bar of hard steel is magnetized, it tends to remain so longer after the force has been removed. This piece is called a permanent magnet. The reason for magnetic retention is that domains in the hard steel are more difficult to influence than those in the iron. This interpretation is confirmed by the observation that a greater magnetizing force, or the necessity for employing a given force a longer period of time to establish a certain level of magnetization in the harder material.

Alloys of nickel and iron tend to remain in a strong magnetic condition, their strength diminishing only slightly in a space of several years. Such magnets are widely used in radio and hi-fi loudspeakers, as well as in the field of instrumentation.

In recent years the two most widely used materials for permanent magnets are *alnico* (an aluminum-nickel-cobalt alloy either cast or compacted and sintered from metal powders) or special ceramic-type mixtures of iron oxide and the oxides of either barium or strontium). These so-called ceramic magnets or *ferrite* magnets are used as core materials for inductors in their soft state. The hard ferrites are used in permanent magnets in applications such as audio speakers. Their shape is determined by molding before baking.

Magnetic Induction. If a common wire nail is brought into contact with

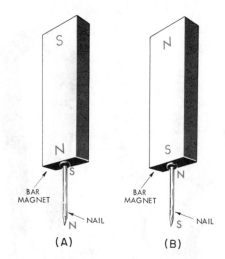

Fig. 5. Illustrating magnetic induction.

one pole of a magnetized bar, Fig. 5(A), the nail will cling to the magnetic pole. If the pole happens to be the N pole of the magnet, an N pole will be found at the lower end of the nail. If the nail is touched to the S pole of the magnet, Fig. 5(B), the lower end of the nail will have an S pole. If taken away from the magnet, the nail will lose its magnetic properties. In other words, the nail is only a temporary magnet.

The lines of force of the magnet cause the domains in the nail to arrange themselves so that their magnetic lines agree with those coming from the magnet. Therefore, an S magnetic pole is formed at the end of the nail in contact with the N pole of the magnet, and an N pole at the lower end. When the nail touches the S pole of a magnet, an N pole is formed at the end in contact with it, and an S pole at the opposite end. This method of inducing magnetism by contact with a magnetized piece is called *magnetic induction*.

Types of Magnets

Magnets are produced in a variety

61

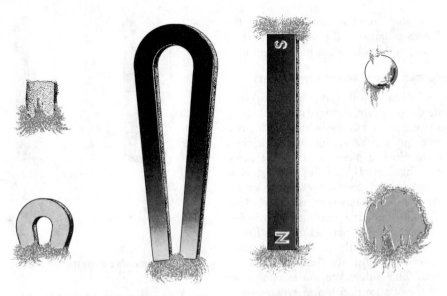

Fig. 6. Some of the various forms of magnets. The iron filings indicate the field patterns.

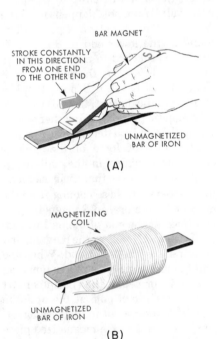

BAR MAGNET

STROKE CONSTANTLY
IN THIS DIRECTION
FROM ONE END
TO THE OTHER END

UNMAGNETIZED
BAR OF IRON

(A)

MAGNETIZING
COIL

UNMAGNETIZED
BAR OF IRON

(B)

Fig. 7. (A) Magnetization of a bar by friction, (B) Magnetization by means of an electric current.

of shapes and sizes as shown in Fig. 6. The pattern assumed by lines of force is shown by iron filings attracted to the poles. These magnets are employed in numerous small devices, in direct-current measuring instruments, and in tachometer generators for indicating speed of rotation. Some of them have accompanied satellites into orbit.

Magnetizing and Demagnetizing Forces

Ferromagnetic substances can be magnetized in any way that encourages domains to form end-to-end combinations. Stroking a piece of unmagnetized iron with a magnet, Fig. 7(A), is one way to accomplish such a result. Mechanical friction sometimes produces a low degree of magnetization, but the common method is that of Fig. 7(B), the unmagnetized metal being placed inside a direct-current magnetizing coil.

Demagnetizing most often occurs through the simple process of aging, do-

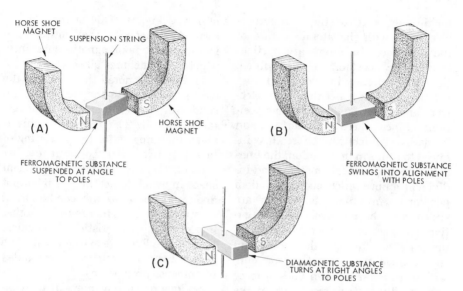

Fig. 8. Comparative reactions of ferromagnetic diamagnetic substances.

mains gradually reverting to mutually neutralizing formations. Mechanical vibrations also will cause domains to relax. A sure quick way to reduce or entirely eliminate magnetization is to heat the material to a moderately high value, heat energy violently disrupting end-to-end alignments.

Diamagnetic Substances

If a rectangular sample of ferromagnetic metal is suspended between the poles of a horseshoe magnet, but at an angle to them, Fig. 8(A), poles will be induced in the object, and it will swing around until directly aligned with the magnetic poles, Fig. 8(B). Movement results, of course, because of attraction between unlike poles, the N pole of the magnet inducing an S pole in the end nearest it, the S pole inducing an N pole.

As explained earlier, these induced polarities result from domain rear-

rangements which are in substantial agreement with magnetic lines from the magnet. This type of response is characteristic only of ferromagnetic substances. Paramagnetic ones show slight magnetic tendencies, others none at all, so that the magnetic lines would have no effect whatever upon a block of wood, for example, that was suspended in place of the ferromagnetic sample.

Certain metals, however, exhibit more or less negative reaction to the presence of magnetic lines, and are said to be diamagnetic. One such is bismuth. When a piece of this material is suspended in a very strong magnetic field, it places itself at right angles to the direction of magnetic lines, as indicated in Fig. 8(C). Yet, this metal shows no magnetic qualities whatever when tested with the most delicate scientific instruments, and it does not have a domain type of structure.

Of the reasons put forth to account

for this odd behavior, the most acceptable deals with the atomic nature of the substance. Its atoms are particularly stable, most of the shells filled, the orbits closely interlaced so that positive-spin and negative-spin electrons are held strongly in relative positions. Since there are no domain formations which can be readily induced to conform in direction, the lines are compelled to work upon myriad individual atoms which occupy random positions. Any magnetic effects are stifled, and the positive-spin or negative-spin electrons, as the case may be, oppose efforts to turn them over and thus change their directions of spin. Violent reaction between lines of force and unwilling electrons result in the whole object turning to the position of Fig. 8(C), where the lines of force have the smallest possible effect.

ELECTROMAGNETISM

Circles of Force About an Electrical Conductor

When current flows through an electrical conductor, magnetic lines of force are set up at right angles to the direction of current flow. As already noted in connection with Fig. 2, these lines close upon themselves. They form circles around the wire and are called *circles*

Fig. 9. Magnetic circles surrounding conductor through which current is flowing.

of force, Fig. 9. This shape will be assumed whenever the entire path is through a nonmagnetic substance, which term includes air.

It should be noted that some of the circles are close to the surface of the conductor, while others are some distance from it. The reason for this is much the same as for the spreading of lines through the air adjacent to a bar magnet. Each circle requires sufficient space in which to act, and, if it cannot find room close to the conductor, it must expand. This fact reveals another important characteristic of magnetic lines: these lines are extremely flexible —they may be stretched or expanded to immense proportions.

So long as current flows in the conductor, the circles of force surround it. When the current ceases to flow, the circles collapse and the magnetism near the conductor disappears, the free electrons which made up the flow of electric current being absorbed by the atomic structure in which the circles of force travel.

Direction in Which Circles of Force Act

A simple way to learn the direction in which circles of force travel or act around a conductor is to use the compass. But the compass needle moves because of a reason different from that noted in connection with the testing of magnetic poles in Fig. 4. There, the compass needle was moved by attraction between unlike magnetic poles. Here, the reason for the movement of the compass needle may be explained somewhat differently.

If the compass needle, Fig. 10(A), is held directly above the conductor, the N pole of the needle swings until it points in the direction along which the

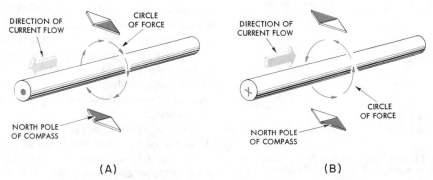

Fig. 10. The indication of a compass depends on the direction of current flow.

circle of force moves. Since the compass needle is a small magnet, with lines of force traveling along its body from the *S* end to the *N* end inside the metal, it turns until the direction of these lines agree with that of the circle of force. The direction of the circle of force at the top, where it touches the needle is from left to right, the compass needle turning so that the *N* pole points in this direction.

Transferring the compass to the bottom of the conductor, Fig. 10(A), results in a reversal of the compass needle because direction of the circle of force at this point is opposite from that at the top. This illustration shows a method often used to indicate direction of electron flow. An arrow may be pictured as a head on view showing the point, and is represented on the end of the conductor by a dot. This indicates that the current is flowing out of a conductor. In Fig. 10(B), the tail may be represented by a cross, and indicates that current is flowing into the conductor.

Determining Direction of Current Flow in a Conductor

Referring back to Fig. 10(A), the di-

rection of the force circle is clockwise and the current flows toward the observer. If the current flows away from the observer, the circle of force is counter-clockwise. In other words, when the direction of the current changes, the motion of the circle of force also changes.

This fact makes it possible to determine direction of current by using a compass. With the compass held above the conductor in which current flows, Fig. 10(A), the *N* pole of the needle points to the right. The compass needle indicates that circles of force surrounding the conductor act in a clockwise direction. Since the circle is clockwise, the direction of current flow in the conductor is toward the observer. If the compass is held beneath the conductor, the *N* pole of the needle is to the left, the direction of a clockwise circle of force at that point. The same result is obtained regardless of the place where the test is made.

Should the compass test give an opposite indication, Fig. 10(B), the circle of force would be counter-clockwise, and current flow away from the observer.

The Left-Hand Rule. A method com-

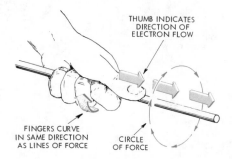

THUMB INDICATES
DIRECTION OF
ELECTRON FLOW

FINGERS CURVE
IN SAME DIRECTION
AS LINES OF FORCE

CIRCLE
OF FORCE

Fig. 11. Left-hand rule for determining direction of magnetic lines set up by electron flow in conductor.

monly used for determining the direction of the magnetic circles is shown in Fig. 11. If the conductor is grasped in the left hand, with the thumb pointing in the direction of current flow, the fingers wrap around the conductor in the same direction as that of the circles of force. If the direction of the circles of force is given, the left hand may be wrapped around the conductor, the fingers representing the circles of force, and the thumb will then indicate the direction of current flow.

Motion of Circles of Force. It has been stated that circles of force have direction, perhaps giving the impression of them in continuous motion. Such is not the case, for although the circle has a tendency to act in a clock-

wise or counter-clockwise direction as the case may be, it is stationary with respect to its source. The circle is a form of energy, and energy is the ability to do work. The energy represented by the circle of force tends to release itself in a certain direction, but it does not revolve.

There is another motion of the circle of force, its radial motion, Fig. 12. When there is no current in the conductor, there are no circles of force surrounding it. When a small current flows, there will be a circle of force at F_1. If the current increases, the circle will increase in size until it occupies the position at F_2. A further increase in current will cause it to expand until it occupies the position at F_3.

As the original circle of force moves from F_1 to F_2 and F_3, new circles take its place at F_1 and F_2. In other words, as current increases, the number of circles of force increases so the outer ones are farther and farther from the conductor.

Should current flow cease, as by opening a switch, the circles shrink quickly until they no longer exist. This growth and reduction in the size of the circles of force are usually termed the *expansion* and *collapse* of the lines of force.

Magnetic Effect of Current Flowing in Wire Loop

In Fig. 13 a piece of copper wire is bent to form a loop. The loop is connected to an electrical generator or other source of current so that current flows as shown by the arrows. Applying the left-hand rule so that the thumb indicates current flow, the fingers will wrap around the wire so that they are on the inside of the loop at all times.

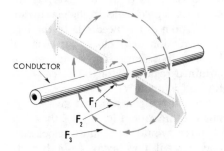

CONDUCTOR

F_1
F_2
F_3

Fig. 12. Showing expansion of lines with increase in current flow.

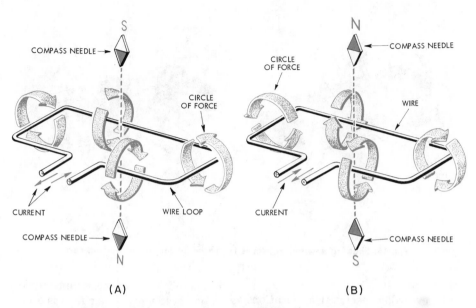

Fig. 13. Using compass needle to show magnetic effect of current flow in wire loop.

Thus, the direction of the circles of force inside the loop will always be downward, as shown in Fig. 13(A).

If a compass is held over the loop in the center of the open space, the N pole of the needle will be attracted. Were the loop of wire a bar magnet, the needle would show the upper part of the loop an S pole. Actually the wire loop has become the equivalent of a bar magnet, the force of all the circles adding up to produce an S pole at the top.

When the compass is held near the center of the lower face of the loop, the S pole of the needle is attracted, Fig. 13(A), indicating that an N pole has been formed there. If current flow is cut off, the compass test will show nothing. The magnetic condition then, is the product of current flow. It is called *electromagnetism,* and the wire loop represents a simple form of electromagnet.

Reversing Direction of Current in Loop. Should the current in the loop be reversed, Fig. 13(B), the left-hand rule show the circles of force moving upward inside it. The S pole of the compass is now attracted toward the upper surface of the loop, showing an N pole there and an S pole underneath.

In Fig. 13(A), where current is in a counter-clockwise direction, an S pole is formed at the upper face of the loop. In Fig. 13(B), where current flow in the upper face is clockwise, an N pole is formed there.

It is best to concentrate on remembering a single relationship, since one follows from the other. If the observer is always in such position that he is looking at that surface of the loop in front of which he has mentally placed himself, the question of whether it is the upper or lower face is unimportant. Under this condition, an easily remembered rule may be stated: *Clockwise direction of current forms an N pole.* Knowing this rule, it is a simple matter to tell at a glance the polarity of an

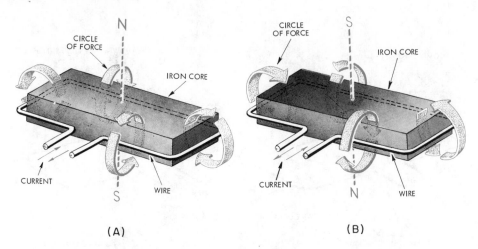

(A)

(B)

Fig. 14. Showing how magnetism is increased by the use of an iron core.

electromagnet.

In everyday electrical applications, the simple electromagnet is met with infrequently, because the magnetic poles formed by the loop of wire alone are too weak for practical purposes.

Effect of Introducing Iron Core

The magnetizing effect of a wire loop may be greatly increased by the use of an iron core. Fig. 14(A) shows a wire loop around a rectangular core of iron. Since the current in the loop flows clockwise when viewed from the top an N pole would be expected there if it were a plain loop. Introduction of the iron core does not alter this fact. An N pole will be formed at the upper face of the iron core, its strength many times as great as it would be without the iron core.

There are two reasons for the tremendous increase in magnetism. The iron provides an easier path for the circles of force, and the magnetic lines induce domains in the iron to assume end-to-end positions.

When the direction of current flow is reversed, Fig. 14(B), polarity of the

iron core is reversed also. Since the direction of the magnetic circles is downward inside the loop, domains arrange themselves so magnetic lines emerge from the lower face of the core forming an N pole there, and an S pole on top. Knowing the direction in which current is flowing when observed from one face or other of the loop, the polarity at the end is readily determined with the rule: *Clockwise direction of current forms an N pole.*

Increasing Strength of Electromagnet. The quantity of magnetism produced in a piece of iron depends upon *permeability,* (magnetic conductivity), of the iron and the strength of the magnetizing force. The easiest way to increase the magnetizing force is to increase current flow. The unit of magnetic force which corresponds to the volt as a unit of electrical force is the *ampere-turn.* By *turn* is meant a complete loop around the core. In Fig. 14, where the wire makes one complete loop around the iron core, it is said to consist of one turn. If 1 ampere of current flows in the loop, the magnetic force established is 1 ampere × 1 turn,

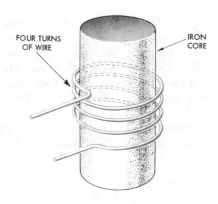

FOUR TURNS OF WIRE

IRON CORE

Fig. 15. Showing an iron core with four turns of wire. Strength of electromagnet is determined by number of ampere turns.

or 1 ampere-turn. With 2 amps flowing through the loop, there are 2 × 1, or 2 ampere-turns. In the same way, 5 amps produce a force of 5 ampere-turns. With an increase in current strength, an increased number of magnetic circles are formed and a greater number of domains are influenced.

The same result may be obtained by increasing the number of turns while the current remains at the original value. A loop of more than a single turn is usually termed a *coil*. The coil surrounding the iron core in Fig. 15 has 4 turns. When current flows in the coil, 4 times as many circles of force result as in the case of 1 turn. The magnetic force for a current of 1 ampere becomes 1 ampere × 4 turns, or 4 ampere-turns. The current of 1 ampere flowing through a coil of 4 turns produces the same quantity of magnetism as 4 amperes flowing through a coil of 1 turn. Eight times as much magnetism will be produced by a current of 2 amperes flowing through 4 turns as by a current of 1 ampere, and 20 times as much by a current of 5 amperes. Electromagnetic coils may have hundreds or thousands of turns.

Factors Determining Strength of Magnetism

Although it is often required to find the value of current, voltage, or resistance in an electrical circuit, equivalent magnetic calculations are seldom made in practice. For the sake of completeness, however, the relationships between magnetic force, resistance, and density of resultant magnetic lines should be understood. First, it should be noted that when a great number of lines of force is involved, the lines are designated, collectively, as *magnetic flux*.

The formula expressing the above relationships is similar to the: $E = IR$ of Ohm's law:

$$IN = kDR_e$$

where

$IN =$ Amps × Number of turns

$D =$ Density (lines per sq in in air)

$R_e =$ Magnetic resistance, or reluctance

$k =$ .313, a multiplying factor

As mentioned before, the term *re-*

luctance is seldom used, the term permeability being more common. *Permeability,* or magnetic conductivity, varies somewhat with density of magnetic lines in a particular sample. For example, the permeability of a certain grade of iron may be 1000 at a density of 20,000 lines per sq in, whereas it may be only 800 at a density of 40,000 lines. Curves from which values of permeability may be read for any given density are prepared by manufacturers.

Determination of Flux Density

When no magnetic core is present, the above formula will read:

$$IN = .313 \times D \times L,$$

where L = length in inches, and R_e = 1, or $IN = .313DL$. When an iron core is present, it reads:

$$IN = \frac{.313DL}{P}$$

where P = permeability, which is unity in air or a vacuum.

The formula may be rearranged to give D, as follows:

$$D = \frac{IN \times P}{.313 \times L}$$

This is the more common form, and will be used to determine the value of flux in the magnetic path represented in Fig. 16. Cross-sectional area of the circuit is 2-in by 4-in, length 6-in, number of amp-turns 100, and permeability given by the manufacturer's data for an assumed probable density of 8000 lines per sq-in as 1500. Therefore:

$$D = \frac{100 \times 1500}{.313 \times 6} =$$

7980 lines per sq-in (approx.)
Total flux = $2 \times 4 \times 7980 =$ 68,830 lines of force.

A probable density of 8000 lines per sq-in was assumed on the basis of experience, but if the result had differed greatly from this amount, a new calculation would have been necessary, using a closer assumption.

Other Aspects of Magnetism

When a nickel alloy bar is magnetized, agitation of atoms within the do-

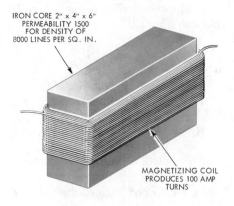

IRON CORE 2" x 4" x 6"
PERMEABILITY 1500
FOR DENSITY OF
8000 LINES PER SQ. IN.

MAGNETIZING COIL
PRODUCES 100 AMP
TURNS

Fig. 16. Flux calculation.

main structures tends to increase its dimensions, making it longer, and increasing its cross-section, both only in small degrees, yet scientifically measurable. This effect is termed *magnetostriction,* a practical example of which will be given in a later chapter.

Another effect is *magnetoresistance,* which results when certain paramagnetic substances are placed in a strong magnetic field, the magnetic lines agi-

tating the atomic structure to the point where resistance increases.

A common industrial application of magnetism makes use of powder patterns to examine steel forgings or castings. Badly disarranged domain groupings on the surface of an object, which has been placed in a strong magnetic field, may indicate the presence of cracks, voids, or other imperfections deep inside.

REVIEW QUESTIONS

1. What activities of the electron are involved in producing magnetic effects?

2. State two characteristics of the line of force emanating from a moving electron.

3. Describe the easiest way for an electron to change its direction of spin.

4. How many positively-spinning electrons will be found in a complete shell of eight?

5. Is iron paramagnetic?

6. Through what kind of material do magnetic lines pass with the greatest ease?

7. Which atomic shells are usually incomplete in ferromagnetic substances?

8. What is a domain?

9. Are domains visible in a pattern of iron filings?

10. Are domains present only at the surface of an iron bar?

11. State the law of magnetic attraction and repulsion.

12. What is the principal difference between a temporary magnet and a permanent one?

13. Explain magnetic induction.

14. How does a diamagnetic substance behave when placed in a strong magnetic field?

15. Is wood a diamagnetic substance?

16. Describe the left-hand rule.

17. If the current is seen to be counter-

clockwise in a loop of wire, what polarity exists at this surface?

18. How much magnetic force is created by a current of 5 amps flowing in a 6-turn coil?

19. Looking down upon the end of an electromagnet, current in the coil circulates clockwise. What is the polarity here?

20. How many ampere-turns are needed to produce a flux of 10,000 lines per sq-in in an air path 1-in long?

Chapter 6

Simple Electrical Generators

6

The term simple, as applied here, refers mainly to lack of moving parts, the principles behind operation of a simple generating device being sometimes rather complicated. Production of static electricity has already been discussed. This chapter deals with a number of means for creating a flow of electrons, some commercially successful, others not yet at this stage of development. The first, and up to this time one of the most practical of them, uses chemical reactions.

PRIMARY CELLS

Elementary Voltaic Cell

Invention of the voltaic cell started man along the road to electrical knowledge. The term *voltaic* was obtained from the name of the Italian scientist, Alessandro Volta, who first devised a

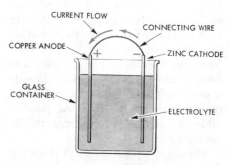

Fig. 1. An elementary voltaic cell.

successful *cell* of this type. Basically, a voltaic cell consists of two dissimilar metals which are immersed in a chemical solution called an *electrolyte*. Fig. 1 represents such a unit.

A glass container holds a quantity of electrolyte, in this case sulfuric acid diluted with water. Sulfuric acid is made up of sulfur, hydrogen, and oxygen and is chemically indicated as H_2SO_4. The positive (+) element, or plate, of the cell is made of copper and is called the *anode*. The negative (−) plate made of zinc is called the *cathode*. When a circuit is connected between the plates, current flows.

Principles of the Elementary Cell

The electrolyte removes electrons from the positive plate and deposits electrons on the negative plate throughout the whole time the cell is functioning. The acid breaks up chemically into two kinds of charged atomic particles, or *ions,* one positive the other negative.

The electrolyte starts working on the plates as soon as the two metals are placed in the solution, and a certain amount of action takes place. But it ceases shortly, and nothing more happens until a wire is connected from one electrode to the other as shown in the figure. At this time, negative charges on the zinc plate repel electrons into the wire, and these free electrons pass along the conductor because of attractive forces on the copper plate, which is in need of electrons. Meanwhile, ions in the electrolyte become activated again, removing electrons from the positive plate and depositing electrons on the negative plate.

Although it presents an over-simplified picture of actual processes, one may look upon the operation as a transfer of electrons from the positive plate directly to the negative plate inside the cell, and a transfer of electrons from the negative plate to the positive plate by way of the external conductor, an enormous quantity of molecules taking part in this action.

Operating Features of the Elementary Cell. The voltage or pressure of this simple cell is approximately 1 volt. This voltage may not be raised by increasing the size of the plates or the amount of electrolyte. Voltaic cells may be made with metals other than copper and zinc. Some of these will form pairs which have a higher voltage than copper and zinc, while others will form pairs having a lower voltage. In any case, the voltage established by any pair is a fixed quantity, for it appears that any two metals create only a certain total amount of electrical force, and this total force is the electromotive force of the cell.

Internal Resistance

The current in a circuit depends upon the total resistance. This resistance consists of two parts: the internal and the external. Internal resistance is that within the cell. Ions encounter opposition to movement through the electrolyte as do electrons in their passage through a conductor.

After the cell has operated for a time,

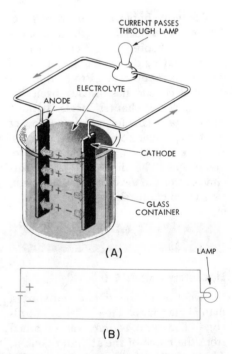

Fig. 2. In (A) a lamp is in series with a cell. In (B) the circuit is shown in schematic form.

its internal resistance increases considerably because of hydrogen gas bubbles that collect on the copper plate. These bubbles prevent the electrolyte from making contact with the anode and, consequently, stop the transfer of electrons from the anode to hydrogen ions. This development is known as *polarization*. It may be counteracted, as will be seen later, by the use of chemical substances which absorb the hydrogen bubbles.

The zinc plate gradually dissolves into the electrolyte as the cell is used and must, therefore, be renewed periodically. The copper plate remains altogether unaffected during operation of the cell.

The external resistance of the circuit is that which is connected across the terminals of the cell, lamps, resistors, and the like.

Fig. 2(A) represents a circuit consisting of a cell and a small lamp. The lamp and the conducting wires make up the total external resistance. A cell is represented in a schematic diagram by a short line and a long line drawn parallel, as in Fig. 2(B). The long line is always the positive terminal, the short line the negative. Drawn in schematic form, the circuit of Fig. 2(A) is shown in Fig. 2(B).

The Leclanché Cell

A cell of commercial importance is the *Leclanché* cell, Fig. 3. The positive electrode is made of carbon, while the negative is of zinc. The electrolyte is a solution of water and ammonium chloride. This cell creates a pressure of approximately 1.5 v. In operation, hydrogen bubbles collect upon the carbon electrode, causing polarization unless steps are taken to prevent it. It is

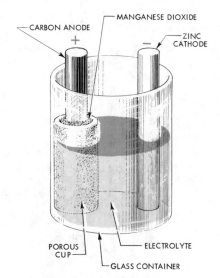

Fig. 3. Leclanché cell.

customary to employ a substance called manganese dioxide. In Fig. 3 a porous cup containing a quantity of manganese dioxide prevents the electrolyte from depositing hydrogen bubbles upon the surface of the electrode.

The Dry Cell

The Leclanché cell provided the basis for the modern *dry cell* which has such a vast field of application. Actually, the cell is not dry but appears so when the top has been sealed with compound, and a cardboard cover placed around it, Fig. 4. The positive carbon electrode is placed in the middle of the unit, the negative zinc electrode forming the outer shell of the cylindrical container. Inside the zinc sheath, a layer of cardboard is soaked with ammonium chloride paste which serves as the electrolyte. Between the carbon rod and the cardboard is a mass of soft material containing manganese dioxide,

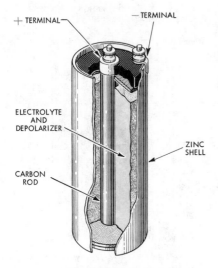

+ TERMINAL

− TERMINAL

ELECTROLYTE AND DEPOLARIZER

CARBON ROD

ZINC SHELL

Fig. 4. The dry cell.

the depolarizing agent, and ammonium chloride mixed with zinc chloride. A terminal lug fastened to the carbon rod, forms the positive (+) connection, another fastened to the zinc shell forms the negative (−) connection.

The voltage of the unit is the same as that of the Leclanché wet cell, or 1.5v. Dry cells are useful for applications where small amounts of current are required for short periods of time. Where required to give even a small amount of current continuously, their useful life is shortened.

In operation, the depolarizing agent gradually loses its effectiveness through absorption of hydrogen gas, and the liquid materials eventually dry out. Both of these occurrences have the effect of increasing internal resistance of the cell and thus lowering its useful output. The processes of deterioration are speeded up by heating. If the cell is required to carry a great deal of current, or is kept in use for long periods of

time, the heat thus generated greatly shortens its life.

Standard Cells

The voltage of an ordinary cell, wet or dry, alters with passage of time, and changes according to conditions of use such as varying temperature. For this reason, they are unsuitable for precise electrical measurements in laboratory operations. The *Clark cell* was developed with a view to providing a unit which maintained a definite voltage that could be utilized as a standard during scientific experimentation.

In this respect, the Clark cell offered considerable improvement over the others, but it too, displayed severe limitations in practice. Weston, profiting by Clark's efforts, devised a *standard cell*, Fig. 5, which has proven itself

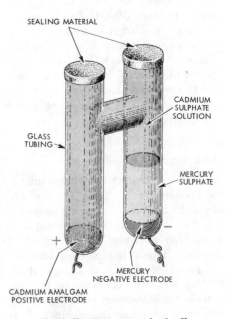

SEALING MATERIAL

CADMIUM SULPHATE SOLUTION

GLASS TUBING

MERCURY SULPHATE

+

−

MERCURY NEGATIVE ELECTRODE

CADMIUM AMALGAM POSITIVE ELECTRODE

Fig. 5. The Weston standard cell.

during a period of more than half a century.

The cell is shaped like a capital H, with the positive electrode, *Cadmium Amalgum* in the bottom of one leg, the negative electrode, *Mercury,* in the other. The electrolyte is *Cadmium Sulphate,* the depolarizer *Mercury Sulphate.* Upper portions of the twin tubes are carefully sealed to exclude outside contamination and to insure against evaporation of liquid. The pressure is slightly less than 1.02 volt at 20°C, and remains constant, but the cell can deliver only the minute current needed for close laboratory work.

Fig. 6. A hydrogen-oxygen fuel cell. This unit has a continuous duty power rate of 1KW with a nominal voltage of 28 volts.
Courtesy Union Carbide Corp.

Fuel Cells

Scientists have been engaged for many years in the attempt to obtain electrical power from fuels without resorting to intermediate apparatus such as steam turbines, or electrical generators. Advances in technology and materials have revived the possibilities of the fuel cell, Fig. 6, as an efficient source of large blocks of electrical power. This device was first developed by Sir William Grove in 1839, but materials available at that time made their use impractical. Recent advances in these units were brought about in the search for an efficient source of power for space travel, and a Hydrogen-Oxygen fuel cell was used for internal power in the Gemini Spacecraft.

The fuel cell has a number of advantages compared to other means of generating electricity; one being that efficiency levels ranging from 50 to 80 percent are being obtained at the present time with an efficiency of 98 percent being theoretically possible. The steam turbine on the other hand has a maximum of about 40 percent.

A cross section of a modern fuel cell is illustrated in Fig. 7. It consists of a rectangular chamber which is divided into two parts, separated by a barrier which is termed an *electrolyte.* This electrolyte may be a ceramic, *a polymer* (synthetic solid with dense molecular structure), a semi-liquid or paste. In this particular unit, the electrolyte is solid, a porous ceramic with a thin coating of metal sprayed on either face. The essential characteristic of all fuel-cell electrolytes is that they must permit ionized particles of gas to diffuse through them from one face to the other, but reject non-ionized particles.

Hydrogen gas, whose molecular symbol is H_2, is supplied to the lower cham-

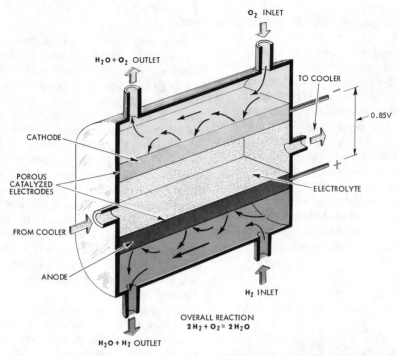

O₂ INLET

H₂O + O₂ OUTLET

TO COOLER

0.85V

CATHODE

POROUS
CATALYZED
ELECTRODES

ELECTROLYTE

+

FROM COOLER

ANODE

H₂ INLET

OVERALL REACTION
2H₂ + O₂ = 2H₂O

H₂O + H₂ OUTLET

Fig. 7. A cross section of a modern fuel cell similar to the one illustrated in Fig. 6.
Courtesy Union Carbide Corp.

ber of the cell under pressure, the number "2" signifying that two atoms of hydrogen combine to form the hydrogen molecule. Oxygen, symbol O_2, is furnished to the upper chamber at a somewhat lower pressure. The hydrogen molecule which, in its normal state, cannot pass through the material, gives up two electrons at the lower face of the electrolyte. The metallic coating thereby acquires a negative charge and becomes the cathode of the cell, while the hydrogen molecule becomes a hydrogen ion whose symbol is H^{++}, the "+ +" meaning that the ionized molecule now has a positive charge of two units.

This hydrogen ion passes through the

ceramic to the opposite face. When it emerges, it draws two electrons from the metallic coating, giving this face a positive charge. The coating thus becomes the anode of the cell, and the hydrogen ion becomes a molecule of H_2 again. Hydrogen molecules combine with oxygen molecules in the right proportions to form molecules of water.

The nickel screen on either side of the electrolyte is termed a *catalyst,* meaning a substance that assists in a chemical reaction without suffering chemical or physical change itself. In the lower chamber the screen helps electrons escape from their atomic orbits. In the upper chamber, the screen helps H_2 molecules to unite with O_2 molecules. The electrical output of this unit, and that of most fuel cells, is approximately 1 volt.

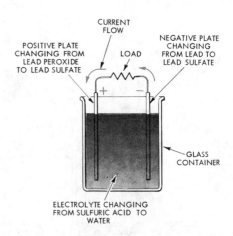

Fig. 8. The chemical changes which occur in a lead cell during discharge.

SECONDARY CELLS

Principle of the Storage Cell

In principle, the storage cell is exactly the same as the voltaic cell. It uses electrodes (plates of different materials) and an electrolyte which carries electrons from the positive to the negative plate. The essential difference between the two types is found in the electrodes. The positive (+) copper electrode of the voltaic cell remains copper so long as the cell exists, and the negative (−) zinc electrode remains zinc, although it must be replaced from time to time as it dissolves.

In the storage cell, the positive (+) electrode changes into a somewhat different metal as it discharges, and the negative (−) plate also changes its nature. When discharged, the plates may be restored to their original condition by a current sent through the cell in the opposite direction. This process is called *charging*. There are, in general, three common types: the lead cell, the nickel-iron, or, Edison cell, and the nickel-cadmium cell.

The Lead Cell Discharge

The lead cell employs sulfuric acid as the electrolyte and electrodes made of lead compound. In a charged state, the positive electrode is lead peroxide, the negative electrode pure lead.

When the cell is connected to a load, current flows through the circuit, and chemical changes take place in the cell as shown in Fig. 8. The positive elec-trode, which was originally composed of lead peroxide (PbO_2), is changed into lead sulfate ($PbSO_4$). The negative electrode, on the other hand, changes from pure lead (Pb) to lead sulfate. And the electrolyte is gradually transformed from sulfuric acid (H_2SO_4) into water (H_2O). In this process the negative electrode gives up electrons and the positive electrode absorbs electrons.

Charging the Lead Cell. The discharged cell has been connected to a generator in Fig. 9. Current now flows through the circuit in the direction opposite to that during discharge, passing from the generator to the negative plate through the liquid electrolyte to the positive plate, and thence to the positive terminal of the generator.

As current flows through the cell, chemical changes again occur, the process opposite to that taking place during discharge. Electrons are forced onto the negative plate by the genera-

79

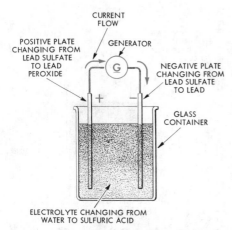

Fig. 9. The chemical changes which occur during the charging of a lead cell are shown here.

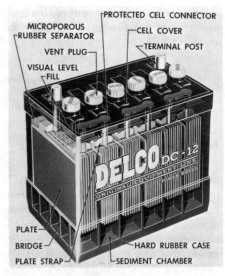

Fig. 10. The internal construction of a lead-cell storage battery.

Courtesy Delco-Remy Div.—General Motors Corp.

tor, and electrons are removed from the positive electrode. In this process, the negative plate is changed from lead sulfate ($PbSO_4$) back to pure lead (Pb). At the same time, the positive plate is changed from lead sulfate ($PbSO_4$) to lead peroxide (PbO_2), and the electrolyte is returning to its originally acid state (H_2SO_4).

Commercial Lead Cell. The commercial lead cell has a number of positive and negative plates, Fig. 10. They are arranged in two closely interwoven groups. The placement is such that a negative plate lies on either side of each positive plate, the two sets being prevented from touching by means of thin wooden or hard rubber sheets called separators. There is one more negative than positive plates.

In one form, the surfaces of both positive and negative plates are ridged with a special tool in order to expose more surface to the electrolyte. Other cells use this type of negative plate but employ a positive unit which has ribbons of lead set into holes in a metallic grid. The automobile battery makes use of a grid type of plate in which the spaces are filled with a soft lead paste, which is somewhat porous. The goal of all these types is to obtain the greatest amount of working area for the electrolyte.

After a certain amount of operation, material which peels off the plates gathers in the bottom of the cell as sediment, to such a depth that it forms a connection between positive and negative plates. For this reason, the plate assembly is elevated from the bottom. The enclosing cases for stationary commercial cells are made of glass; for portable applications, such as in automobiles, they are hard rubber.

Operating Facts about the Lead Cell. The voltage of the lead cell is approximately 2 v, being slightly greater than this when fully charged, and slightly less when discharged. The state of charge of the cell is not judged, however, by its voltage, for this is not a good indicator. The best method is to check the density of the electrolyte. By *density* is meant the weight of the liquid as compared to the weight of an equal volume of water. As the electrolyte gives up its sulfate content to the plates of the cell on discharge, it decreases in density. It is not necessary to weigh the liquid in order to determine this fact. Instead, a device known as a *hydrometer* is employed.

A hydrometer, Fig. 11, consists of an air-filled glass bulb with a lead weight in the bottom. When placed in the cell, it sinks into the liquid to a distance which depends upon the density of the liquid. The electrolyte of a fully charged cell, will read, perhaps, 1275. When the cell is fully discharged, it will read approximately 1150. The hydrometer is usually placed inside a syringe so that a quantity of the electrolyte can be drawn up into it.

The reading thus obtained is called the *specific gravity* of the electrolyte. It affords an accurate indication of the general state of charge of the cell, the density slowly changing between full charge and discharge. When a low specific gravity reading of 1150 is reached, it is not safe to discharge the cell further. After this value has been reached, further discharging will result in permanent *sulfation* of the plates. This is when the lead sulfate becomes crystalline and is no longer chemically active.

Capacity of Cell. The ability of a storage cell to hold an electrical charge is expressed in *ampere-hours.* A 100 ampere-hour cell will supply 12.5 amps for a period of 8 hours (12.5 × 8 = 100). Hence, an ampere-hour is simply 1 amp × 1 hour, or ½ amp × 2 hours. Storage cells are usually rated on an 8-hour basis. Thus, a 200 ampere-hour cell would put forth a current of 25 amps steadily for a period of 8 hours. This ability of the cell to supply electricity is termed its *capacity.*

Storage Batteries. Storage cells are not often used singly. They are invariably assembled into batteries which consist of two or more cells. The 6-volt automobile battery consists of three separate cells connected in series. The 12-volt battery consists of six cells connected in series. In commercial applications where stationary batteries are required, as in stand-by service for power stations, they may be grouped into batteries of 50 or more cells connected in series.

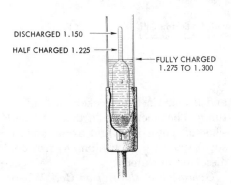

DISCHARGED 1.150

HALF CHARGED 1.225

FULLY CHARGED 1.275 TO 1.300

Fig. 11. The density of a battery's electrolyte is checked with a hydrometer.

The Edison Cell

Fig. 12 shows a cutaway view of the Edison Cell. The positive plates are made up of a number of nickelled tubes which contain the active material, nickel oxide. In order for the electrolyte to contact the active material, the tubes

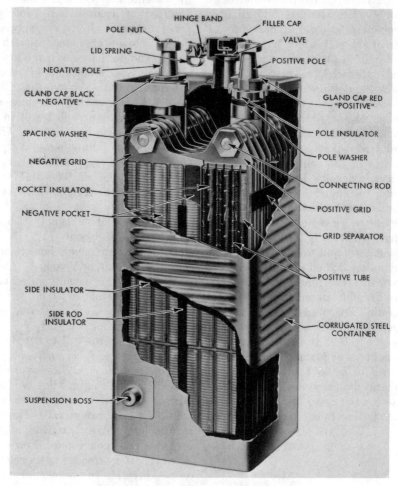

HINGE BAND
FILLER CAP
POLE NUT
VALVE
LID SPRING
POSITIVE POLE
NEGATIVE POLE
GLAND CAP BLACK "NEGATIVE"
GLAND CAP RED "POSITIVE"
SPACING WASHER
POLE INSULATOR
NEGATIVE GRID
POLE WASHER
POCKET INSULATOR
CONNECTING ROD
NEGATIVE POCKET
POSITIVE GRID
GRID SEPARATOR
POSITIVE TUBE
SIDE INSULATOR
SIDE ROD INSULATOR
CORRUGATED STEEL CONTAINER
SUSPENSION BOSS

Fig. 12. A cutaway view of the Edison Cell.
Courtesy Thomas A. Edison, Inc.

are thoroughly perforated. To obtain improved electrical conductivity the nickel oxide is alternated with layers of nickel flake at the time it is tamped into the tube.

The negative plate is of similar construction except that finely divided iron oxide is used as the active ingredient

and is contained in pockets instead of tubes. The electrolyte in this cell consists of potassium hydroxide and a small quantity of lithium hydroxide, dissolved in water.

Operation of the Edison Cell. When the cell is charged, the iron oxide is changed into iron, and the positive plate is nickel peroxide. As the cell delivers current to a load, the positive plate changes from nickel peroxide to nickel oxide, and the negative plate changes from iron to iron oxide.

Charging reverses this process, namely, the positive plate again becomes nickel peroxide, and the negative plate becomes iron. The electrolyte does not take part in these changes, and the density varies only slightly during these changes.

Facts about the Edison Cell. It is not possible to determine the state of charge of an Edison cell by means of a hydrometer reading because density of the electrolyte varies but slightly between charge and discharge. The voltage is somewhat less than that of the lead cell, being approximately 1.4 v when fully charged and 1 v when discharged. The capacity of this cell, although rated in ampere-hours like the lead cell, is usually based on a 5-hour discharge period. It is more rugged than the lead cell, and it will stand considerable abuse. It may be roughly handled; it may be overcharged and overdischarged without permanent damage. Edison cells can be frozen solid, yet operate without difficulty when thawed out.

Nickel-Cadmium Cell

The *nickel-cadmium cell* resembles the Edison in many respects, both in general appearance and operating principles. Its positive plate is of the same material as the Edison's, but the negative plate is a cadmium-iron mixture. The alkaline electrolyte contains a preservative chemical that protects the elements from deterioration. As in the *nickel-iron unit,* density of the electrolyte does not change during charge and discharge cycles, while oxygen is being transferred from one plate to the other.

The construction is extremely rugged, permitting rapid charge and discharge, overcharge, or even short-circuiting without permanent damage. These cells are made in sizes large enough to handle starting load of huge diesel engines, and small enough for use in electric hand tools, or even electric razors. The razor battery is less than 2-in long, and $\frac{1}{2}$-in in diameter.

Cells Now in Process of Development

The caesium-diode cell, whose function is to convert heat energy directly into electrical energy, holds great promise if certain technical difficulties are overcome. One form of this unit has a tantalum electrode, called the emitter, and a copper electrode, called the collector. Caesium metal in the bottom of the enclosing vessel is exposed to great heat, such as from an atomic reactor, and is vaporized thereby. The caesium vapor, largely ionized, condenses on the two electrodes, more on the copper than on the tantalum, setting up a potential difference between the two elements. Electrical pressures somewhat less than 5 volts are presently obtainable, with a maximum operating efficiency of 10 percent.

The bacterial cell is still undergoing laboratory experimentation. A type currently under test is in the form of a U-tube similar to that of the Weston Standard Cell. Contrasting types of bacteria occupy solutions in either vertical member, while osmotic membranes seal off the horizontal member which contains an electrolyte. When further improvements have made the device commercially practicable, the unit will take its place in the category of fuel cells.

OTHER MEANS OF GENERATING ELECTRICITY

Thermoelectricity

When two dissimilar metals are welded together, and the junction between them is heated or cooled, an electrical pressure is created. Fig. 13 shows such a union between iron and copper wires. It appears that the joining process disturbs atomic orbits at this point so that outer electrons in both metals are but loosely held. Under this condition any small addition or subtraction of energy will set them free.

In Fig. 13(A), the heat of the flame

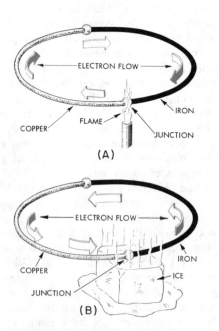

(A)

(B)

Fig. 13. The operation of a thermocouple is illustrated here. In (A), heat creates a current flow. In (B), cooling results in a current flow opposite to that created by heat.

provides additional energy, electrons from copper atoms being repelled into the conductor. Loss of electrons by copper atoms creates a positive force that draws electrons from the iron, the process continuing as in any other type of generating device.

When the junction is cooled, Fig. 13(B), the lowered temperature subtracts energy, affecting atomic vibrations, and so weakening the hold of iron nuclei upon outer electrons that flow starts in the opposite direction from before.

Other pairs of metals than iron and copper form *thermocouples*. Nickel-iron alloys are commonly paired with dissimilar metals, each particular combination producing its own characteristic temperature-voltage curve of values. A number of thermocouples are usually grouped together so that all the junctions may be heated at the same time. When the free ends are properly joined so that the tiny voltages may add up, the unit is called a thermopile.

Fig. 14 shows two common applications of the thermopile. In View (A), the thermopile is connected to a sensitive electric meter whose scale is marked, or calibrated, to indicate degrees of temperature inside an electric furnace. The junction end of the thermopile is exposed to severe heat there, the volume of electron flow depending upon intensity of the heat.

The common *fail-safe pilot flame* thermal device used in connection with residential gas furnaces is shown in Fig. 14(B). If the pilot flame is lighted, sufficient voltage is generated by the device to open the gas valve when the switch closes. If the pilot flame goes out, lack of generated voltage renders the valve inoperative.

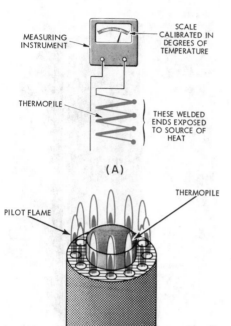

(A)

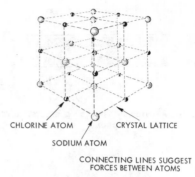

CHLORINE ATOM · SODIUM ATOM · CRYSTAL LATTICE

CONNECTING LINES SUGGEST
FORCES BETWEEN ATOMS

Fig. 15. The atomic arrangement in the crystaline structure of salt.

(B)

Fig. 14. Two applications of the thermopile. (A) illustrates its use in measuring furnace heat, (B), its use in a fail-safe pilot flame for gas furnaces.

Piezoelectricity

Forces exist between atoms inside a given sample of material because of interlaced electronic orbits and mutual repulsion between positively charged nuclei. Stability or equilibrium can be maintained only through certain arrangements characteristic of each substance. Many elements and compounds have what is known as a *crystalline structure*, the atoms being assembled in layers or patterns. X-ray and spectroscopic investigations show that such arrangements are always the same for any given kind of material.

Fig. 15 shows the crystalline structure of common salt, atoms of sodium and chlorine falling into a cubical design in which atoms of sodium are adjacent to atoms of chlorine in all directions. Countless numbers of such cubes unite to form larger crystals that may be seen with the aid of a laboratory microscope.

Individual atoms vibrate, and their electrons follow interlacing orbits, but they stay in these same relative positions. In most of these substances, the formations resists any change by heat, pressure, or other external influences. A few, however, like *tourmaline, quartz,* and *rochelle salt,* are pressure sensitive, releasing electrons when their crystals are compressed along a certain line or axis.

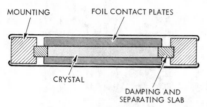

MOUNTING   FOIL CONTACT PLATES

CRYSTAL

DAMPING AND
SEPARATING SLAB

**Fig. 16. A cutaway view of a crystal micro-
phone.**

**Fig. 17. A solar power pack made up of silicon
cells. The small size of the unit may be judged
by comparison with the pencil in the photograph.**
Courtesy International Rectifier Corp.

This principle is applied in the crystal microphone, Fig. 16, voice vibrations being transmitted to a thin disk of rochelle salt by means of a flexible metallic diaphragm. As the crystalline material is alternately compressed and released, electron flow is established in the circuit of which the sensitive disk is a part.

Photoelectricity

When a ray of light strikes upon certain materials, electrons are set free. Those which show this quality to a marked degree are termed *photoelectric* substances. It is likely that all materials are more or less photoelectric although the metals are comparatively far more sensitive than non-metals. Zinc, of the common metals, is quite responsive, caesium and selenium more so. This subject will be discussed in some detail in a later chapter.

Solar Batteries

Man's attempts to harness energy from the sun have met with comparatively little success. Parabolic collectors have been devised for the purpose of driving small steam engines, but variable weather conditions, clouds, and the fact that the collector must be geared to follow the sun make such efforts impractical for most applications.

On a smaller scale, silicon solar batteries, Fig. 17, are practical for a number of applications. The *Telstar* and *Courier* satellites made use of solar cells to provide operating power for a number of their devices. The telephone companies are also utilizing these devices to provide a part of the operating current for the system. Mounted on the top of telephone poles where they will receive the maximum amount of sunlight, they also keep storage batteries charged for use during those days when the sun does not shine.

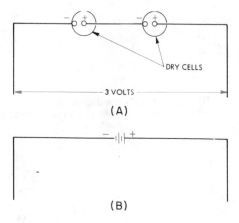

(A)

(B)

Fig. 18. (A) illustrates the series connection of cells. In (B), the equivalent circuit is shown in schematic form.

CELLS AND BATTERIES IN CIRCUITS

Series Combination

A single unit is properly termed a *cell,* a group of cells, a *battery.* Many

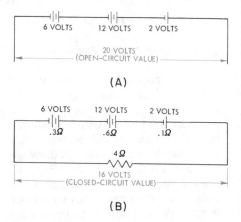

(A)

(B)

Fig. 19. Series connection of cells with different voltages.

applications require a voltage greater than that of a single unit, and a battery must be employed. The series arrangement of Fig. 18(A) is much like that of series resistors except that attention must be given to polarities, the positive terminal of one cell being attached to the negative terminal of the other. Any number of cells may be so connected if the polarity sequence is observed.

Cells in the automobile battery, already mentioned, are grouped in this manner, three being in series for a 6-volt battery, six for a 12-volt one. Any type of generating device may be treated in this manner. The separate thermocouple of the thermopile, for example, were in series, and the two generators of the three-wire circuit discussed earlier.

Fig. 18(B) shows the manner in which cells are shown in schematic drawings, only two or three pairs of short and long vertical lines being used regardless of the number of cells they represent.

Voltage and Current

The voltage obtainable from the series combination is equal to the number of units times the voltage of one. Thus, if the cells are each rated at 1.5 volts, the voltage of twelve in series would be: 12×1.5 volts, or 18 volts. It is not necessary for voltages of series units to be the same. In Fig. 19, a 6-volt, a 12-volt, and a 2-volt unit are connected in this way to provide a total of 20 volts, the individual voltages being merely added together.

This is the open-circuit voltage shown in Fig. 19(A), that is, the reading of a voltmeter when no current is being delivered. The internal resistance of large rotating generators with voltages of 120, 240, or 600, as the case may

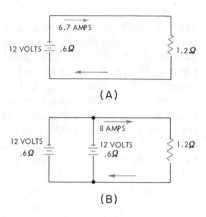

(A)

(B)

Fig. 20. Circuits illustrating the effect of paralleling batteries to provide greater current capacity.

be, is so low that it need not be considered in estimating current flow. Where small units are involved, whose resistances are relatively large, the factor must be taken into account.

The resistances in Fig. 19(B) are: .3-ohm, .6-ohm, and .1-ohm respectively, adding up to 1-ohm. With a 4-ohm resistor in circuit, the current is:

$$\frac{20 \text{ volts}}{4 \text{ ohms} + 1 \text{ ohm}} = 4 \text{ amps.}$$

The voltage drop in the cells is:
$$4 \text{ amps} \times 1 \text{ ohm} = 4 \text{ volts,}$$

and, the closed circuit voltage is:
$$20 \text{ volts} - 4 \text{ volts} = 16 \text{ volts}$$

Parallel Combination

A 12-volt battery with an internal resistance of .6-ohm, Fig. 20(A), supplies a 1.2-ohm load. Current flow equals:

$$\frac{12 \text{ volts}}{1.8 \text{ ohms}} = 6.7 \text{ amps.}$$

If another 12-volt battery with internal resistance of .6-ohm is connected in parallel with the first one, as in Fig. 20(B)

the two negative terminals being attached to the upper supply wire, the two positive terminals to the lower, the combined resistance of the two batteries is one-half that of either alone. Current flow is:

$$\frac{12 \text{ volts}}{1.5 \text{ ohms}} = 8 \text{ amps.}$$

where the maximum safe output from a single unit is not sufficient for load demands. If the load requires 20 amps, and the output of a single battery is limited to 10 amps, a second one may be connected in parallel as shown.

Importance of Battery Characteristics

Several precautions are necessary in paralleling cells. They must have the same voltage and the same internal resistance. If a battery with a small voltage is used in parallel to one with a large voltage, the smaller will draw current from the larger. In effect, the smaller voltage cell will act as a load resistance. One may go further and say that the two cells should be of the same age. If one becomes weak, the load will be forced onto the stronger.

REVIEW QUESTIONS

1. Describe the simple voltaic cell.
2. What term denotes the collection of hydrogen bubbles on the copper plate?
3. What term is applied to the negative plate?
4. Describe a dry cell.
5. What is the voltage of a dry cell?
6. What is meant by the term *standard cell*?
7. What metals are used for anode and cathode of the Weston cell?
8. Name a successful type of solar battery.
9. What form of electrolyte is used in a modern fuel cell?
10. What kind of atoms pass readily through the fuel cell electrolyte?
11. Of what material is the positive plate composed in a charged lead cell?
12. Of what material is the negative plate in the charged cell?
13. What is the composition of the positive plate in a discharged lead cell?
14. How do plates and electrolyte of the Edison cell differ from those of the lead cell?
15. How does the negative plate of the nickel-cadmium cell differ from that of the Edison cell?
16. Describe the flow of electrons from the heated junction of an iron-copper thermocouple.
17. What is piezoelectricity?
18. What is meant by the term *crystalline structure*?
19. How is the voltage of a series group of generating devices determined?
20. Under what conditions may batteries be successfully connected in parallel?

Chapter 7

Direct Current Generators

7

GENERAL PRINCIPLES

Effect of Magnetic Lines on Moving Conductor

Magnetic lines of force are a form of energy. They have power to act upon metallic substances so as to free electrons from planetary orbits when relative motion exists. In Fig. 1, for example, when a copper wire moves across the magnetic field established by pole N, electrons are set free and a flow of electrons is induced. The direction of

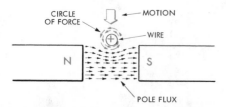

Fig. 1. A conductor moving downward through the flux field of the two poles has a current induced in the direction shown and circles of force in a counter-clockwise direction.

flow, if the ends of the wire are connected to form a complete circuit, is such that magnetic circles will form about the conductor as shown, the circles moving counter-clockwise in this particular case.

Two facts should be noted here. First the direction of the magnetic circles of force is downward on the left of the conductor, the arrows indicating the same direction as the lines of force established by pole N. These magnetic circles therefore tend to crowd into space already occupied by the lines from N. The second point to note is that mutual repulsion exists between magnetic lines which flow in the same direction, just as repulsion exists between like charges of static electricity.

The N lines thus resist the intrusion of lines circling the conductor, trying to repel them, and to push the conductor upward, physical force being required to overcome this magnetic resistance so long as the conductor is in motion. When the conductor stops moving, the opposing force dies out because electrons are no longer set free within the conductor, and the circles of force cease to exist.

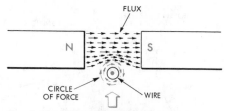

Fig. 2. Reversing the direction of the conductor through the magnetic field also reverses the direction of the current and the circles of force.

The direction of current in the conductor may be determined by means of the left-hand rule already learned. It will be seen that current tends to flow downward, or "into the paper."

Reversing Direction of Motion. If the conductor is moved upward across the N pole, as in Fig. 2, current will be established in the opposite direction. The circles of force will be clockwise, current flow toward the observer, or "out of the paper", and the force of repulsion will be downward. The essential fact to notice is that reversing direction of motion reverses direction of current.

Reversing Magnetic Polarity. Fig. 3 is similar to Fig. 1 except that magnetic polarity has been changed from N to

S. It will be seen that the direction of circles of force surrounding the conductor is opposite from that in Fig. 1. Current flow in the conductor, according to the left-hand rule is toward the observer, or "out of the paper". Hence, reversing the magnetic polarity, direction of motion remaining the same, reverses direction of current in a conductor.

Elementary Generator

The elementary electromagnetic generator consists of a loop of wire which revolves between two magnetic poles, Fig. 4. In the figure, the black conductor moves across the N magnetic pole of the generator, while the white conductor moves across lines from the right, or S magnetic pole. The parts of the loop which are in the same direction as the flux from the poles do not cut magnetic lines of flux since they are moving parallel to them.

Direction of Current. In Fig. 4, the direction of current in the black conductor is seen to be from front to back. In the white conductor the direction is from back to front. The current flows

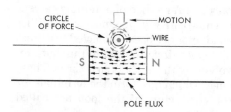

Fig. 3. Reversing the polarity of the magnetic flux reverses the direction of the current and the circles of force.

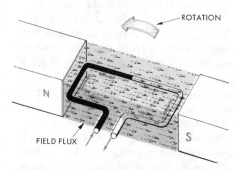

Fig. 4. An elementary electromagnetic generator. In this position the loop is cutting the maximum number of lines of flux and will generate a flow of current.

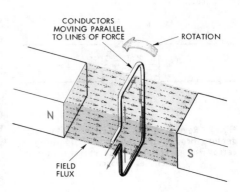

CONDUCTORS
MOVING PARALLEL
TO LINES OF FORCE

ROTATION

N

S

FIELD
FLUX

Fig. 5. The generator loop is outside the magnetic flux field and is not cutting lines of flux. No current will be generated in this position of the loop.

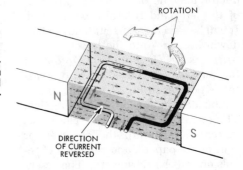

ROTATION

N

S

DIRECTION
OF CURRENT
REVERSED

Fig. 6. Rotation of the loop reverses the position of the black and white conductors from that shown in Fig. 4. The current flow is still clockwise through the loop although its direction through the conductors is reversed.

clockwise from the white to the black conductor, thus completing the circuit.

A quarter turn later as the loop continues to revolve, it reaches the position shown in Fig. 5. Here, the conductors of the loop are no longer cutting lines of force from the magnetic poles but are again moving parallel to them, so that no voltage is generated in them. Current at this instant, therefore, is zero.

Another quarter turn, Fig. 6, finds the loop in a position similar to that of Fig. 4, but with the position of the black and white conductors interchanged. Comparison with Fig. 4 shows that while the direction of current is

still clockwise in the loop, the direction of the current through the black and white conductors has been reversed.

Obtaining Current from the Elementary Generator. Electricity generated in a closed loop would serve no useful purpose. A method of obtaining current is shown in Fig. 7. One end of either conductor has been connected to a brass or copper ring attached to the shaft, revolving with the shaft as it turns. These rings, which now form the terminals for the conductors of the loop are called *collector rings*.

The collector rings are contacted by two stationary pieces of brass which are

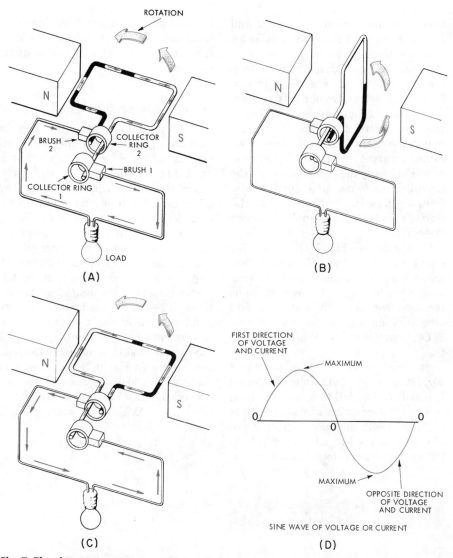

Fig. 7. The elementary generator with brushes and collector rings. Drawings (A), (B), and (C), show current flow through the circuit for half a revolution of the loop. In (D), the output waveform of current and voltage from the generator is illustrated.

termed *brushes*. As the loop and collector rings turn, they are always in contact with the brushes which are connected to the load.

At the instant shown in Fig. 7(A), current flows in the white conductor from back to front, to collector ring 1, through brush 1 to the load. From the load it returns through brush 2 and collector ring 2 to the black conductor loop, completing the circuit.

A quarter turn later, when the loop has reached the position shown in Fig. 7(B), conductors are not cutting lines of force and the current is zero. During the next quarter turn, Fig. 7(C) current in the white conductor is now front to back, current through the loop clockwise. Following the current through the circuit, it is found that current now flows out through collector ring 2, and through the load in a counter-clockwise direction.

Comparison of Fig. 7(C) with 7(A) shows that direction of current flow through the load reverses while the loop makes a half turn. It reverses again as the loop continues through a second zero position similar to that of Fig. 7(B), and back to its original position.

As shown in Fig. 7(D), voltage and current increase from zero to maximum in one direction, fall to zero again, then reverse, following an identical outline in the opposite direction. This pattern of voltage or current, discussed fully in the next chapter, is termed a *sine* wave.

THE DIRECT-CURRENT ARMATURE

Alternating Current and Direct Current

With this elementary generator, current in the load reverses direction twice each revolution and is called an *alternating current*. Alternating current is suitable for the greater majority of tasks met with in practice, but is not desirable for certain types of loads. One example is the charging of a stor-

age battery. A current which flows continuously in one direction, like that from a storage cell is called a *direct current*.

Generating Direct Current

The elementary generator may be made to produce direct current. A device called a *split ring commutator* is used, Fig. 8, which is basically a collector ring sawn in half with the two parts insulated from each other. Referring to the generator in Fig. 9, brushes are located so that each is in contact with one segment of the commutator, and is connected to one end of the load.

As shown in Fig. 9(A), current is clockwise from the black conductor, through the white conductor, the right hand portion of the commutator, the brush, and clockwise through the load.

When the loop has moved to the position shown in Fig. 9(B), current flow through the load has ceased because the conductors are not cutting magnetic lines. At this time each brush is

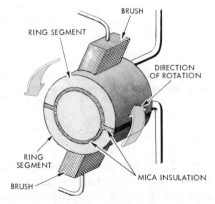

Fig. 8. A split-ring commutator.

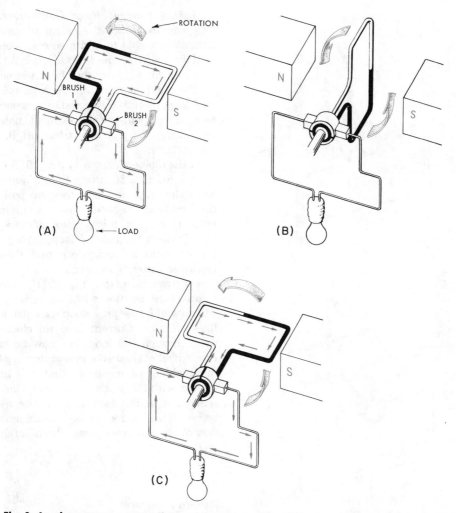

Fig. 9. An elementary generator with commutator and brushes for generating direct current. The loop is shown through half a revolution to illustrate the fact that current flow through the load remains in the same direction regardless of the loop position.

in contact with both halves of the commutator.

When the loop reaches the position of Fig. 9(C), the black conductor and its commutator segment are now in contact with the right hand brush. If

the circuit is traced it is seen that current again flows through the load in a clockwise direction as it did in Fig. 9(A).

The essential point here is that current through the load flows in the same

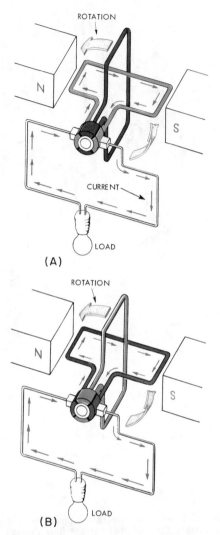

Fig. 10. The addition of a second loop to the generator produces a more continuous flow of direct current through the load.

Making Current Flow Continuous. One way to reduce the extent of zero current values is to add a second loop at right angles to the first, Fig. 10(A). A second loop (grey), has been placed on the shaft at right angles to a colored loop, its ends connected to a second pair of segments; each segment now comprising only one-quarter of the commutator.

At the moment shown in Fig. 10(A), the colored loop is cutting lines of force while the grey loop is in the zero position midway between the magnetic poles. Current in this position is clockwise through the colored loop, through the commutator and brush and then clockwise through the load.

A quarter turn later, Fig. 10(B), the colored loop has moved into the zero position while the grey loop is cutting lines of force. Current is again clockwise, through the loop, but now flows through the shaded commutator segment which has moved so that it is in contact with the brush on the right. Current through the load is clockwise as before. Thus, a reasonably continuous flow of direct current passes through the load.

Modern Armatures

In practice, the rotating member which supports the loops is called an *armature,* Fig. 11. Essentially, it consists of a shaft upon which are mounted a slotted, sheet-steel magnetic *core* and a *commutator.* The *armature winding* has a great number of wire loops, termed *coils,* which are inserted in the *slots.* Fiber wedges are driven between the coils and overhanging slot sides or *teeth* to hold the wires in place. The commutator has a number of copper

direction at all times. In other words, it is a direct current. The flow is not actually continuous, because the loop passes through two zero positions in each revolution; but its direction remains unchanged.

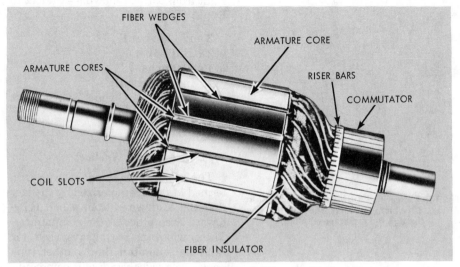

FIBER WEDGES

ARMATURE CORE

ARMATURE CORES

RISER BARS

COMMUTATOR

COIL SLOTS

FIBER INSULATOR

Fig. 11. A generator armature and its associated parts.

segments, commonly known as *bars,* which are insulated from one another and from the shaft by strips of mica or other suitable material.

An Armature Winding

Fig. 12 presents a developed illustration of an armature winding, core slots being shown radially, along with an end view of the commutator bars. Individual coils usually have a great number of loops or turns, but are shown here by a single turn. Two half-coils are installed in each slot, the top half being shown in solid color lines, the bottom half in dashed color lines. Coil ends are attached to commutator bars in such a way as to provide a closed circuit through the winding. It is possible to start at any commutator bar, such as *A,* and trace a path that includes every

conductor before returning to the starting point.

As the armature revolves between

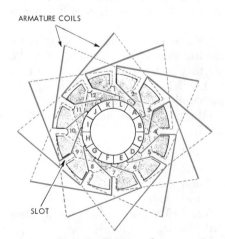

ARMATURE COILS

SLOT

Fig. 12. Schematic diagram of an armature winding.

97

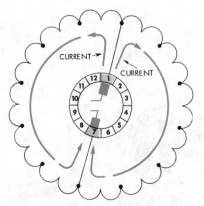

Fig. 13. A simplified drawing of the schematic of Fig. 12 which indicates the current flow through the armature winding.

magnetic poles, the wires cut lines of force, each generating a voltage which adds to that generated by other conductors in series with it between the brushes. The total voltage is the amount generated by only one-half the conductors, because there are two parallel paths from positive to negative *brushes.* Fig. 13 shows a simplified schematic of the circuit through the armature.

The schematic may be verified by tracing through the winding, Fig. 12, starting at the positive brush in contact with bar A. Moving to the left, the path follows from bar A through top conductor 1 to bottom conductor 5, to bar B;

2 and 6 to bar C;
3 and 7 to bar D;
4 and 8 to bar E;
5 and 9 to bar F;
6 and 10 to bar G, which is in contact with the negative brush.

The second path begins toward the

right from bar A, following 4 and 12 to bar L;

3 and 11 to bar K;
2 and 10 to bar J;
1 and 9 to bar I;
12 and 8 to bar H;
11 and 7 to bar G which is in contact with the negative brush.

Types of Armature Windings

The type of winding shown here, termed *lap,* is characterized by the fact that the lead wires from a coil come toward each other at the commutator. These windings are always used for two-pole generators, but another type known as a *wave winding* is commonly employed with *multipolar* generators, those having more than two poles. The lap winding has as many paths as poles, but the wave winding has only two paths regardless of the number of poles. Other than the sequence of attaching coil leads to commutator bars, the construction of the wave-connected armature is exactly the same as the lap.

Induction

Before going on to the important subject of commutation, it is necessary to understand the meaning of induction. It was shown earlier that circles of force surround a conductor in which a current flows, spreading out farther from the conductor as the value of current is increased. Fig. 14 shows a conductor in which current flows as indicated by the arrow. Applying the left-hand rule, circles of force have directions indicated by small arrows in the figure.

When the current is small, circles of force surround the wire at a distance

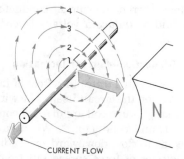

Fig. 14. Expanding circles of force generated by current flow through a conductor. The numbered circles show progression of the circles as they expand to their maximum position.

Fig. 15, this movement being similar to that of a N magnetic pole which is moved in the direction shown by the large arrow.

Effect of Current Change upon Adjacent Circuit

Fig. 16(A) shows two wire loops, $L1$ and $L2$. Loop $L1$ contains a battery and a switch. $L2$ is a closed circuit. When the switch contacts are closed, current begins to flow around loop $L1$ in a counter-clockwise direction so that in wire M it is upward. As will be seen shortly, the value of current does not reach its highest point immediately, but must

shown by circle 1 in Fig. 14. As current increases slightly, circles spread out as far as circle 2. Further increase brings them to position 3, and a still greater increase brings them to position 4. While current is increasing in this way, the spreading of circles of force on the right side of the conductor is similar to the movement of an N magnetic pole in a direction shown by the large colored arrow. When the current decreases, the circles of force shrink, or collapse, toward the conductor as in

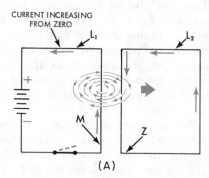

(A)

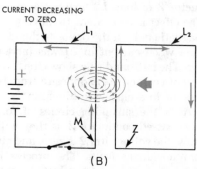

(B)

Fig. 16. Magnetic induction between adjacent conductors.

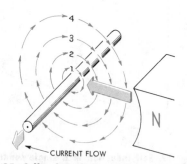

Fig. 15. Collapsing circles of force around a conductor as current flow decreases.

build up gradually from zero. While current is building up, circles of force spread out from conductor M toward conductor Z of L2.

The left-hand rule shows that direction of the circles of force about conductor M is clockwise. As they continue to spread out and to cut across conductor Z, the effect is the same as if a N magnetic pole were moved across the conductor. The voltage induced in Z forces current to flow in a downward direction in the wire and to circulate in the loop in a counter-clockwise direction. As the current in conductor M rises to maximum value, circles of force stop spreading and remain at their respective distances from conductor M.

Under this condition, magnetic flux no longer cuts across conductor Z, voltage and current in L2 dying to zero. There will be no further voltage induced in L2 so long as the switch in loop L1 is closed. If the switch is now opened, the current flowing in L1 will cease. But, it cannot do so upon the instant, for it takes time to die out in much the same way that it required time to build up.

While current in conductor M dies out, the magnetic circles of force shrink toward the conductor, Fig. 16(B). Those which have passed beyond conductor Z of loop L2 now cut across it. The effect is equivalent to moving a N magnetic pole in the opposite direction from when current in L1 was increasing. The left-hand rule shows the direction of current in L2 opposite from before, or in a clockwise direction.

After the shrinking circles of force have passed conductor Z as they withdraw, the voltage in L2 is zero, and current no longer flows. The process by which current in one conductor creates a voltage in another is called *induction*.

Here, current flowing in conductor M is said to *induce* a voltage in conductor Z.

Self-Induction—Current Increasing

Fig. 17 illustrates the effect of induction upon the circuit itself. The wires in the single-turn coil are marked conductor 1, and conductor 2. When current begins to flow downward in conductor 1, circles of force spread out from the wire, their direction being determined by the left-hand rule. As the circles of force from conductor 1 cut across conductor 2, a voltage is induced in it just as if an N magnetic pole was moved across it as indicated by the large arrow. The voltage induced in conductor 2 is in a direction to oppose the flow of current. It is referred to as *counter-electromotive-force*, (*cemf*), and is indicated by the dotted arrow E1.

At the same time, current flowing in conductor 2 creates circles of force which cut across conductor 1, as shown by the large arrow pointing to the left in Fig. 17. The voltage induced in con-

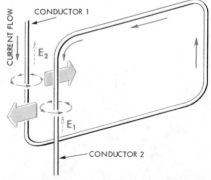

Fig. 17. Self induction in a single conductor with current increasing. The voltage induced opposes the current flow in the conductor.

ductor *1* opposes current flow as shown by the dotted arrow *E2*. Current rising in each conductor induces in any adjacent conductor a voltage opposing the flow of current in the circuit. This induced force continues until current flow in the circuit reaches a steady value.

Self-Induction—Current Decreasing

Fig. 18 shows the effect of self-induction with decreasing current. When magnetic circles of force around conductor *1* start to shrink, they cut across conductor *2* as indicated by the large arrow pointing toward conductor *1*. The voltage induced in conductor *2* is in such a direction as to assist, or continue, the flow of current in the circuit, rather than opposing as was the case with increasing current. This induced voltage is shown by dotted arrow *E3*. In the same way, the shrinking circles of conductor *2* induce a similar voltage in conductor *1*, as shown by the dotted arrow *E4*. This induced force continues so long as current is decreasing until it falls to zero.

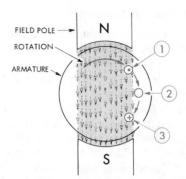

Fig. 19. Simplified diagram of the commutation process.

Purpose of Commutation

Fig. 19 shows in diagrammatic form, three positions of coil *Z* as the armature rotates in a clockwise direction between *N* and *S* poles. In position (*1*), the top coil side is cutting lines from the *N* pole, the bottom coil lines from the *S* pole. At position (*2*), both top and bottom coils are passing through the neutral zone, and are cutting no lines of force. Continued rotation brings coil *Z* to position (*3*) where the top coil side cuts lines of the *S* pole, the bottom side, lines from the *N* pole, the voltage generated in it being opposite from that of position (*1*).

Electron flow in the coil must change direction, too. In position (*1*), current from one-half the armature winding passes through the coil on its way to the brushes. In position (*3*), current from the other half of the winding flows through the coil, but in the opposite direction. This reversal of current, while the coil travels from one side of the neutral zone to the other, is termed *commutation.*

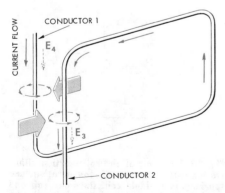

Fig. 18. Self induction with current decreasing. The induced voltage is in the same direction as the current in the conductor.

The Problem of Commutation

The altered direction of voltage generated in the coil when it approaches position (3) aids in this process, but the force of self-induction opposes it. The three positions of the coil are presented in detail by Fig. 20. Coil *M* is connected to commutator bars *A* and *B*, the brush contacting only bar *B* in position (1). Current from the upper half of the winding passes through *M* in a clockwise direction to reach bar *B* and the brush. Rotation brings *M* to position (2) where the brush contacts both *A* and *B*, thereby short-circuiting the coil. Although no voltage is being generated in *M* at the instant, self-induction results in continuing flow of current, as shown by the arrows, in a clockwise direction.

When *M* reaches position (3), it joins the lower half of the winding. If the force of induction has not dissipated, the induced current in coil *M*, color arrow, continues in a clockwise direction, opposing the current in the lower half of the winding. These opposing currents result in a spark which will jump from bar *B* to the tip of the brush as shown in the illustration. This *sparking* aggravates the original condition, burning and roughening the commutator surface so that the trouble becomes progressively worse.

Improving Commutation

Design engineers pay considerable attention to this problem, reducing coil inductance as much as possible, adjusting the width of neutral spaces, altering brush resistance, and so on. There are limits to design precautions, but once the generator has been placed in operation, sparking may be reduced by shifting the brushes forward somewhat

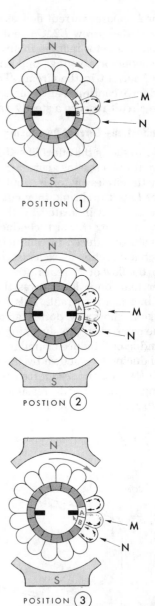

POSITION ①

POSTION ②

POSITION ③

Fig. 20. Diagrams of the coil positions illustrating the sequence of commutation and sparking in a single coil. Only one coil and brush is used in the explanation for clarity, but the process shown also occurs at the brush on the left.

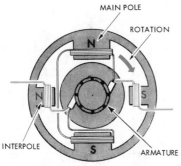

Fig. 21. Generator with interpoles.

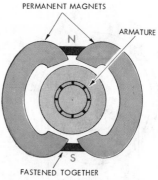

Fig. 22. Permanent magnet field structure.

in the direction of rotation so the coil short-circuited by the brush cuts magnetic flux from the pole ahead, and thus generates a voltage which helps oppose the force of self-induction. This method is not always practical, however, especially in those generators and motors where armature magnetism seriously affects flux distribution, requiring a new brush position for each change in load.

The method illustrated in Fig. 21 is widely used to avoid the necessity of shifting brushes. Two small poles, called *interpoles,* or *commutating poles,* are set midway between the main poles, their polarities being the same as that of the pole ahead in the direction of rotation. Here, the armature turns clockwise, and an S pole is needed at the right, an N pole at the left. The coil cuts the interpole flux, the result being the same as that accomplished by shifting the brushes.

DIRECT-CURRENT FIELD STRUCTURE

Permanent Magnetic Field

The armature of the commercial generator is completely surrounded by a magnetic field structure. The purpose of the field structure is to provide magnetic flux for armature conductors to cut. The flux provided by the magnetic poles is called the *field* of the generator. The term *field flux* is often used in this regard. A complete and balanced field structure might be formed by two permanent magnets bent and fastened together as in Fig. 22. The two N magnetic poles, together, form the N field pole of the generator, while the two S magnetic poles form the S field pole. When the armature turns, voltages are generated in the conductors.

Electromagnetic Field Structure

In practice, permanent field magnets are employed only in auxiliary devices, field excitation for generators being obtained from electromagnets.

In Fig. 23, the field frame, or yoke, is made of cast iron or soft steel. N and S are the magnetic poles. A field coil K surrounds the N pole, coil M the S pole. When the battery is connected to the field coil circuit, current flows from the negative battery terminal through coil M, then to coil K, and returns to the positive terminal.

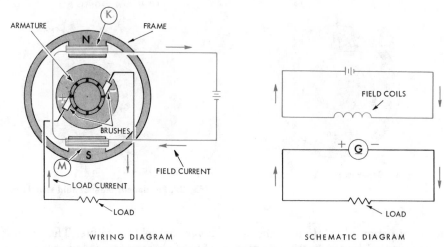

WIRING DIAGRAM   SCHEMATIC DIAGRAM

Fig. 23. Wiring diagram and schematic of a separately excited generator.

Coil *M* creates an *S* pole, *K*, an *N* pole. The flux established exists only so long as current is supplied by the battery. Since the material of the field structure is soft iron, it loses its magnetism when current flow ceases.

Separately Excited Generator. The unit shown in Fig. 23 is known as a *separately excited generator,* because its field coils are excited by means outside the generator. Here, the separate source is a battery. Quite often, however, the separate source is another, smaller, generator.

Separately excited, direct-current generators are not frequently met with in practice. It is customary to excite field coils with current taken from the armature of the generator itself. There are three general types of self-excited generators.

Shunt Generator

A *shunt generator,* also termed a *shunt-wound generator,* is illustrated in Fig. 24. The two lead wires which connected to the battery in Fig. 23 now connect to the brushes so that part of the current from the generator flows through the coils. Current from the negative brush passes through the coils and back to the positive brush as the armature revolves. If the armature stops turning, current ceases, and flux decreases almost to zero.

A small amount of magnetism still remains in the field. It is called *residual magnetism* because it remains for an indefinite period, some of the magnetic domains holding their new positions. Such magnetism is comparatively weak, but sufficient to produce a low voltage in the armature winding when the armature turns. This voltage, however small, supplies some current to the field coils, the resulting excitation adding to the flux of residual magnetism and producing a higher voltage in the armature winding. The current in the field coils increases again so that an

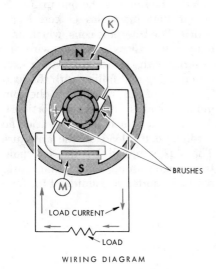

WIRING DIAGRAM

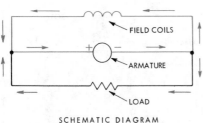

SCHEMATIC DIAGRAM

Fig. 24. Diagrams of a shunt generator.

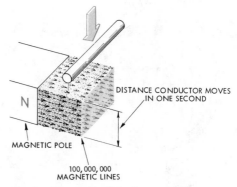

Fig. 25. The generation of one volt in a conductor moving in a magnetic field.

even greater amount of flux is produced. This process, called *building up,* continues until the generator is operating at full rated voltage. The load is disconnected before attempting to excite the generator, thus enabling the voltage to build up faster.

Strength of Generated Voltage. The voltage generated in a conductor is measured by the rate at which it cuts magnetic lines of force. One volt of electrical pressure is generated when 100,-000,000 magnetic lines are cut by a conductor in the space of one second,

as illustrated in Fig. 25. In the figure, the magnetic flux from pole *N* is made up of 100,000,000 lines. When the conductor moves across this flux in the space of one second, one volt of electrical pressure is generated.

Increasing the value of the field flux to 200,000,000 lines results in the voltage being doubled. Decreasing the value of field flux to 50,000,000 results in the voltage decreasing to ½ volt. In other words, increasing the flux increases the voltage, and decreasing flux decreases voltage. It may be stated in the form of a rule: *Generator voltage varies directly as the field flux.*

Field Resistor. In practice, an adjustable resistance called a *field rheostat* is connected in series with the field coil. In Fig. 26, the rheostat is connected between field coil *M* and brush B_1. By turning a knob, the value of the resistor may be increased or decreased at will. The usual way of indicating a rheostat in a schematic diagram is as shown in Fig. 26. If the resistance in a circuit is increased,

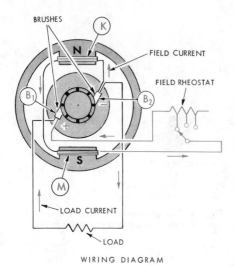

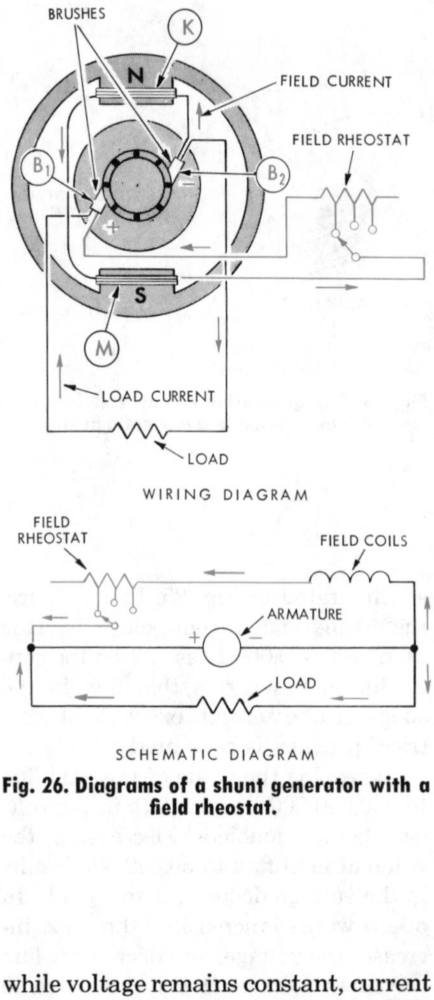

WIRING DIAGRAM

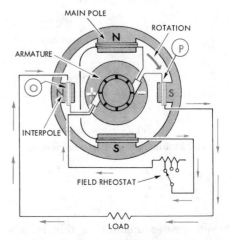

exactly the same as the one in Fig. 26 except that interpoles O and P are added. The interpole coils, which have only a few turns of heavy wire, are in series with the armature, their excitation rising and falling with the load. The force of self-induction in the commutated coil is in direct proportion to the current, which in turn varies according to the load on the armature. The flux supplied by the interpoles thus creates an opposing voltage automatically suited to counteract the force

SCHEMATIC DIAGRAM

Fig. 26. Diagrams of a shunt generator with a field rheostat.

while voltage remains constant, current flowing in the circuit decreases, and if resistance decreases, current increases.

When current in the field circuit decreases, the flux decreases, and generated voltage in the armature decreases. On the other hand, when current in the field circuit increases, the voltage increases. Thus a field resistor may be used to control the voltage generated in the armature of the shunt unit.

Shunt Generator with Interpoles

The shunt generator in Fig. 27 is

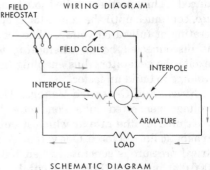

SCHEMATIC DIAGRAM

Fig. 27. A shunt generator with interpoles.

of self-induction, and eliminates the tendency to spark at the brushes. Interpoles may be used with any type of direct-current generator.

Compound Generator

A *compound generator,* also termed a *compound-wound generator,* has two field coils on each pole, Fig. 28. *K* and *M* are the usual type of shunt field

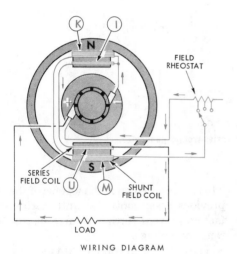

WIRING DIAGRAM

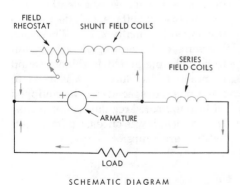

SCHEMATIC DIAGRAM

Fig. 28. Diagrams of the compound-wound generator.

coils and are connected in circuit with a field resistor. *I* and *U* are called *series coils,* or *series field coils.* One of the two remaining leads is connected to a brush and the second to the load. The other terminal of the load connects to the remaining armature terminal. The series field coils are connected in series with armature and load.

Flux created by the series coils adds to that created by the shunt coils, so that armature voltage is increased. If the load is disconnected, the field flux is only that from the shunt coils. When the load comes on again, the voltage rises.

The increase in voltage, which thus occurs automatically in the compound generator when the load comes on, supplies the voltage lost in the resistance of connecting wires between generator and load. If the load draws little current, the voltage lost in the wires is of a low value, but, if the load draws considerable current, the voltage loss may be quite high. When series field windings are accurately proportioned, voltage may be kept fairy constant in spite of the fact that the load may vary from moment to moment.

Series Generator

A *series,* or *series-wound,* generator has but one set of field coils which are much like the series coils of the compound generator. They are connected in circuit between armature and load, Fig. 29. This generator resembles the compound generator of Fig. 28 except that the shunt coils are omitted. The voltage generated depends upon the amount of current taken by the load. If its resistance increases, the current through the field coils decreases. This

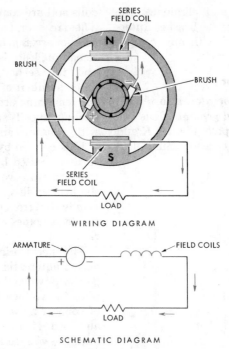

WIRING DIAGRAM

SCHEMATIC DIAGRAM

Fig. 29. Diagrams of the series-wound generator.

decrease in field current results in a decrease in field flux and the decrease in field flux brings about a decrease in armature voltage, which causes a still smaller current to flow to the load.

This process of decreasing current and decreasing voltage continues until certain factors prevent current and voltage from decreasing further. If the resistance of the load is now decreased, a greater current flows through it. Since this same current flows in the series field coils, the flux is increased. The armature voltage rises, and the current through the load becomes even greater.

This process, which is the reverse of the previous case, continues until certain factors prevent further change in current and voltage.

The decrease in voltage with increase in load resistance is considerable, and the increase in voltage with decrease in resistance is equally great. Therefore, the voltage of a series generator is said to be unstable and it is seldom found in practice. In Europe, it is used to some extent in special circuits equipped with complicated regulating apparatus. In America, it was once used for street-lighting arc lamps.

REVIEW QUESTIONS

1. A conductor moves from right to left under an N pole. In what direction does current tend to flow in the conductor?

2. State the left-hand rule for conductors.

3. Could current produced by a loop and two slip-rings be used to charge a storage battery?

4. Name the essential parts of an armature.

5. If each armature coil has ten conductors, how may wires occupy each armature slot?

6. How many parallel paths in a two-pole lap armature winding?

7. How many paths in a four-pole wave armature winding?

8. What is the purpose of commutation?

9. What is the principal difficulty encountered during commutation?

10. What is the simplest operational procedure for improving commutation?

11. What is residual magnetism?

12. Describe how a generator "builds up."

13. What are interpoles?

14. How do interpoles assist commutation of a coil?

15. What is the relative polarity of generator interpoles?

16. What is a separately-excited generator?

17. How are field coils of a shunt generator excited?

18. Define the volt in terms of magnetic lines cut by a conductor.

19. How is field flux created in a compound-wound generator?

20. Describe how the voltage of a series generator varies with changing load.

Chapter 8

Alternating Current Principles

GENERATION OF ALTERNATING CURRENT

Nature of Single-Phase Current

A revolving loop equipped with two *collector rings* and a pair of brushes is an elementary *single-phase generator:* the term single-phase meaning a single generating circuit. Commercial alternating-current generators, which are called *alternators*, have a number of loops connected in series, so that a single-phase armature may be defined as one that has a single winding terminating in a pair of collector rings.

Alternating current, flowing first in one direction, then the other, is designated not only by voltage of the circuit, but also by the number of current reversals or *alternations* per second. A pair of alternations, that is a clockwise flow plus a counterclockwise flow, is termed a *hertz*. Until recent years, this term for a pair of alternations was called a *cycle*. One cycle per second (cps) is the same as one hertz. Current produced by the elementary loop, it will be recalled, went through one complete hertz in each revolution, passing first through a zero position to a maximum, then through another zero position and again to a maximum during the time the loop makes a single revolution.

The form of alternating current presently used throughout most of the United States is termed *60-hertz,* because it goes through 60 complete hertz in one second. And, since the number of hertz is called the *frequency,* it is customary to say that current is supplied to the American home at a frequency of 60 hertz. Many frequencies have been employed at one time or another, such as 25, 40, 50, and 133 hertz, but the 60-hertz frequency has supplanted practically all of them. It may be observed at this time, that the armature of a two-pole alternator must rotate at a speed of 60 revolutions a second, or 3600 revolutions per minute, to produce 60-hertz current.

Polyphase Current

Two additional forms of alternating current have been commonly used, *two-*

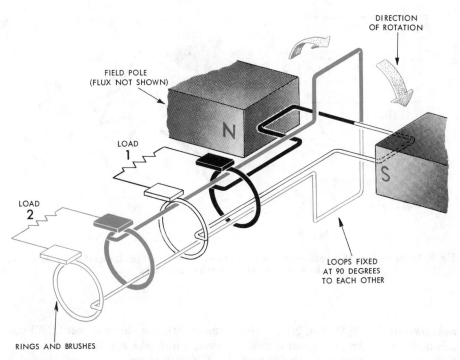

DIRECTION
OF ROTATION

FIELD POLE
(FLUX NOT SHOWN)

N

LOAD
1

LOAD
2

S

LOOPS FIXED
AT 90 DEGREES
TO EACH OTHER

RINGS AND BRUSHES

Fig. 1. An elementary two-phase generator.

phase and *three-phase,* both of which come under the heading of *polyphase* current. The two-phase elementary generator of Fig. 1, has two loops at right-angles to one another, each with a pair of collector rings. At the instant shown in the figure, the colored and white is in the neutral zone, hence generating no voltage, but the black and white loop is cutting directly across lines from both *N* and *S* poles. A quarter-turn later, the colored and white conductors move across the poles while the black and white lie in the neutral zone.

The two-phase unit actually produces two distinct single-phase currents whose flow cycles begin at different

instants. It was developed primarily to supply large induction motors which, as will be seen later, do not start easily on single-phase current. There is relatively little two-phase current generated today, three-phase having monopolized the field.

The three-phase generator of Fig. 2 has three loops spaced at equal distances from one another, and it would seem that six collector rings would be necessary. However, when loop terminal lead wires are properly grouped in what is known as a three-phase *Wye,* or *star* connection only three collector rings are used. In practice the star connection is made inside the winding

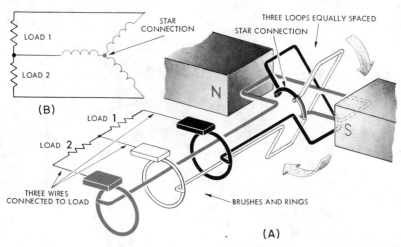

Fig. 2. An elementary three-phase generator is shown in (A). Inset (B) illustrates the details of the star connection in schematic form.

as shown in Fig. 2(A). Fig. 2(B) is the schematic of the armature star connection. As represented here, the output of the unit is the same as three distinct single-phase loops in which flow cycles begin at different times.

Mechanical and Electrical Degrees

The circle, Fig. 3(A), contains 360 degrees, each degree representing the size of a wedge whose area is 1/360th that of a full circle. Angles are stated in terms of how many such wedges they contain, thus measuring the distance between base radius 0-1 and a second radius whose outer end moves around the circumference in a counter-clockwise direction. At point (2), the angle between the radii is 30 degrees; at point (3), 45 degrees; at (4), 60 degrees; at (5), 90 degrees, or one-quarter circle; at (6), 120 degrees, or one-third of a circle; at (7), 180 degrees, or a half-circle; at (8), 270 degrees, three-quarters of a circle;

and at (1), the starting point, 360 degrees, which is a whole circle.

Fig. 3(B) represents an end view of a revolving loop, only the two conductors being indicated. Conductor (a) is directly opposite the middle of the N pole, while (b) is opposite the middle of the S pole. If the loop moves counterclockwise a quarter-turn until (a) and (b) lie in the *neutral zone*, the rotation is 90 degrees. Another quarter of a circle brings (a) to the middle of the S pole, (b), the N pole, and the distance is equal to 180 degrees. A third quarter-turn places the conductors in the neutral zone again, 270 degrees from the starting point, and the final quarter-turn finishes the revolution at 360 degrees.

In the two-pole unit, *electrical degrees* and *mechanical degrees* are exactly alike, a full revolution constituting both 360 geometrical or mechanical degrees and 360 electrical degrees. The two values differ, however, where more

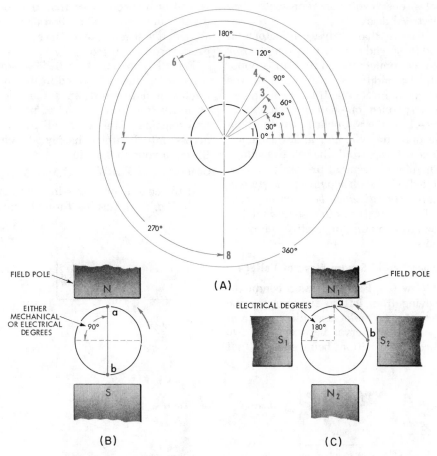

Fig. 3. Comparison between mechanical and electrical degrees.

than two poles are involved. Note that, in Fig. 3(B), the distance from the middle of an N pole to the middle of the next S pole amounts to 180 electrical degrees. Likewise, the distance from the middle of this S pole to the middle of the next N pole, in the direction of rotation, is 180 electrical degrees.

Refer now to Fig. 3(C), which shows a four-pole generator. The distance from pole N_1 to pole S_1 is only 90 *mechanical* degrees, but since it is the dis-

tance from the middle of an N pole to the middle of the next S pole, it is 180 *electrical* degrees. Continuing rotation, conductor (a) passes through 180 electrical degrees from the middle of S_1 to the middle of N_2, 180 electrical degrees from N_2 to the middle of S_2, and finally 180 electrical degrees to the starting point at N_1.

When a generator has four poles, therefore, a conductor moves 720 electrical degrees in a single revolution of

113

the armature. If the generator has six poles, one revolution represents 1080 electrical degrees.

Observe that active conductors (a) and (b) spread or span only one-quarter of a circumference, so that (a) is opposite the middle of an N pole, (b) the middle of an S pole. If the two conductors spanned one-half the circumference, as in the two-pole unit, both would lie opposite N poles, and the generated voltages would be self-cancelling. Therefore, a general principle may be stated: *Coil span should equal approximately the distance between adjacent pole centers.* It is often somewhat less than this amount, in practice, but never more.

Generation of Square Wave of Voltage

View (A), Fig. 4, shows a conductor moving from left to right across the face of an N pole at a uniform rate of speed, the interval between any two successive positions being equal to that

between any other pair. If the pole flux is evenly distributed so that the conductor cuts across 500 million lines per second, it will generate 5 volts each and every second as it moves.

Fig. 4(B) shows the so-called "square-wave" of voltage generated in the conductor, a flat-topped wave that rises abruptly from 0 to 5 volts at point 1, and remains at this value all the way across until it drops sharply at point 6 from 5 volts to 0 again.

Generation of Sine Wave of Voltage

If the conductor moves in a circular path, Fig. 5(A), turning through equal

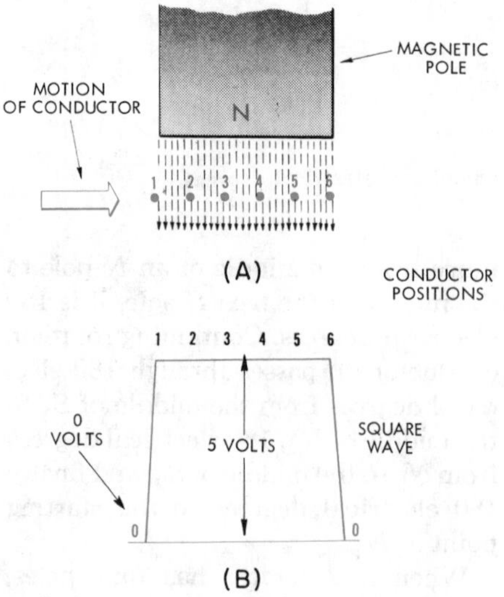

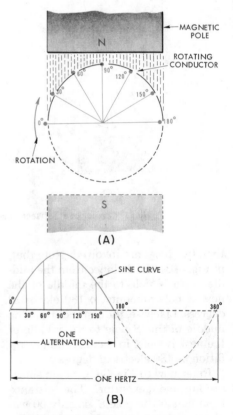

Fig. 4. The generation of a square wave of voltage.

Fig. 5. The development of a sine wave.

angular distances in equal spaces of time, it no longer cuts across magnetic lines at a uniform rate of speed. This observation follows because the motion of the conductor is partly vertical and partly horizontal. When moving vertically, it is parallel to the magnetic lines, hence not cutting them, and only the amount of horizontal movement is effective in producing voltage. The resulting voltage alters from instant to instant, as already observed in connection with the revolving loop of the preceding chapter.

At the 30° point on the circle, horizontal motion is 50 percent of maximum value so that only 250 million lines are cut in one second, and 2.5 volts are generated. At the 60° point, the horizontal motion is 86.6 percent so that 433 million lines are cut per second, and 4.33 volts are generated. In the 90° position, horizontal motion is at maximum value, and 5 volts are generated.

As rotation continues, and the conductor starts downward from its maximum position, the percentage of horizontal motion decreases, and the voltage falls to 4.33 at the 120° point, 2.5 at the 150° point, and 0 at 180°. These values are plotted in Fig 5(B), where the base line is divided into six 30° spaces, and the heights of vertical risers at each point are proportional to the voltage generated there. When a smooth line is drawn through the upper ends of the risers, it takes a characteristic shape known as a sine wave.

The solid outline in Fig. 5(B) represents an alternation. If the conductor is now rotated downward from the 180° position to cut across magnetic flux of an *S* pole, voltage will be generated in the opposite direction, as indicated in dotted outline, from 180° to 360°. The two alternations represent one complete hertz of voltage, or of current flow resulting therefrom.

Effective Value of Voltage or Current Wave

Since current in the sine wave changes continually from zero, through the various steps to the highest or *maximum* point, then decreases to zero again, it may be asked how the current in an alternating circuit can be specified. In Fig. 6, the zero and maximum points are marked, along with two others.

The *average value* is obtained by taking the sum of a great number of instantaneous values, and dividing this sum by the number so taken. It is found to be .636 times maximum current, so that for a maximum of 10 amps, the average is: .636 × 10 = 6.36 amps. When this amount of current passes through a given resistor, however, the heat generated is much greater than from a direct current of 6.36 amps.

Tests show the amount of generated heat equal to that from a direct current of 7.07 amps, and this is said to be the *effective value* of an alternating current which has a maximum rating of 10 amps, or .707 times maximum rating. And, since a 10-volt maximum wave produces a current of 7.07 amps in a

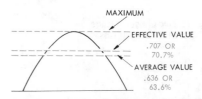

Fig. 6. The average and effective voltage values of a sine wave.

1-ohm resistor, the effective value of such a voltage wave is also taken as .707 times maximum.

Representation of Two-Phase Voltage or Current

Although two-phase current is of comparatively minor importance to-day, its sine-wave diagram may serve to introduce the more complicated three-phase one. Loops of the elementary two-phase generator that was seen in Fig. 1 are at right-angles to each other, voltages in them 90 electrical degrees apart.

Sine waves of Fig. 7 follow the voltage outlines as each loop starts from the neutral zone and makes a full revolution. The zero point on the base line marks the position of Loop 1 when it lies in the neutral zone. At this instant, Loop 2 is generating maximum voltage as it cuts directly across flux from the two poles, having started from zero 90 electrical degrees earlier. When voltage in Loop 1 reaches a maximum, that in Loop 2 will have come to the base or zero line, 90 degrees from 0, and 180 degrees from its own starting point.

Representation of Three-Phase Voltage or Current

It is clear that voltages in the elementary three-phase generator of Fig. 2 are 120 electrical degrees apart because the loops are spaced a distance of one-third circumference. Starting points of the three curves in Fig. 8(A), therefore, are placed at 120° intervals along the base line. At the instant chosen here, the colored loop is adjacent to the poles as shown in position (1), and its voltage is maximum as shown on the waveform. Voltage of the white loop (dotted line) is rising toward a maximum in the negative direction while that of the black loop has passed maximum, and is decreasing in the positive direction. As the loops rotate further, the white loop will next be adjacent to the poles and voltages will be as shown in position 2.

Reference was made earlier, to the fact that only three collector rings were required for the elementary three-phase generator. Vertical dashed lines (1), (2), and (3) show why. The sum of voltages in two conductors on one side of the base line are equal, exactly, to that of one on the opposite side, and the cur-

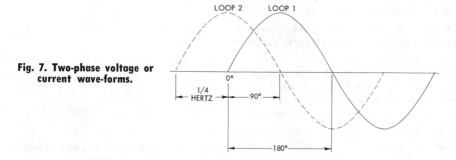

Fig. 7. Two-phase voltage or current wave-forms.

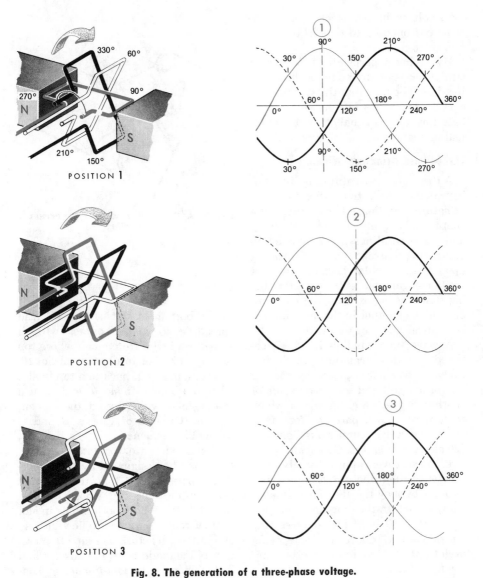

POSITION 1

POSITION 2

POSITION 3

Fig. 8. The generation of a three-phase voltage.

rents established by these voltages will bear this same relationship. At (*1*) for example, the voltage in the colored loop is the same but opposite in direction to that in the white and black loops; at (*2*) the voltage in the white loop is numerically equal but opposed to that in both the colored and black loops and at

117

(3), voltage in the black loop is the same but opposed to that in the white and colored loops.

By laying a perpendicular ruler or straight-edge at any point on the drawing, it will be seen that the sum of voltages in two loops on one side of the base line always equals that of the remaining loop.

Three-Phase Armature Winding

A three-phase armature winding is illustrated in Fig. 9. In the direct-current armature winding of the preceding chapter, a lap type was shown. Both lap and wave windings are employed in alternating current units, depending upon the particular design, the most easily recognizable feature of either being that lead wires from the coils approach one another in the lap winding, but spread apart in the wave.

A wave type winding has been selected for this armature which has twelve slots, each occupied by the top side of one coil and the bottom side of another. Since the generator of which the armature is a part has four field poles, one full circumference takes in 720 electrical degrees. *Coil span*, therefore, must cover approximately one-quarter of the armature circumference in order to obey the 180° span rule already stated. In the diagram, coil span is exactly 180 electrical degrees, the top side of one coil, for example, lying in slot *1*, the bottom in slot *4*, others following a like pattern.

There are three distinct windings, Phase *A* starting with main lead wire A_1 and terminating at A_2, Phase *B* starting with B_1 and terminating at B_2, Phase *C* starting with C_1 and terminating at C_2, each consisting of a single path, or circuit. That of Phase *A* passes from A_1 to the top coil side in slot *1*,

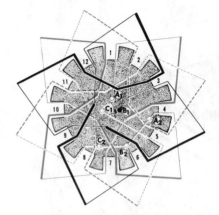

Fig. 9. A three-phase wave type armature winding.

to the bottom in slot *4*, the coil lead spreading to make contact with a spread lead wire from a coil whose top lies in slot *7*, then to a bottom in slot *10*, a bottom in slot *1*, back to a top in slot *10*, and a bottom in slot *7*, ending at a top in slot *4*, which goes to the A_2 connection. Circuits of Phases *B* and *C* follow like sequences.

Certain facts are worth noting. First, coils may be designated by *top-to-bottom* numbering so that the notation *1-4* identifies the coil whose top lies in slot *1*, and whose bottom lies in slot *4*. If current is assumed to flow into A_1 and out A_2, its clockwise direction creates *N* magnetic poles in coils *1-4* and *7-10*. Then, a *bottom-to-bottom* connection sends current through *10-1* in a counter-clockwise direction, forming *S* poles there and in *4-7*. Thus, coil lead wires are arranged to produce alternate *N* and *S* poles around the circumference of the armature, four altogether.

When current is assumed to flow through the other phases in the same manner, four armature poles are produced in each case. A total of twelve poles does not result, however, but only four, because current of a given instantaneous direction does not flow into A_1, B_1, and C_1, at the same time. Instead, it flows in the sequence illustrated in Fig. 8(A), A_1, B_1, and C_1, being spaced around the circumference in such order that three adjacent coils always produce the same polarity at any selected moment.

The proper arrangement of main lead wires is arrived at by spacing them at intervals of 120 electrical degrees. If Phase A is represented by the colored loop in Fig. 8, Phase C, represented by the black loop, must start 120 electrical degrees from Phase A, or two-thirds of the distance between top and bottom coil sides of *1-4*, which is slot *3*. In the same way, Phase B, represented by the white loop, must start two-thirds of the distance between top and bottom coil sides of *3-6*, which is slot *5*.

The final operation is to attach main lead wires to three collector rings. Although the winding illustrated in Fig. 9 could be placed in an actual generator, its total of twelve slots allows only one for each phase, for each pole, which amounts to one slot per pole per phase. The usual design has a number of slots per pole per phase. In such case, each phase winding encircles the armature more than one time. The principles involved, however, are exactly the same as with the single-slot type.

A lap connection is also commonly used for alternators, the coils being arranged to form alternate N and S poles around the circumference. The *single-circuit* wye or star connection is employed here, but other arrangements are possible. They will be discussed, along with voltage relationships, in Chapter 10.

Revolving Field Magnets

Except in small sizes, alternators are of the *revolving-field* type shown in Fig. 10. The winding is placed in slots on

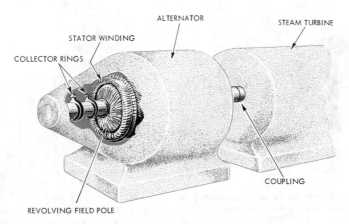

ALTERNATOR

STEAM TURBINE

STATOR WINDING

COLLECTOR RINGS

COUPLING

REVOLVING FIELD POLE

Fig. 10. A revolving field alternator.

the inner circumference of a stationary armature which is termed a *stator*. It is the same as the armature winding of Fig. 9, but the main leads are connected directly to load supply wires rather than to collector rings. Collector rings are needed, however, on the rotating element, or *rotor,* which carries the field poles. Direct current is supplied to the collector rings for exciting the field coils, this current being furnished by a small direct-current generator termed an *exciter.*

Eddy Currents

When an iron core cuts across magnetic lines, or is cut by them, a voltage is generated in it, and current flows, Fig. 11 (*top*), the resulting power loss appearing as heat which rapidly destroys insulation on armature conductors. Such wasteful flows are called *eddy currents.* They may be reduced to harmless proportions by *laminating* armature and stator cores, that is, constructing them of iron sheets which are insulated from one another by a thin coating of lacquer, or oxide, Fig. 11 (*bottom*). Magnetic properties are unaffected by laminating, but the core is made highly resistant to the flow of eddy currents.

INDUCTANCE IN ALTERNATING-CURRENT CIRCUITS

Normal Voltage and Current Curves

Curve *E*, Fig. 12, is a sine curve of voltage. *I* is the current curve. The current at any instant is determined by Ohm's Law, depending upon voltage and resistance of the circuit. At zero point on the voltage curve, the current is also zero. At maximum on the voltage curve, current is maximum. The two curves remain in step during the entire

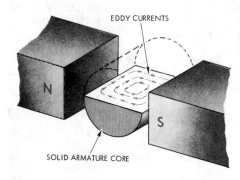

SOLID ARMATURE CORE

EDDY CURRENTS

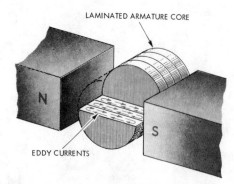

LAMINATED ARMATURE CORE

EDDY CURRENTS

Fig. 11. Eddy currents in an unlaminated armature core, *top,* and a laminated core, *bottom.*

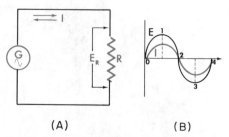

(A)   (B)

Fig. 12. The relationship of current and voltage in a non-reactive circuit.

cycle and are said to be *in phase*. A circuit which contains only resistance is said to be non-reactive, current always being in phase with the applied voltage. Usually, circuits do have some inductance so that the relationship of current and voltage curves will differ from that of Fig. 12.

Current Lag in a Pure Inductive Circuit. Fig. 13 illustrates the relationship between voltage and current curves in an alternating-current circuit which has pure inductance. As the voltage curve, *E*, rises from zero, it is opposed by the voltage of self-induction. At point *1*, the circuit voltage is at maximum and the value of current represented by *I* is zero.

At point *1*, the circuit voltage is greater than the voltage of self induction so that current flows in the circuit. The circuit voltage reaches maximum value at point *1*, but the current does not reach maximum until point *2*. When the voltage passes through zero, point *2* in the figure, the value of current is at maximum. From point *2* to *3*, circuit voltage tries to force current in the negative direction but the voltage of self induction maintains flow in the positive direction.

Circuit voltage reaches a maximum in the negative direction at point *3*. The current, however, does not reach its maximum value until point *4*, and the voltage becomes zero again at point *4*, while the current does not reach zero until point *5*. It should be noted that the normal shape of the sine wave, *I*, which represents current is not altered by inductance whose only effect is to shift the current curve some distance to the right of the voltage curve. Therefore, it is said that current lags behind the voltage by 90 degrees in a pure inductive circuit.

Reactance of Inductive Circuit. The strength of induced voltage depends on the physical characteristics of the circuit such as number and spacing of turns of wire, and proximity to iron objects. The term used to describe this quality is called *inductance* and is expressed in *henrys*. In some respects inductance acts like resistance. For this reason, *inductive reactance* as it is called, is always expressed in ohms. Thus a certain coil which has an inductance of 1/100 henry offers as much opposition to the flow of 60 hertz current as does a resistance of 3.77 ohms.

The inductive reactance of a circuit is indicated by the term X_L, and its value is determined by the following formula.

$$X_L = 2\pi f L$$

$X_L =$ inductive reactance in ohms.

$\pi =$ the constant 3.1416

$f =$ frequency of the supply current in hertz

$L =$ value of the inductance in henrys

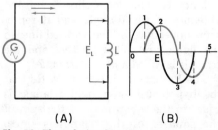

(A)   (B)

Fig. 13. The relationship of current and voltage in a pure inductive circuit.

To determine the inductive reactance of the coil which has an inductance of 1/100 henry when placed in a 60-Hz circuit, substitute these values in the formula;

$$X_L = 2\pi f L$$
$$X_L = 2 \times 3.1416 \times 60 \times \frac{1}{100}$$

Multiplying $2 \times 3.1416 \times 60$, we have 376.8 (377 approximately), so,

$$X_L = 377 \times \frac{1}{100}$$
$$X_L = 3.77 \text{ ohms}$$

Note that the frequency of the voltage is an important factor in determining reactance, and that reactance varies in direct proportion to frequency. For example, the coil with an inductance of 1/100 henry offers less opposition to the flow of 50-Hz current because:

$$X_L = 2\pi f L$$
$$X_L = 2 \times 3.14 \times 50 \times \frac{1}{100}$$
$$X_L = 314 \times \frac{1}{100}$$
$$X_L = 3.14 \text{ ohms}$$

Determining Impedance in a Resistive-Inductive (RL) Circuit

Comparison of the curves in Figure 13 illustrates that resistive and inductive opposition to current flow are not in step (in time phase) with each other. The effect is somewhat like two forces resisting a moving body from two different directions.

We can always find an equivalent force with the combined effects of the two resisting forces. The same triangular and vector methods of combining forces are used in finding combined effects of resistance and inductance in RL circuits. Impedance values are analogous to the magnitude of the forces, and the time phase relationships are analogous to angular relations. It is necessary to combine the resistance and the inductive reactance to obtain the actual resistance of an RL circuit.

This total opposition to current flow, the combination of both resistance and inductive reactance, is called the *impedance*. It is represented by the letter Z. There are two methods of determining impedance. One is mathematical, using the Pythagorean theorem. The other, an alternate method, utilizes graph paper.

Determining Impedance in a Series Resistive Inductive (RL) Circuit Using Pythagorean Theorem. The mathematical method of calculating impedance in a series RL circuit is with the use of the *Pythagorean theorem.* In the Fig. 14 (C) a right triangle is drawn with one side representing inductive reactance X_L and the other side representing resistance R. The dotted line is the hypotenuse of the triangle and it represents impedance Z. The arrowed lines in the figure are called *phasors* and the triangle is called a *phasor triangle*.

The Pythagorean theorem states that the hypotenuse of a right triangle is equal to the square root of the sum of the squares of the triangle's sides. ($H = \sqrt{S^2 + S^2}$). If we substitute electrical terms for the parts of this equation, we create the impedance formula $Z = \sqrt{X_L^2 + R^2}$. The calculation of impedance in the Fig. 14 involves the following:

$$Z = \sqrt{X_L^2 + R^2}$$

$$Z = \sqrt{4^2 + 3^2}$$

$$Z = \sqrt{16 + 9} = \sqrt{25}$$

$$Z = 5 \text{ ohms}$$

This particular calculation has been done with small numbers. If larger values of inductive reactance and resistance were used (and this is generally the case) square and square root tables or a slide rule would have to be used. As an example:

$$Z = \sqrt{X_L^2 + R^2}$$

$$Z = \sqrt{32^2 + 24^2}$$

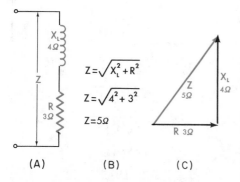

(A)   (B)   (C)

Fig. 14. Determining impedance in a series RL circuit using the Pythagorean theorem.

$$Z = \sqrt{1024 + 576} = \sqrt{1600}$$

$$Z = 40 \text{ ohms}$$

A square root table is provided in Appendix E. Also see Appendix D for an alternate and much simpler solution using trigonometry tables or slide rule.

Determining Impedance in a Series Resistive Inductive (RL) Circuit Using Alternate Method. This special method involves the right angle method of addition, and can be performed easily with the help of graph paper. This paper, which may be obtained at stationery stores, has horizontal and vertical lines which form small squares. The ¼-inch size used here is a common variety.

In Fig. 15(A) the inductive reactance in the circuit is 4 ohms, the resistance 3 ohms. It is desired to find the impedance of the circuit. On a piece of graph paper draw two lines, H horizontal, and V vertical as in Fig. 15(B). From the point of their intersection, mark off 3 squares to the right to indicate resistance. At the end of this line, draw a vertical line upward, 4 squares long to represent inductive reactance of 4 ohms.

Next, draw line Z, dotted line in Fig. 15(B), from the left end of the resistance line to the upper end of the reactance line, thus completing a right triangle. If Z is measured with a ruler as in Fig. 15(C), it will be found to be 1¼ inches long, this distance being equal to 5 of the ¼-inch squares, and thus indicating an impedance of 5 ohms.

This method is analogous to drawing a map to scale. Any distance may be obtained if we know how many miles are represented by one inch. Likewise, if we know how many ohms are represented by one graph unit, or any unit of length, along one of the known legs of the triangle, we may apply the same scale to the hypotenuse.

Determining Impedance in a Parallel Resistive Inductive (RL) Circuit. Impedance in a parallel RL circuit is found using Ohm's law. This method assumes that the applied voltage and total current are known. Calculation procedures are as follows:

$$Z = \frac{E}{I_T}$$

$$Z = \frac{100V}{5A}$$

$$Z = 20 \text{ ohms}$$

Determining Total Current in a Resistive Inductive (RL) Circuit

Determining Total Current in a Series (RL) Circuit. In a series RL circuit, current stays the same throughout the circuit. The phase angles mentioned previously have to do with the relationship of generator (supply) voltage and its current.

Determining Total Current in a Parallel (RL) Circuit. In a parallel RL circuit current is mathematically calculated with the use of the *Pythogorean theorem.* In the Fig. 16 (C), a right triangle is drawn with one side (phasor) representing current through the inductor I_L and the other side (phasor) representing current through the resistor I_R. The dotted line is the hypotenuse of the triangle and it represents total current I_T. Calculation of the values in the circuit of Fig. 16 are as follows:

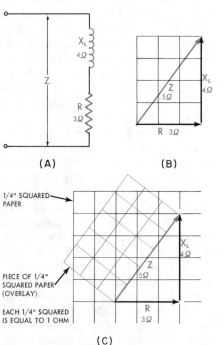

(A)   (B)

1/4" SQUARED PAPER

PIECE OF 1/4" SQUARED PAPER (OVERLAY)

EACH 1/4" SQUARED IS EQUAL TO 1 OHM

(C)

Fig. 15. Determining impedance in a series RL circuit using alternate (graph paper) method.

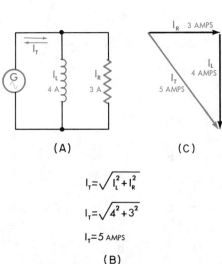

(A)   (C)

$$I_T = \sqrt{I_L^2 + I_R^2}$$

$$I_T = \sqrt{4^2 + 3^2}$$

$$I_T = 5 \text{ AMPS}$$

(B)

Fig. 16. Determining total current in a parallel RL circuit.

$$I_T = \sqrt{I_L{}^2 + I_R{}^2}$$

$$I_T = \sqrt{4^2 + 3^2}$$

$$I_T = \sqrt{16 + 9} = \sqrt{25}$$

$$I_T = 5 \text{ amps.}$$

Determining Total Voltage in a Resistive Inductive (RL) Circuit

Determining Total Voltage in a Parallel (RL) Circuit. In a parallel RL circuit, voltage stays the same throughout the circuit. The phase angles mentioned previously have to do with the relationship of generator (supply) voltage and its current.

Determining Total Voltage in a Series (RL) Circuit. In a series RL circuit, voltage is mathematically calculated with the use of the *Pythagorean theorem.* In the Fig. 17 (C), a right triangle is drawn with one side (phasor) representing voltage across the inductor X_L and the other side (phasor) representing voltage across the resistor R. The dotted line is the hypotenuse of the triangle and it represents total applied voltage E_A. Calculation of the values in the circuit of Fig. 17 are as follows:

$$E_A = \sqrt{V_L{}^2 + V_R{}^2}$$

$$E_A = \sqrt{4^2 + 3^2}$$

$$E_A = \sqrt{16 + 9} = \sqrt{25}$$

$$E_A = 5 \text{ volts}$$

Determining Inductive Phase Angles

When an electric force (voltage) is applied to electrons, electron flow is not always immediate. If a delay exists in the electron response to the applied voltage, the current is said to differ in time phase with the source voltage.

The phase relationship between the current and voltage in a pure resistive (non-reactive) circuit is in phase. See Fig. 12. That is, the current and the voltage are in phase in a circuit that has no inductance.

In a pure inductive circuit such as that shown in Fig. 13, the current lags the voltage by 90 degrees. That is, the current is at maximum, 90 degrees or one quarter cycle after the voltage arrives at maximum. Since the cycle represents time, then it can be stated that the current lags the voltage in a pure inductive circuit by one quarter cycle in time. This relationship between cur-

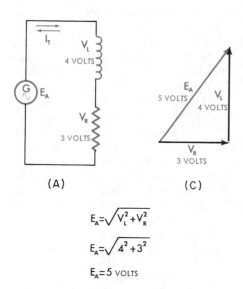

(A)   (C)

$$E_A = \sqrt{V_L^2 + V_R^2}$$

$$E_A = \sqrt{4^2 + 3^2}$$

$$E_A = 5 \text{ VOLTS}$$

(B)

Fig. 17. Determining total voltage in a series RL circuit.

rent and voltage across any inductor is always the same.

However, when an inductor is inserted in a circuit with another device such as a resistor, the circuit becomes reactive and a phase relationship exists between the generator voltage and its current. The relationship we speak of is the *phase angle*. This phase angle is normally known by the Greek letter theta and is represented by the symbol θ.

Determining Inductive Phase Angle in a Series (RL) Circuit. The phase angle in a series RL circuit can be calculated from the impedance triangle shown in Fig. 18. Use of the values of resistance and inductive reactance, along with the simple trigonometry formula tangent $\theta = \dfrac{X_L}{R}$, will provide us with the *phase angle* relation between the generator voltage and its current.

The tangent function of an angle θ (or tan θ for short) is a trigonometric relation between the legs of a right triangle. The tangent of the angle is defined as the ratio of the leg opposite the angle to the leg adjacent to same angle. In the Fig. 18, the opposite side is the inductive reactance X_L and the adjacent side is resistance R. Calculation of the phase angle in the circuit in Fig. 18 is as follows:

$$\text{tangent } \theta = \frac{X_L}{R}$$

$$\tan \theta = \frac{4}{3}$$

$$\tan \theta = 1.333$$

$$\text{angle } \theta = 53 \text{ degrees}$$

From the trigonometry table in Appendix D, the angle that has a tangent of 1.333 is 53 degrees. Therefore, the phase angle of the circuit is 53 degrees and line current lags generator voltage by that amount in this particular circuit.

In any series combination of X_L and R, current will lag generator (supply) voltage by some value between 0 and 90 degrees. The larger the X_L is, the larger the phase angle will be. Since this is a series circuit, the larger the X_L is, the larger the voltage drop will be across X_L. Current will remain the same throughout the circuit.

It must be repeated here that current through the resistor is in phase with the voltage drop across the resistor. Similarly, current through the inductor lags the voltage drop across the inductor by 90 degrees. The current-to-voltage phase angle relates to the generator (supply) voltage. See the Fig. 19 for the phase

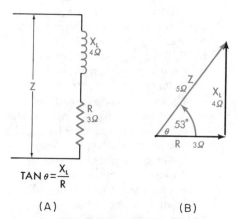

TAN $\theta = \dfrac{X_L}{R}$

(A)   (B)

Fig. 18. Determining inductive phase angle in a series RL circuit.

relationship of all voltages and current in a series RL circuit.

Determining Inductive Phase Angle in a Parallel RL Circuit. The phase angle in a parallel RL circuit can be calculated from the current triangle shown in Fig. 20. Use of the values of current through the resistor and the inductor along with the simple trigonometry formula tangent $\theta = -\frac{I_L}{I_R}$ will provide us with the *phase angle* relation between the line current and the generator voltage. In the figure, the current through the inductor is the opposite side of the triangle and the current through the resistor is the adjacent side of the triangle. Calculation of the phase angle in the circuit in Fig. 20 is as follows:

$$\text{tangent } \theta = -\frac{I_L}{I_R}$$

$$\tan \theta = -\frac{4}{3}$$

$$\tan \theta = -1.333$$

$$\text{angle } \theta = -53 \text{ degrees}$$

From the trigonometry table in Appendix D, the angle that has a tangent of 1.333 is 53 degrees. Therefore, the phase angle of the circuit is −53 degrees and line current lags generator

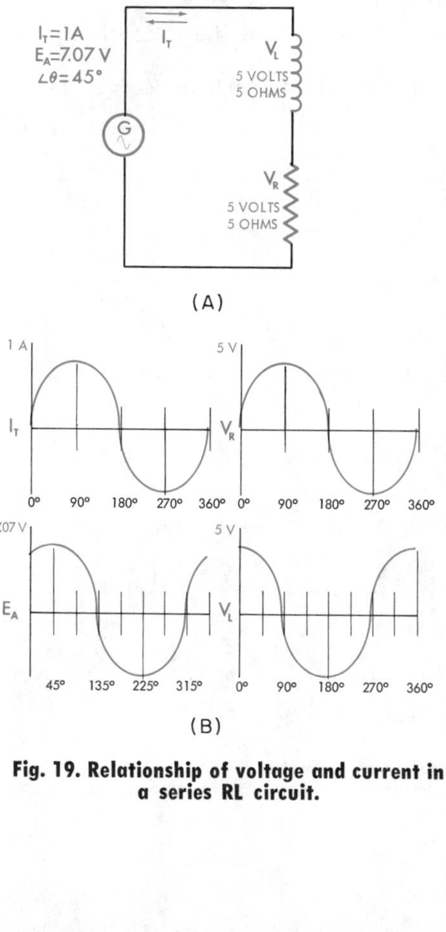

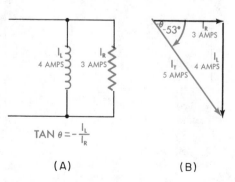

I_T=1A
E_A=7.07 V
∠θ=45°
I_T
G
V_L
5 VOLTS
5 OHMS
V_R
5 VOLTS
5 OHMS
(A)
1 A
I_T
0° 90° 180° 270° 360°
5 V
V_R
0° 90° 180° 270° 360°
7.07 V
E_A
45° 135° 225° 315°
5 V
V_L
0° 90° 180° 270° 360°
(B)

Fig. 19. Relationship of voltage and current in a series RL circuit.

I_L
4 AMPS
I_R
3 AMPS
TAN θ = - I_L / I_R
(A)
θ
-53°
I_R
3 AMPS
I_T
5 AMPS
I_L
4 AMPS
(B)

Fig. 20. Determining inductive phase angle in a parallel RL circuit.

voltage by that amount, in this particular circuit.

You will notice that this angle is negative because the I_L lags the parallel voltage by 90 degrees and also I_R by the same amount. Refer to Fig. 21 for the phase relationship of all currents and voltage in a parallel RL circuit.

Series RL Circuit Electrical Summary

The following is provided as a summary of the electrical characteristics of a series RL circuit:

a. Current stays the same throughout the circuit. ($I_T = I_L = I_R$)

b. Voltage applied is the square root of the sum of the squares of the inductor volts and the resistor volts:

$$E_A = \sqrt{V_L{}^2 + V_R{}^2}$$

c. Impedance is the square root of the sum of the squares of the inductive reactance and the resistance ($Z = \sqrt{(X_L)^2 - R^2}$

d. Voltage across the inductor leads the voltage across the resistor by 90 degrees.

e. Phase angle is calculated by the ratio of the inductive reactance and the resistance:

$$\tan \theta = \frac{X_L}{R}$$

Parallel RL Circuit Electrical Summary

The following is provided as a summary of the electrical characteristics of a parallel RL Circuit:

a. Current total is the square root of the sum of the squares of the current through the inductor and the resistor. ($I_T = \sqrt{I_L{}^2 + I_R{}^2}$)

b. Voltage is the same throughout the circuit. ($E_A = V_L = V_R$)

c. Impedance is the ratio of applied voltage and total current. $Z = \dfrac{E_A}{I_T}$

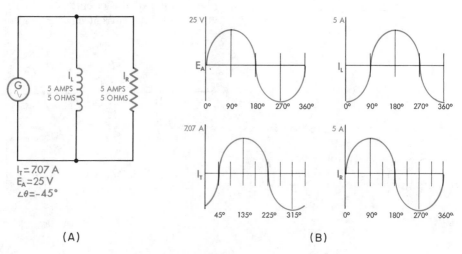

(A)
(B)

Fig. 21. Relationship of voltage and current in a parallel RL circuit.

d. Current through the inductor lags the current through the resistor by 90 degrees.

e. Phase angle is calculated by the ratio of the inductor current and the resistor current:

$$\tan \theta = -\frac{I_L}{I_R}$$

CAPACITANCE IN ALTERNATING-CURRENT CIRCUITS

The Capacitor

A *capacitor* is primarily a device for storing electricity and capactors in a-c circuits are as important as inductances.

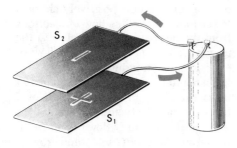

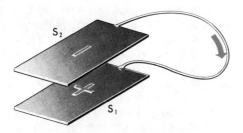

Fig. 22. The charge and discharge of a capacitor.

The manner in which a capacitor acts is illustrated in Fig. 22. It consists of two metal plates, S_1 and S_2, which are separated by a small air space or by an insulating material. When a dry cell is attached to the plates by connecting wires, current flows to the negative plate. That is, the electron flow is from the positive plate into the dry cell, resulting in a shortage of electrons on this plate. The negative plate on the other hand gains a surplus of electrons.

If the dry cell is removed the unit will store these charges for a considerable length of time. But if a wire is connected between the two plates, Fig. 22(B), the excess electrons on S_2 rush through the connecting wire as a current of electricity to neutralize the positive charge on S_1.

It must be noted here that *current cannot flow through the dielectric of a capacitor.* However, the charge and discharge of the capacitor develops current flow in the circuit the capacitor is tied to. Furthermore, if the current applied is direct in nature, the capacitor will initially charge to maximum, then block any DC current of the same magnitude. If a capacitor is to react (charge and discharge) the applied current must be pulsating direct current (varying DC) or alternating current (AC).

Unit of Capacitance. The amount of charge a capacitor will hold depends on its physical dimensions as well as on the nature of the insulating material between the plates. This quality of a capacitor is termed its *capacitance* or its *capacity*. The latter term, though not exactly correct, is often used. The unit of capacitance is the *farad*. This unit is much too large for expressing the quantity of capacitance ordinarily found in electrical circuits, and the term *microfarad* is more common. A

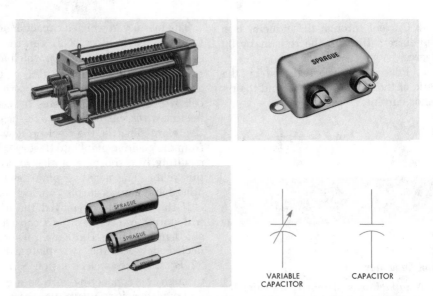

Fig. 23 Types of capacitors. *Top left*, is a variable capacitor. *Top right*, is a bathtub type, and *lower left*, shows paper capacitors. Schematic symbol is shown at the *lower right*.

microfarad is simply one-millionth (1/1,000,000) part of a farad. These terms need not be dwelt upon, however, because the *effect* of capacitance, like that of inductance, is always expressed in ohms. Several capacitor types are illustrated in Fig. 23.

Effect of Capacitance in Circuit

In Fig. 24(A), an alternating-current generator, G, is connected in series with a resistor, R, and capacitor C, the upper terminal of G being negative, the lower terminal positive. Electrons are attracted from the lower plate of C to the positive terminal of the generator, their passage through the lower wire forming a current of electricity. At the same time, electrons move from the negative terminal of the generator to the upper plate of C to supply atoms

whose electrons have been drawn out of orbit by the now positively-charged lower plate of C. This flow of electrons from the negative terminal of G to the upper plate of C constitutes a flow of electric current in the upper wire.

It should be observed that no current flows directly from the upper plate of C through the intervening space to the lower plate. The only flow of current in the circuit is from G to C, and from C to G.

As a positive charge grows on the lower plate of C, and a negative charge increases on the upper plate of C, electrical pressure (voltage) is built up in C. This voltage tries to force electrons from the upper plate of C through the generator in a counter-clockwise direction to the lower plate of C. The generator voltage opposes this force, since G is trying to send current in the opposite, or clockwise, direction. But, as the

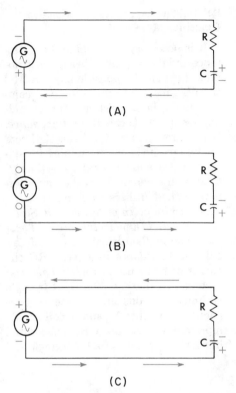

(A)

(B)

(C)

Fig. 24. Alternating current in a capacitive circuit.

generator voltage decreases toward zero on the sine curve, the voltage of the capacitor finally exceeds that of the generator and manages to force current in the counter-clockwise direction.

At the instant shown in Fig. 24(B), when the generator voltage is passing through zero, the flow of current established by the voltage on C is quite high. When the generator voltage starts to increase again, but in the below-the-line direction, it tends to establish a current which flows in a counter-clockwise direction as shown in Fig. 24(C). The capacitor has already started a flow of current in this direction. A capacitor is said, therefore, to force cur-

rent to lead the voltage of a circuit. This effect is directly opposite to that of inductance which forces current to lag behind the voltage.

Current Lead in a Pure Capacitive Circuit. Fig. 25 shows the sine curves of voltage and current in a circuit which contains pure capacitance. At point *1*, when the voltage E of the generator is zero, current is at maximum. At point *2*, current is at zero while voltage has increased to maximum.

The negative alternation is exactly like the positive alternation, current reaching its maximum and zero values 90 degrees ahead of the voltage. Thus, while current lags behind voltage in an inductive circuit, it leads voltage by 90 degrees in a pure capacitive circuit.

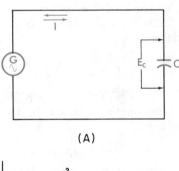

(A)

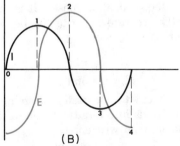

(B)

Fig. 25. The relationship between current and voltage in a pure capacitive circuit.

Determining the Reactance of a Capacitor

The capacitor, like the inductor, resists the flow of alternating-current. This opposition is known as *capacitive reactance* and is measured in ohms. The symbol for capacitive reactance is X_C and its value determined by the formula;

$$X_c = \frac{1}{2\pi fC}$$

X_c = capacitive reactance in ohms
π = 3.1416
f = frequency of the supply
    current in hertz
C = capacity in farads

An example will show how this is done. A 665-microfarad capacitor is connected in a 60-Hz alternating-current circuit. To express the capacitance in farads it is necessary to divide microfarads by 1,000,000. This is done by moving the decimal point 6 places to the left. Therefore, $665 \div 1,000,000 = 0.000665$. Substituting in the formula we have

$$X_c = \frac{1}{2\pi fC}$$

$$X_c = \frac{1}{2 \times 3.1416 \times 60 \times 0.000665}$$

Multiplying the numbers in the denominator,

$$X_c = \frac{1}{0.250572}$$

For all practical purposes, it is sufficiently accurate to write the denominator as 0.25, which gives,

$$X_c = \frac{1}{0.25}$$
$$X_c = 4 \text{ ohms}$$

To avoid the necessity of changing microfarads into farads, the above formula may be written:

$$X_c = \frac{1,000,000}{2\pi fC}$$

Determining Impedance in a Resistive-Capacitive (RC) Circuit

It is necessary to combine the resistance and the capacitive reactance to obtain the actual opposition of an RC circuit. This total opposition, the combination of both resistance and capacitive reactance, is called the *impedance*. It is represented by the letter Z. There are two methods of determining impedance. One is mathematical, using the Pythagorean theorem. The other, an alternate method, utilizes graph paper.

Determining Impedance in a Series RC Circuit Using Pythagorean Theorem. The mathematical method of calculating impedance in a series RC circuit is with the use of the *Pythagorean theorem*. In the Fig. 26 a right triangle is drawn with one side representing capacitive reactance X_C and the other side representing resistance R. The dotted line is the hypotenuse of the triangle and

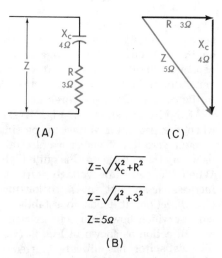

$$Z = \sqrt{X_c^2 + R^2}$$
$$Z = \sqrt{4^2 + 3^2}$$
$$Z = 5\Omega$$

(B)

Fig. 26. Determining impedance in a series RC circuit using the Pythagorean theorem.

it represents impedance Z. The arrowed lines in the figure are called *phasors* and the triangle is called a *phasor triangle*.

The Pythagorean theorem states that the hypotenuse of a right triangle is equal to the square root of the sum of the squares of the triangle's sides. ($H = \sqrt{S^2 + S^2}$). If we substitute electrical terms for the parts of this equation, we create the impedance formula $Z = \sqrt{X_C^2 + R^2}$. The calculation of impedance in the Fig. 26 involves the following:

$$Z = \sqrt{X_C^2 + R^2}$$

$$Z = \sqrt{4^2 + 3^2}$$

$$Z = \sqrt{16 + 9} = \sqrt{25}$$

$$Z = 5 \text{ ohms}$$

This particular calculation has been done with small numbers. If larger values of capacitive reactance were used (and this is normally the case) square and square root tables or a slide rule would have to be used. As an example:

$$Z = \sqrt{X_C^2 + R^2}$$

$$Z = \sqrt{32^2 + 24^2}$$

$$Z = \sqrt{1024 + 576} \quad \sqrt{1600}$$

$$Z = 40 \text{ ohms}$$

A square root table is provided in Appendix E. Also see Appendix D for an alternate and much simpler solution using trigonometry tables or slide rule.

Determining Impedance in a Series Resistive Capacitive (RC) Circuit Using Alternate Method. Fig. 27 shows a circuit which contains a 3 ohm resistor and a capacitor with a reactance of 4 ohms. The impedance of the group may be obtained by the use of graph paper in a way similar to the method for determining impedance in the inductive circuit.

Two lines are drawn at right angles, see Fig. 27. As before, the horizontal line represents resistance of the circuit. The vertical line represents capacitive reactance. It should be noted that the vertical line is downward from the resistance line instead of upward.

Mark off three spaces to the left to indicate 3 ohms resistance and four spaces downward to indicate 4 ohms of capacitive reactance. The length of the Z line, drawn from the end of the vertical line to the end of the horizontal line, is measured in the same way as it was done in Fig. 15. The impedance is found to be 5 ohms.

Determining Impedance in a Parallel RC Circuit. Impedance in a parallel RC

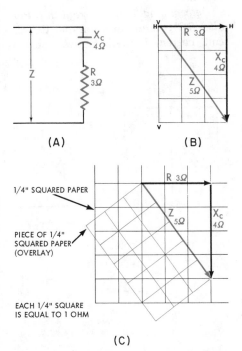

Fig. 27. Determining impedance in a series RC circuit using alternate (graph paper) method.

circuit is found using Ohm's law. This method assumes that the applied voltage and total current are known. Calculation procedures are as follows:

$$Z = \frac{E}{I_T}$$

$$Z = \frac{100V}{5A}$$

$$Z = 20 \text{ ohms}$$

Determining Total Current in a Resistive Capacitive (RC) Circuit

Determining Total Current in a Series RC Circuit. In a series RC circuit, current stays the same throughout the circuit. The phase angles mentioned previously have to do with the relationship of generator (supply) voltage and its current.

Determining Total Current in a Parallel RC Circuit. In a parallel RC circuit, current is mathematically calculated with the use of the *Pythagorean theorem.*

In the Fig. 28 a right triangle is drawn with one side (phasor) representing current through the capacitor I_C and the other side (phasor) representing current through the resistor I_R. The dotted line is the hypotenuse and represents total current I_T. Calculation of the values in Fig. 28 are as follows:

$$I_T = \sqrt{I_C{}^2 + I_R{}^2}$$

$$I_T = \sqrt{4^2 + 3^2}$$

$$I_T = \sqrt{16 + 9} = \sqrt{25}$$

$$I_T = 5 \text{ amps.}$$

Determining Total Voltage in a Resistive Capacitive (RC) Circuit

Determining Total Voltage in a Parallel RC Circuit. In a parallel RC circuit, voltage stays the same throughout the circuit. The phase angles mentioned previously have to do with the relationship of generator (supply) voltage and its current.

Determining Total Voltage in a Series RC Circuit. In a series RC circuit, voltage is mathematically calculated with the use of the *Pythagorean theorem.*

In the Fig. 29 a right triangle is drawn with one side (phasor) representing voltage across the resistor R. The dotted line is the hypotenuse of the triangle and it represents total applied voltage E_A. Calculation of the values in the circuit of Fig. 29 are as follows:

$$E_A = \sqrt{V_C{}^2 + V_R{}^2}$$

$$E_A = \sqrt{4^2 + 3^2}$$

$$E_A = \sqrt{16 + 9} = \sqrt{25}$$

$$E_A = 5 \text{ volts}$$

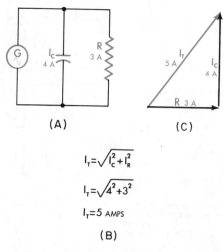

(A)

(C)

$$I_T = \sqrt{I_C{}^2 + I_R{}^2}$$

$$I_T = \sqrt{4^2 + 3^2}$$

$$I_T = 5 \text{ AMPS}$$

(B)

Fig. 28. Determining total current in a parallel RC circuit.

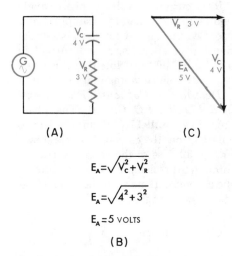

$$E_A = \sqrt{V_C^2 + V_R^2}$$

$$E_A = \sqrt{4^2 + 3^2}$$

$$E_A = 5 \text{ VOLTS}$$

(B)

Fig. 29. Determining total voltage in a series RC circuit.

Determining Capacitive Phase Angles

In a pure capacitive circuit such as that shown in Fig. 25, the current leads the voltage by 90 degrees. That is, the current is at maximum 90 degrees or one quarter cycle before the voltage arrives at maximum. Since the cycle represents time, then it can be stated that the current leads the voltage in a pure capacitive circuit by one quarter cycle in time. This relationship between current and voltage across any capacitor is always the same.

However, when a capacitor is inserted in a circuit with another device such as a resistor, the circuit becomes reactive and a phase relationship exists between the generator voltage and its current. The relationship we speak of is the *phase angle*. This phase angle is normally known by the Greek letter theta and is represented by the symbol θ.

Determining Capacitive Phase Angle in a Series (RC) Circuit. The phase angle in a series RC circuit can be calculated from the impedance triangle shown in Fig. 30. Use of the values of resistance and capacitive reactance, along with the simple trigonometry formula:

$$\text{tangent } \theta = - \frac{X_C}{R}$$

will provide us with the *phase angle* relation between generator voltage and its current.

Tangent (or tan) is a trigonometric function of a right triangle. The tangent is the ratio of the opposite side to the adjacent side of any right triangle. In the Fig. 30, the opposite side is the capacitive reactance X_C and the adjacent side is resistance R.

Calculation of the phase angle in Fig. 30 is as follows:

$$\text{tangent } \theta = - \frac{X_C}{R}$$

$$\tan \theta = - \frac{4}{3}$$

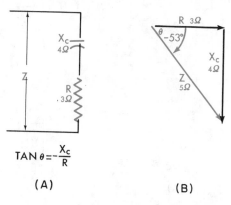

$$\text{TAN } \theta = - \frac{X_C}{R}$$

(A)

(B)

Fig. 30. Determining capacitive phase angle in a series RC circuit.

$$\tan \theta = -1.333$$

$$\text{angle } \theta = -53 \text{ degrees}$$

From the trigonometry table in Appendix B, the angle that has a tangent of 1.333 is 53 degrees. Therefore, the phase angle of the circuit is 53 degrees and current leads generator voltage by that amount, in this particular circuit.

In any series combination of X_C and R, current will lead generator (supply) voltage by some value between 0 and 90 degrees. The larger the X_C is, the larger the phase angle will be. Since this is a series circuit, the larger the X_C is, the larger the voltage drop will be across X_C. Current will remain the same throughout the circuit.

It must be repeated here that current through the resistor is in phase with the voltage drop across the resistor. Similarly, current through the capacitor leads the voltage drop across the capacitor by 90 degrees.

The current-to-voltage phase angle relates to the generator (supply) voltage. See the Fig. 31 for the phase relationship of all voltages and current in a series RC circuit. Note that this angle is negative since voltage lags current.

Determining Capacitive Phase Angle in a Parallel RC Circuit. The phase angle in a parallel RC circuit can be calculated from the current triangle shown in Fig. 32. Use of the values of current through the resistor and current through the capacitor along with the simple trigonometry formula:

$$\text{tangent } \theta = \frac{I_C}{I_R}$$

will provide us with the *phase angle* relation between the line current and the generator voltage.

In Fig. 32, the current through the capacitor is the opposite side of the triangle and the current through the resistor is the adjacent side of the triangle.

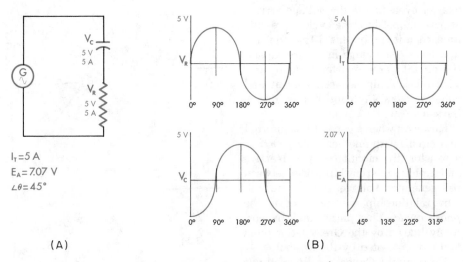

$I_T = 5 \text{ A}$
$E_A = 7.07 \text{ V}$
$\angle \theta = 45°$

(A)   (B)

Fig. 31. Relationship of voltage and current in a series RC circuit.

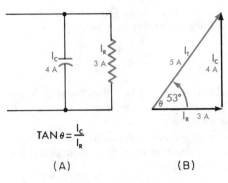

$$TAN\theta = \frac{I_C}{I_R}$$

(A)   (B)

Fig. 32. Determining capacitive phase angle in a parallel RC circuit.

Calculation of the phase angle in the circuit in Fig. 32 is as follows:

$$\text{tangent } \theta = \frac{I_C}{I_R}$$

$$\tan \theta = \frac{4}{3}$$

$$\tan \theta = 1.333$$

angle $\theta = 53$ degrees

From the trigonometry table in Appendix D, the angle that has a tangent of 1.333 is 53 degrees. Therefore, the phase angle of the circuit is 53 degrees and line current leads generator voltage by that amount, in this particular circuit. Refer to Fig. 33 for the phase relationship of all currents and voltage in a parallel RC circuit.

Series RC Circuit Electrical Summary

The following is provided as a summary of the electrical characteristics of a series RC circuit:

a. Current stays the same throughout the circuit. ($I_T = I_C = I_R$)

b. Voltage applied is the square root of the sum of the squares of the capacitor volts and the resistor volts. ($E_A = \sqrt{V_C^2 + V_R^2}$

c. Impedance is the square root of the sum of the squares of the capacitive

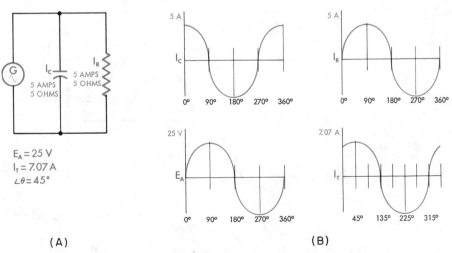

$E_A = 25$ V
$I_T = 7.07$ A
$\angle\theta = 45°$

(A)   (B)

Fig. 33. Relationship of voltage and current in a parallel RC circuit.

reactance and the resistance. ($Z = \sqrt{(X_C)^2 + R^2}$

d. Voltage across the capacitor lags the voltage across the resistor by 90 degrees.

e. Phase angle is calculated by the ratio of the capacitive reactance and the resistance. ($\tan \theta = -\dfrac{X_C}{R}$)

Parallel RC Circuit Electrical Summary

The following is provided as a summary of the electrical characteristics of a parallel RC circuit:

a. Current total is the square root of the sum of the squares of the current through the capacitor and the resistor.

$I_T = \sqrt{I_C^2 + I_R^2}$

b. Voltage is the same throughout the circuit. ($E_A = V_C = V_R$)

c. Impedance is the ratio of applied voltage and total current:

$$(Z = \frac{E_A}{I_T})$$

d. Current through the capacitor leads the current through the resistor by 90 degrees.

e. Phase angle is calculated by the ratio of the capacitor current and the resistor current:

$$(\tan \theta = \frac{I_C}{I_R})$$

Circuit Containing Resistance, Inductive, and Capacitive Reactance

Fig. 34(A) shows a circuit which

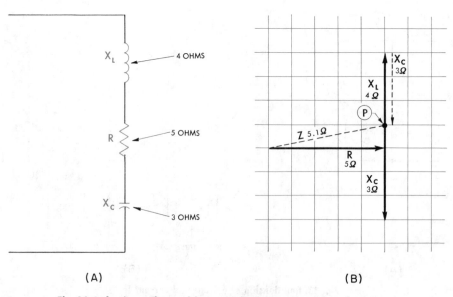

(A)   (B)

Fig. 34. Inductive and capacitive reactances in series with resistance.

contains a resistance of 5 ohms, an inductive reactance of 4 ohms, and a capacitive reactance of 3 ohms. The impedance of this group may be determined as shown in Fig. 34(B). A horizontal line is drawn to represent circuit resistance. A vertical line is drawn which extends both above and below the resistance line and at right angles to it to represent inductance and capacitance. Resistance is marked off on the horizontal line to represent 5 ohms. The inductive reactance of 4 ohms is marked off upward on the vertical line. The capacitive reactance of 3 ohms is marked off downward on the vertical line.

Since inductive reactance forces current to lag behind the voltage, and capacitive reactance forces current to lead the voltage, the two forces are in opposition. And, since the value of the reactance represents the effectiveness of inductance or capacitance in the circuit, the resultant effect on the circuit is equal to the difference between the two. If the inductive reactance is greater than the capacitive, then the second quantity is subtracted from the first. If capacitive reactance is greater the inductive reactance is subtracted from it.

To do this on the graph, transfer the shorter of the two vertical lines, the capacitive reactance, and place it by the longer line of the inductive reactance. The shorter line is marked off from the tip of the longer. In Fig. 34(B), the capacitive reactance arrow is indicated with a broken arrow pointing downward. This measurement brings us to point P. By this means the value of capacitive reactance is subtracted from the inductive reactance.

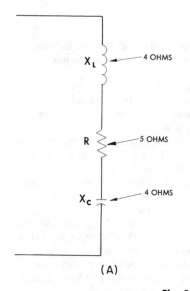

(A)

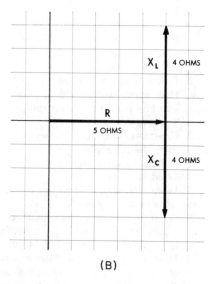

(B)

Fig. 35. A series resonant circuit.

Drawing the Z line from the origin of the resistance line to point P, the impedance measures 5.1 ohms. Current in the circuit is determined by dividing circuit voltage by this impedance. It is well to mention that if capacitive reactance is greater than the inductive, the impedance line will be below the resistance line.

Where inductive reactance is greater, current lags the voltage, which is represented by the fact that point P lies above the resistance line. If capacitive reactance is greater, current leads the voltage, which is indicated by point P lying below the resistance line.

Resonance

Fig. 35(A) shows a series circuit containing a resistance of 5 ohms, an inductance of 4 ohms, and a capacitive reactance of 4 ohms. It is seen that the value of $X_L = X_C$. Fig. 35(B) illustrates how these quantities may be represented on graph paper. The arrows for R, X_L and X_C are indicated, but if either of the reactances are subtracted from the other the result is zero.

Under this condition, when $X_L = X_C$, the circuit is said to be *resonant*, or *in resonance*. The current in the series resonant circuit is found by dividing the voltage by the circuit resistance.

REVIEW QUESTIONS

1. What term denotes a single reversal of alternating current?

2. What is a hertz?

3. Define the term *frequency*.

4. How many mechanical degrees are represented by the armature circumference in an 8-pole alternator?

5. How many electrical degrees are represented in the alternator?

6. Define the term impedance.

7. What is a sine wave?

8. What is the ratio of effective to maximum value of a sine wave?

9. What is the maximum value of a single-phase current wave whose heating effect is equal to a d-c current of 14 amps?

10. How many electrical degrees separate three-phase voltage waves in a 4-pole alternator?

11. What outward difference in appearance is noticeable between lap and wave windings?

12. State the rule for armature coil span.

13. A four-pole, three-phase armature has 48 slots. How many slots are there per pole per phase?

14. What name is given to the stationary member of a revolving-field alternator?

15. What are eddy currents?

16. The inductance of a device used in a 60-hertz circuit is .02 henry. What is its inductive reactance?

17. The capacitance of a device used in a 60-hertz circuit is 500 microfarads. What is its capacitive reactance?

18. If the two devices are placed in series with a 6-ohm resistor on a 60-hertz circuit, what is the total impedance?

19. What conditions must exist for a circuit to be resonant?

20. How much power is consumed in a balanced three-phase circuit which carries 20 amps at a pressure of 240 volts, under a power factor of .65?

21. Calculate impedance in a series RC circuit with a reactance of 10 ohms and a resistance of 8 ohms.

22. Calculate impedance in a parallel RL circuit with a reactance of 12 ohms and a resistance of 9 ohms.

23. Calculate the phase angle of a series RL circuit with a reactance of 16 ohms and a resistance of 12 ohms.

24. Calculate the phase angle of a parallel RC circuit with a reactance of 15 ohms and a resistance of 9 ohms.

25. What is the phase relationship of current and voltage in a pure inductive circuit?

Chapter 9

Electric Motors

PRINCIPLES OF OPERATION

Producing Simple Motion

When current passes through a conductor which lies under a magnetic pole, the conductor will move in a direction at right-angles to the direction of flux. In Fig 1 (*left conductor*), cur-

rent flows into a wire that is surrounded by flux from an *N* pole. The circles of force act in a clockwise direction, crowding against flux lines at the bottom and nullifying those at the top. Resulting magnetic force causes motion upward as indicated by the large arrow.

If the current is reversed, Fig. 1 (*right conductor*), crowding of magnetic lines occurs on the top side, and nullifying of lines at the bottom, so that movement is downward.

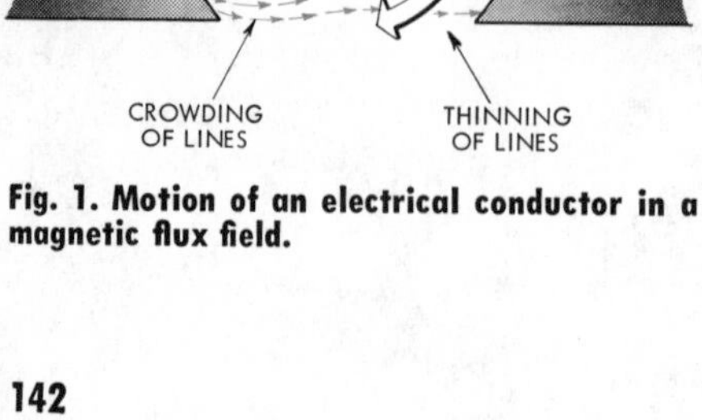

Fig. 1. Motion of an electrical conductor in a magnetic flux field.

Armature Rotation

If the brushes of a direct-current generator are supplied with current, the conductors near the poles will be repelled and the armature will start to revolve, other winding conductors taking the places under the poles of those carried into the neutral zone. The commutator, meanwhile, serves to reverse current flow in conductors that pass under the brushes, the combination of changed current flow and changed magnetic polarity resulting in continuous clockwise or counter-clockwise rotation, as the case may be.

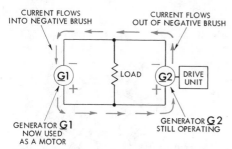

Fig. 2. Shunt generator used as a motor.

Comparison Between Armature Rotation as Generator and as Motor

Two shunt generators, Fig. 2, have been operating in parallel to supply load L. Generator G_1 has been uncoupled from its driving unit, and is to be used as a motor, its negative terminal still connected to the upper wire, its positive terminal to the lower one. Generator G_2 remains attached to its driving engine, furnishing power to load L and to motor G_1. Investigation discloses that the armature of G_1 now rotates in the same direction as when used as a generator.

The seeming contradiction is quickly resolved upon analysis of the facts. Although the armature connections of G_1 have not been disturbed, current flow through the windings has been reversed. When furnishing power, current flowed out of the windings into the negative brush, and then to the negative load conductor. When taking power from the supply wires, on the other hand, current flows from the negative wire into the negative brush, and through the windings in the oppo-

site direction from generated current. At this point, an elementary but important observation is in order: Current flows *out* of the negative brush of a generator, but *into* the negative brush of a motor.

Motor Force Developed in Generator Armature

Fig. 3 illustrates a generator armature which is being driven clock-wise, one of the conductors passing across a pole face with current flowing away from the observer. If this were a motor, the direction of current would force the conductor to move upward, or counter-clockwise. There is in fact, a motor force present in the conductor which attempts to do just that and opposes the rotation of the conductor as shown by the colored arrow in Fig. 3. This motor force is not sufficient to stop rotation of the generator, but does create opposition to the driving force of the engine which turns the armature. This explains why more steam would be necessary for a drive turbine, or more fuel for a gas engine, as the amount of current drawn from the generator increases.

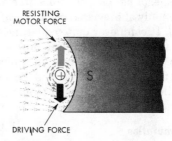

Fig. 3. Driving and resisting forces in a generator armature. The color arrow indicates the direction of the opposing motor force.

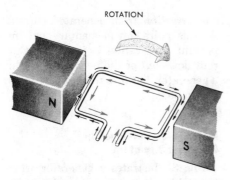

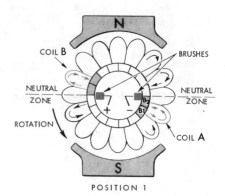

Fig. 4. The counter-voltage generated in a motor armature is shown by the color arrow.

POSITION 1

Counter Voltage Generated in a Motor Armature

Fig. 4 illustrates a conductor of a motor armature which is rotating clockwise with current flow as shown by black arrows. The movement of the conductor through the magnetic field creates a generator action which produces a voltage opposing the current of the motor armature, and attempts to send current through the conductor in the opposite direction as shown by the colored arrows in Fig. 4. That is, voltage generated in the motor armature opposes the flow of current which makes rotation possible.

Such pressure is termed counter-voltage, or *counter-electromotive force,* commonly abbreviated *counter-emf* and *cemf.* It can never quite equal the supply voltage, since current flow would then cease. The manner in which it affects motor operation will be taken up shortly in the study of motor speed characteristics.

Commutation

An important feature of direct current machines is illustrated with the

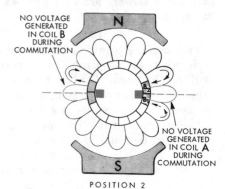

POSITION 2

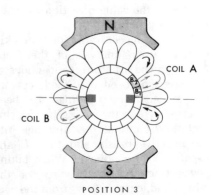

POSITION 3

Fig. 5. The sequence of the commutation process in a motor armature.

help of Fig. 5 which shows a section of ring-wound armature, this type of winding being chosen here, instead of the familiar barrel type, to simplify explanation. Two coils, A and B are illustrated, but only one will be referred to in the explanation, as the process is similar for both coils. Coil A is illustrated in three positions while the armature turns in a counter-clockwise direction as indicated by the large arrow. When coil A occupies position 1, electron flow through it is shown by the small arrow marked there. When coil A lies in position 2, and the neutral zone, its ends are short-circuited by the brush resting on commutator bars B_1 and B_2 electrons from the lower portion of the armature winding passing down a lead wire to bar B_1 and those from the upper portion of the winding passing down a lead wire to bar B_2, and none through coil A, itself. When the coil has reached position 3, electrons from the upper half of the winding flow through it to bar B_1. It should be noted that in passing from position 1 to position 3, electron flow through the coil has been reversed. This process is termed *commutation,* and coil A is said to have been commutated during the interval.

Fig. 6 shows the magnetic state in a motor which is carrying no load, the magnetic flux from the field poles spreading symmetrically over the surface of the armature core whose neutral, or no-flux area is marked by line N-N. When the motor is loaded, however, electrons flowing in the armature winding magnetize the armature core, establishing poles on it. Interaction between field and armature fluxes causes the flux issuing from the field poles to assume a distorted position illustrated in Fig. 7 so that the neutral or no-flux position has shifted to line G-G.

In such case, coil A will be commutated outside the neutral zone, and undesirable sparking will be present during operation, as illustrated in Fig. 8. If the coil shown in color is cutting magnetic lines from the N pole as shown in Fig. 8, the voltage generated

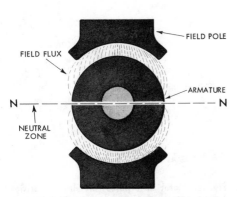

Fig. 6. The undistorted flux field of a motor under no-load conditions.

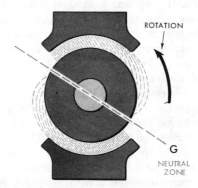

Fig. 7. Flux field of a motor under load showing flux distortion and shifting of the neutral zone. Note that the flux shift is backward against the direction of rotation.

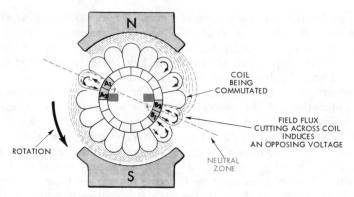

Fig. 8. Sparking at the brushes caused by commutation outside the neutral zone.

in it opposes the flow of electrons in the lower portion of the winding before it is short circuited by the brush. Some of the electrons, therefore, choose the easier path through the air from commutator bar B_1 to the tip of the brush, the resulting spark burning the copper surface.

For this reason it becomes necessary to shift the brushes backward against the direction of rotation in order that the coil may be commutated while in a neutral position. A method employed in commercial machines to avoid necessity for shifting brushes is to use small commutating poles, usually referred to as interpoles, as in Fig. 9. These small poles are excited by coils which are in series between brushes and load, their magnetic strength being directly proportional, therefore, to the strength of the armature magnetic poles. They are carefully designed to nullify that portion of the field magnetic flux which attempts to shift into neutral zone $N\text{-}N$. Under this condition, coil A may be safely commutated in the true neutral position for any value of load.

Flux shift in a motor, shown in Fig. 7, is opposite from that in a generator. Motor commutating poles perform an identical purpose to those in a generator, but must be of opposite polarity.

For clockwise rotation, the motor interpole is of the same polarity as the main pole *behind* it with respect to the direction in which the armature turns.

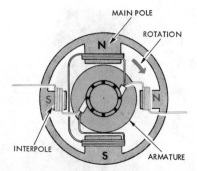

Fig. 9. Motor interpoles showing polarities for counter-clockwise rotation. Polarities of the interpoles must shift the field flux in the direction of rotation.

This relationship is easy to remember, since the right-hand interpole, for example, simply furnishes excitation which could be obtained by shifting the brushes backward, to place the commutated coil under the influence of the pole it was about to leave.

Commercial Direct-Current Motor

The cut-away view of a direct-current motor in Fig. 10 shows armature core and winding, field coils, commutator, and brushes. It also displays certain mechanical features of the commercial unit, such as armature bearings, cooling fan, and mounting.

Motor Speed

The speed of a motor is usually determined by the number of times it rotates or revolves in a period of one minute. This is known as the r.p.m. (revolutions per minute) of the motor.

The *load* applied to the motor is that device that the motor is turning or operating. It may be something easy to turn or it may be difficult to turn. If the device is easy to turn, it is known as

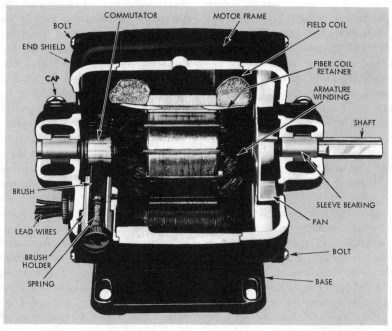

Fig. 10. Cutaway view of a direct-current motor showing its various parts.

a *light load;* and if it is difficult to turn it is known as a *heavy load.* Now, in most cases, a motor will turn more slowly under a heavy load than it will under a light load, but the amount of difference in speed of a motor between turning a light load and a heavy load depends upon the individual type of motor. As we will see, different types of motors will react differently to various loads, and the manner in which the motor reacts to a load is known as the *speed characteristic* of the motor.

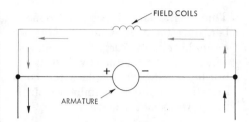

Fig. 11. Schematic of a shunt motor.

Shunt-Motor Speed Characteristics

The manner of exciting the fields in direct-current motors is quite similar to that used in direct-current generators. Except in special devices, motor field coils are never separately excited. The various types of direct-current motors are noted by the type of field coils employed, for each field winding produces a motor which has different speed characteristics than the others. We will consider each of these arrangements separately.

The *shunt-wound* motor, which is more commonly known as a *shunt motor,* has the field coils connected in parallel with the armature, as shown in the schematic diagram of Fig. 11. Since the field coils are connected directly to the supply lines, the current flowing through the coils is constant (with a constant supply voltage) and, accordingly, the field excitation in the shunt motor remains practically the same regardless of the nature of the load on the motor. With a constant field flux, the counter-emf is directly proportional to the speed of the armature, for the faster the armature revolves the more counter-emf will be developed. In a like manner, the slower

the armature rotates, the less counter-emf is generated.

Suppose the supply voltage is 115, armature resistance .2 ohm, no-load armature current 5 amperes, and no-load speed 1500 rpm (revolutions per minute). Since the armature draws a current of 5 amperes in order to rotate with no load connected to its shaft, the counter-emf cannot be greater than the difference between the supply voltage and the voltage drop in the armature. Here, 5 amperes flowing through an armature resistance of .2 ohm causes a voltage drop of 1 volt ($5 \times .2 = 1$). The counter-emf developed in the armature must be the difference between 115 volts and 1 volt, or 114 volts.

When full load is applied to the armature shaft, it is found that the armature draws a current of 25 amperes in order that it may continue to revolve. The voltage drop in this case, is equal to: $25 \times .2$, or 5 volts. The counter-emf, under this condition, must be equal to 115 minus 5, or 110 volts. And, since the armature speed at no-load is 1500 rpm with a counter-emf of 114 volts, the armature speed must be 1447 rpm at full-load with a counter-emf of 110 volts. This result follows because the counter-emf is directly proportional to speed when the field strength remains constant, and

$$\frac{110 \text{ volts}}{114 \text{ volts}} \times 1500 \text{ rpm} = 1447 \text{ rpm}$$

Between no-load and full-load, then, the speed of the motor decreases only 53 rpm (1500 minus 1447). The values obtained here apply only to the particular motor tested, but similar results will be found in connection with any other shunt motor. For this reason, the shunt motor is said to be a constant-speed unit.

Compound-Motor Speed Characteristics

Fig. 12 shows the diagram of a compound motor. The supply voltage is 115 v, as before, the armature resistance is 0.2 ohm, no-load current is 5 amps, and the speed at no load is 1,500 rpm. The resistance of the series field winding is so low that it may be disregarded here. This winding is designed so that it increases the field flux $\frac{1}{10}$, or 10 percent, at full load. Suppose this armature draws 25 amps at full load. As in the case of the shunt motor, the voltage drop will be 5 v, so the counter-emf must be 110 v.

If the strength of the field flux were unchanged between no-load and full-load, the speed of the armature at full load would be exactly the same as that of the shunt motor, or 1,447 rpm, but the strength of the field flux has been increased by 10 percent, so that for every revolution the armature will produce 10 percent more counter-emf than before. The speed of the armature must be lower than 1,447 rpm. It must be

$$\frac{100\% \; shunt \; flux}{110\% \; compound \; flux} \times 1447 \; rpm$$
$$= 1316 \; rpm$$

The drop in speed between no load and full load is equal to 1,500 minus 1,316, or 184 rpm. It is evident that the speed of the compound motor is not so constant as that of the shunt motor. As a matter of fact, the full-load speed of a compound motor depends upon the strength of the series field as compared to that of the shunt field. Here, the series field produces, at full load, 10 percent of the excitation of the shunt field. If it had produced 20 percent as much excitation, the speed at full load would have been even lower than 1,316 rpm. If it had produced only 5 percent as much excitation as the shunt field, the speed at full load would have been higher than 1,316 but it would not have been equal to that of the shunt motor.

The compound motor also has good speed regulation, but it is not so good as the shunt motor.

Series-Motor Speed Characteristics

Fig. 13 is the diagram of a series motor. The supply voltage is 115 v, and the armature resistance is 0.2 ohm. Since there is no shunt field winding, the strength of the field flux will depend upon the current flowing through

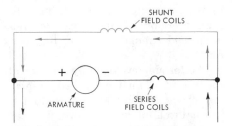

Fig. 12. Schematic of a compound motor.

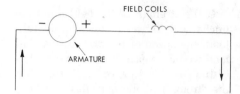

Fig. 13. Schematic of a series motor.

the armature and the number of turns of wire in the field coils. If this winding is designed to produce the same field strength at full load as that of the corresponding shunt motor, it will have full field strength when the armature current is 25 amps. Let us suppose that the speed at this time is 1,450 rpm.

Now, since it requires approximately 5 amps to supply armature losses at no load, so that the armature may continue to rotate, the field strength of the motor will be only ⅕ that at full load. The armature will have to run considerably faster, in this case, in order to generate a counter-emf of 114 v, which is the value necessary to limit armature current to 5 amps. As a matter of fact, the armature will require somewhat more than 5 amps to keep it rotating at the higher speed. Let us say that 10 amps are required.

The armature drop is now 10 amps times 0.2 ohm, or 2 v, so that the counter-emf must be 115 v minus 2 v, or 113 v. With full-load field strength, the speed would be $^{113}/_{114}$ times 1,500 rpm, or 1,487 rpm but the field strength with 10 amperes flowing is only $^{10}/_{25}$ times normal value, or 40 percent. In order to generate the required counter-emf, the armature must run at a speed 2½ times as fast as full-load speed, or 3,717.

The speed variation from no load to full load, therefore, will amount to 3,717 rpm minus 1,487 rpm, or 2,230 rpm. For this reason, the series unit is said to be a varying speed motor, its speed changing drastically with change in load.

Although the maximum speed attained by the motor at no load, here, was only 3,717 rpm, a motor with a lower armature resistance might attain a speed of several thousand revolutions per minute. Such a speed is often sufficient to tear the windings out of armature slots. For this reason, series motors are usually applied to loads in which it is impossible for the motor to be without load. In very small motors, the pressure of the brushes on the commutator is sufficient to limit the speed to a safe value. Larger sizes are usually applied to geared loads, or direct-connected loads, so that it is unlikely for dangerous speeds to occur.

Torque of Motor

The term *torque* means *turning effort*. It is usually expressed as the turning effort in pounds at a distance of 1 foot from the center of the armature shaft. Fig. 14 illustrates this term. Now, the torque of a motor armature depends upon two factors: the strength of the magnetic flux in the field poles, and the strength of the armature poles. The strength of the armature poles depends upon the value of the current flowing in the armature winding. It may be said, therefore, that torque depends upon field strength and armature current.

The torque developed by a motor at the instant of starting is called its *starting torque*. In direct-current motors, it is usual to limit starting current to 150 percent, or 1½ times, its full-load run-

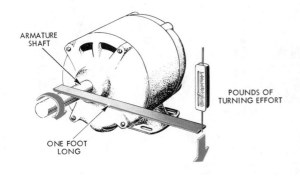

ARMATURE
SHAFT

POUNDS OF
TURNING EFFORT

ONE FOOT
LONG

Fig. 14. The method used to determine motor torque.

ning current. The manner in which the current is limited will be considered shortly. In a shunt motor, whose field strength is the same at all times, the field strength may be said to be 100 percent, or 1. With a current of 1½ times normal full-load value, and a field strength of 1, the starting torque of the motor will be 1½ × 1, or 1½ times normal full-load, or running torque.

Now, the torque required at starting is usually greater than required after the driven machine has started to rotate. In some cases it is several times running torque. Therefore, it is desirable to have a motor whose starting torque is fairly high. In the compound motor, the armature current flows through the series field winding. If this winding is so proportioned that it provides 10 percent, or ⅒, more field flux at full load, it will provide 15 percent, or 3⁄20, more field strength at starting when 1½ times normal current flows. Here, the starting torque will be approximately 1¾ that of the running value. If the series field coils are wound with a greater number of turns, the starting torque will be correspondingly

greater. Because the compound motor has the higher torque, it is often preferred to the shunt motor even though its speed will vary more during operation.

The series motor provides the greatest starting torque of the three, for, if normal running current provides a field strength which is 100 percent, a field current which is equal to 1½ times the normal value will provide a field strength of 1½ times the normal value. This follows from the fact that the same current flows through the series field coils as flows through the armature. The starting torque of the series motor, under this condition, is 1½ times 1½, or 2¼ times normal running torque. Thus, the starting torque of the series unit is much greater than that of the other types for the same amount of current.

Horsepower. Electric motors are rated in horsepower (abbreviated hp). A motor of 1 hp which runs at a speed of 1,000 rpm will exert force of 10.52 pounds at the outer surface of a pulley 12 inches in diameter. The electrical equivalent of 1 horsepower is 746 watts.

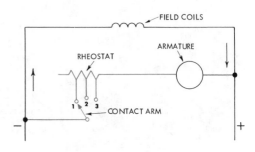

Fig. 15. A starting rheostat connected in the circuit with a shunt motor.

Limiting Starting Current

The armature conductors are not cutting across the field flux before the armature begins to rotate. The counter-emf of the armature at this instant is, therefore, zero, and the current which flows in the armature circuit may be determined by Ohm's law. Fig. 15 shows the *starting rheostat* connected in series with the armature of a shunt motor. It will be noted that the field coils are connected directly to the supply wires. The rheostat consists of a resistor and a movable contact arm. The resistor is connected to three contact-buttons, *1, 2,* and *3*. Contact-button *3* is connected to one brush of the armature. The contact arm is also connected to a supply wire.

At starting, the arm makes contact with button *1* at the end of the starting resistor. The value of the resistor is approximately 2.8 ohms. This resistance added to that of the armature, which is 0.2 ohm, totals approximately 3 ohms. With a supply voltage of 115 v, the current is about 37.5 amps, which is 1½ times running value. As the armature starts to rotate and to generate a counter-emf, the current decreases;

and the arm is moved to button 2. The current increases once more as the resistance is thus decreased, but decreases again as the armature builds up counter-emf. Then the arm is moved to button *3* which connects the armature directly across the supply wires.

This method of starting may be used equally well in connection with the compound or the series motors. Starting rheostats are employed for motors larger than 1 hp. In the smaller sizes, the armature resistance is large enough to limit starting current to a safe value without using a resistor.

Speed Control

If a resistor is placed in series with the armature of a direct-current motor, the voltage drop in the resistor prevents the armature from receiving full supply voltage. In this case, the counter-emf of the armature is less than when the armature is subjected to full voltage. Since the counter-emf is less than normal, the speed of the armature is less than normal. That is, the insertion of a resistor in the armature circuit reduces the full-load speed of the armature.

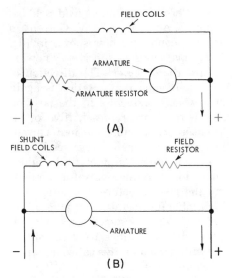

Fig. 16. Shunt motors using armature and field resistors for speed control.

In Fig. 16(A) a resistor of 2.3 ohms is connected in series with the armature of a shunt motor, which draws 25 amps at full load. By applying Ohm's law, it is found that the voltage consumed in the resistor is 57.5 v, which is one-half the supply voltage. The armature is, therefore, subjected to only 57.5 v, and its counter-emf will be approximately ½ normal running value. Since the counter-emf is ½ normal value and the field strength constant, the speed of rotation must be ½ normal speed.

The speed of compound and series motors may be reduced below normal in this way. If various speeds are required, the resistor may be arranged in the form of a rheostat so that its value may be altered to suit the desired speed. It is to be noted, however, that any speed variation obtained by means

of a series resistor is below normal operating value.

Some applications require that the speed be varied upward from normal. This result may be accomplished, in the case of a shunt motor, by inserting a resistor in series with the field coils, Fig. 16(B). The resistor reduces the shunt field current, which results in decreased field flux. The armature is, therefore, forced to run at a higher speed than normal in order to generate the required counter-emf. The amount of the speed increase may be adjusted by making the resistor variable. In this case, an increase in resistor value results in an increase in speed, and a decrease in resistor value results in a decrease, the speed in any case being higher than normal value.

A field resistor may be used in connection with a compound motor, but the results are not as satisfactory as with the shunt motor because of the presence of series field coils. It is impractical to use a field resistor to increase the speed of a series motor. In one special type of series motor, a resistor is connected in parallel with the field coils so that the speed may be increased somewhat; in general, this is seldom attempted.

ALTERNATING-CURRENT MOTORS

Series Motor

The shunt direct-current motor cannot be used successfully on alternating current because of high inductive reactance in the field circuit as compared to that of the armature. Field current, for this reason, is badly out of phase with armature current so that magnetism of field poles and current flow in armature conductors reach maximums at different instants. Little torque, therefore, is developed.

In the series motor, the same current flows through both field coils and armature winding, field-pole strength and armature current increasing and decreasing in step, so that excellent torque results. And, even though current alternates several times per second, rotation carries on in one direction.

Fig. 17 illustrates why this is so. At the instant shown in (A), the left field pole is N polarity, the right one S. Current in the single armature conductor flows away from the observer indicating clockwise rotation. During the

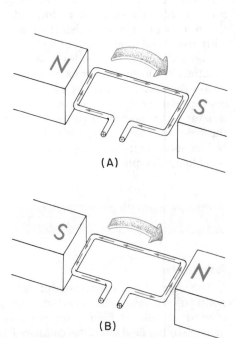

(A)

(B)

next alternation, the left field pole becomes S, and the right N. Current in another conductor now shown in Fig. 17(B) has reversed, and is now flowing toward the observer. The direction of current is such that rotation is again clockwise. Conductors in the position shown here, are representative of all armature conductors because current reverses in all when it reverses in one.

Except for certain relatively minor alterations, and necessity for laminating the field structure, the series unit is suitable for operation on alternating current. When arranged for service on either direct or alternating current at the will of the user, it is called a *universal motor*.

The Squirrel-Cage Motor

The *induction motor* Fig. 18 has a wound stator like that of a revolving-field alternator. Its rotating element, or *rotor*, consists of a *laminated core* which is slotted to receive conducting bars. The assembly of bars and end rings to which they are attached, resembles a *squirrel-cage,* hence the name.

If a stator with a single-phase winding is connected to an alternating-current circuit, N and S poles are created at the inner circumference of the stator core, their polarities changing rapidly with each alternation, or at a rate twice the frequency of the supply.

The flux, sweeping through stator and rotor cores, induces current in the rotor much as though the bars were caused to move across magnetic poles. Reaction between these stator poles and the rotor current, gives the rotor a tendency to revolve. Before mechanical inertia, the tendency to remain at rest, can be overcome, stator polarities change, inducing rotor currents in the

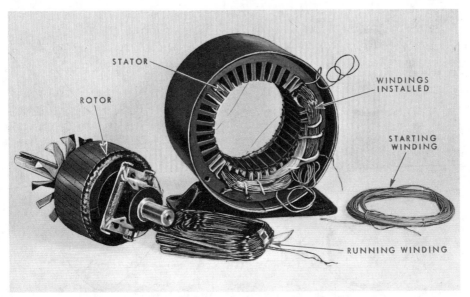

Fig. 18. Disassembled view of a squirrel-cage induction motor.

opposite direction, and thus reversing the force of rotation. As in the case of the single-phase loop discussed earlier, the rotor is prevented from turning in either direction so that it merely vibrates.

Flux Rotation in the Two-Phase Stator

Early experimenters conceived the idea that if stator poles could be forced to move, they would draw the rotor around with them. Two-phase current, developed for this purpose, proved the theory correct. Fig. 19 illustrates the process of *flux rotation*. At a particular instant when current is at a maximum in Phase *A*, zero in Phase *B*, coils of the stator winding create an *N* polarity whose center lies even with the maximum point of wave *A*, this line marked *A* in both views of the figure.

One-quarter alternation later, when current strength in Phase *A* has decreased to one-half maximum, while that in Phase *B* has risen from zero to one-half maximum, stator coils of both phases are equally active in the production of magnetic flux, and the *N* polarity shifts to the line *AB*. The following quarter-alternation finds the current in Phase *A* at zero, and that in Phase *B* maximum, the center of *N* polarity having moved to the line *B*.

Only four positions of flux movement are shown, but there are innumerable progressions between them as currents in the two phases change gradually by small amounts with respect to one another. That is, the stator magnetic poles no longer remain stationary in space, but circulate around the circumference. A single *N* polarity was followed in the

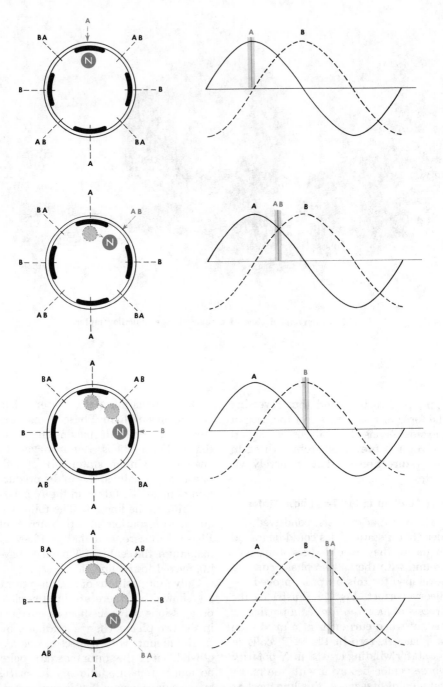

Fig. 19. Two-phase flux rotation. The drawing on the left illustrates the successive pole positions with respect to the indicated value of the current wave on the right.

discussion, but a shifting S polarity would be found directly opposite in the core of a two-pole stator. If the stator winding were arranged as four-pole, there would be a second N pole and two S poles traveling around the inner circumference at the same time.

Flux Rotation in the Three-Phase Stator

Three-phase flux rotation is illustrated in Fig. 20, showing various positions of a shifting polarity in one complete revolution, and movement of the stator pole with respect to current waves in the stator windings. Designations in both views are based upon above-the-line, or N, polarity. As mentioned in connection with the two-phase stator, there are countless intervening positions which give rise to continual rotation of magnetic flux rather than a succession of distinct leaps.

Speed of Flux Rotation

If the stator winding is arranged for two poles in each phase, and the supply frequency is 60 hertz, the shifting N polarity will return to its original position A in Fig. 20, at the end of $\frac{1}{60}$ second. That is, the flux travels 360 electrical degrees in this space of time, or since mechanical and electrical degrees are the same in a two-pole unit, 60 revolutions in one second. The number of revolutions in a minute is 60 times this amount, and the *synchronous* speed of flux rotation in a two-pole, 60-hertz induction motor is said to be 3600 rpm. The term synchronous refers to the fact that flux rotation is in step with, or in synchronism with, the supply frequency.

In a four-pole stator, the time for flux rotation to move through 360 electrical degrees, that is from the middle

of an N pole to the middle of the next N pole, remains $\frac{1}{60}$ second. But the distance traveled is only one-half the stator circumference, and two $\frac{1}{60}$ second intervals are needed to make a complete revolution. The synchronous speed of a four-pole, 60-hertz motor, then, is 30 revolutions per second or 1800 rpm.

Formula for Synchronous Speed

The synchronous speed of any induction motor can be obtained by using the formula:

$$rpm = \frac{frequency \times 60}{pairs\ of\ poles}.$$

the 60 being the number of seconds in a minute. The formula may also be written:

$$rpm = \frac{frequency \times 120}{number\ of\ poles},$$
$$or\ rpm = \frac{f \times 120}{P}$$

Example: What is the synchronous speed of a 25-hertz, two-pole induction motor?

$$rpm = \frac{25 \times 120}{2} = 1500$$

Example: What is the synchronous speed of a 60-hertz, six-pole induction motor?

$$rpm = \frac{60 \times 120}{6} = 1200$$

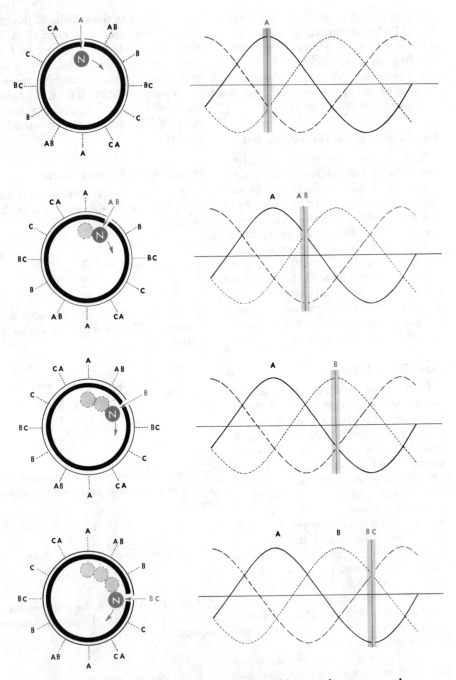

Fig. 20. Flux rotation in a three-phase stator. Pole positions on the stator are shown on the left in relation to the changing phase currents in the waveform on the right.

In the same way, the synchronous speed of a 60-hertz eight-pole induction motor is found to be 900 rpm, that of a twelve-pole 600 rpm.

Effect of Shifting Stator Flux Upon the Rotor

The shifting stator polarity, Fig. 21(A), is represented by the conventional N pole and an arrow. As the flux sweeps across the conductor, it sets up a voltage which tends to force current away from the observer. This causes conductor motion clockwise, in the same direction as the shifting polarity.

Current in the rotor bars flows to the end rings, where it divides, half one way, half the other, to form poles on the rotor surface, Fig. 21(B). Whether or not one chooses to think such rotor po-

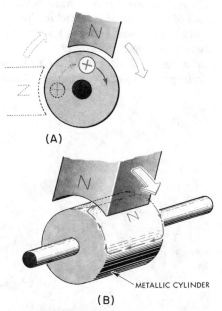

(A)

(B)

METALLIC CYLINDER

Fig. 21. (A) shows the effect of shifting stator flux on the rotor. In (B), the color arrows show the current induced in the rotor.

larities aid in maintaining rotation, it is worthwhile noticing that the position of the N rotor pole is such that it is repelled by the shifting N pole, while the S rotor pole is attracted toward it, both forces acting in the direction of rotation. Nevertheless, the effect of the shifting stator flux upon the rotor is exactly the same, regardless of whether stator excitation is two-phase or three-phase.

Rotor Slip

The rotor of an induction motor does not revolve at synchronous speed, but at somewhat less than this amount, depending upon the mechanical load applied to the shaft. *Torque* results from interaction between current in rotor conductors and the shifting magnetic poles. Although the strength of the revolving poles is constant, current in rotor bars varies according to the rate at which they cut lines of force. If the rotor traveled at synchronous speed, the bars would maintain relative positions with respect to stator poles, and no magnetic lines would be cut. Consequently, no voltage would be generated in them, and no current would flow to support rotation.

Therefore, the rotor must slide backward gradually during each flux revolution. That is, it must turn at a speed somewhat less than synchronous value. Thus, a four-pole, 60-hertz induction motor may turn at 1790 rpm instead of 1800 rpm when there is no load on it, this decrease of 10 revolutions per minute generating enough current to provide torque required at no-load. When load is applied, however, the mechanical drag on the rotor slows it down, thereby generating enough current to maintain rotation against this added physical resistance. The decrease in speed, with this particular motor, amounts to 40 revolutions, making a total of 50. The difference between synchronous speed and full-load speed is termed *slip*. Here, the slip is 50 revolutions per minute, or approximately 3 percent of synchronous speed, the motor turning at the rate of : 1800 − 50 = 1750 rpm.

Flux Shift Produced by Rotation

Earlier discussion revealed that the single-phase induction motor is not self-starting because flow of current, by itself, does not result in a shifting stator flux. If the single-phase rotor is made to turn at a speed of approximately two-thirds synchronous, however, its motion will continue, and it will drive a load.

The reason is presented in Fig. 22(A), where a single-phase rotor and two stator poles are indicated. At the instant shown, flux whose value is constantly changing under a sine wave of current through the coils, induces rotor voltages which tend to create rotor poles directly opposite those of the stator, along line *a-b*. Current in the

rotor bars is 90° out of phase with rotor voltage, for reasons that will be taken up in the next chapter.

Rotation carries line *a-b* to the position shown in Fig. 22(B), while the *N* flux of the upper pole passes through zero, and the flux created by the *out-of-phase* rotor current induces an *N* stator pole in this location. By the time the rotor-created magnetism dies away, the lower stator pole has altered to *N* polarity, with the result that a synthetic, but effective, kind of flux shift is produced in the stator.

Split-Phase Motor

An early method for starting the single-phase motor was to jerk the belt which drove the load pulley, in much the same way that an outboard gasoline engine is started by pulling on a rope. This procedure was not always convenient because all loads were not belt-driven. Various schemes were devised, some of which are illustrated in Fig. 23.

The stator was equipped with two-phase windings, shown here in schematic form, *R* denoting the *running winding*, *S* the *starting winding*. *R* was

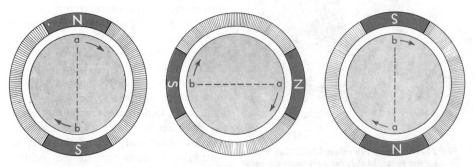

Fig. 22. Shift in rotor poles and rotation caused by the rotating stator flux.

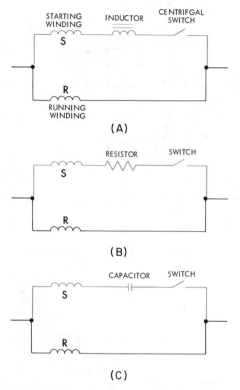

Fig. 23. Schematic diagrams showing various methods for starting split-phase motors. Resistors are used in (A), an inductor in (B), and a series capacitor in (C).

connected directly across supply wires, Fig. 23(A), but S was placed in series with an inductor and a centrifugal switch. When energized by the supply circuit, current through S lagged behind that in R, because of the series inductance, and a rudimentary form of shifting flux took place in the stator core. It was sufficient to start rotation, and when rotor speed reached approximately two-thirds synchronous value, the centrifugal switch opened the circuit through S, the motor continuing to

function on R, alone. This type is called a *split-phase* motor because one phase, in effect, is "split" into two.

The initial arrangement was varied, a resistor substituted for the inductor, Fig. 23(B), current through R, in this case, lagging behind that in S. The same result was accomplished, starting torque was almost as great as before, and the resistor was less bulky than the inductor. In time, this resistance was incorporated in winding S by using smaller size wire than in R, and reducing the number of turns so that its normal inductance was lower than R's. This form of split-phase motor became quite popular, and many are in use today.

Within the past few years, a series capacitor, Fig. 23(C), has been included in circuit with winding S. Its value of reactance approximates that of R, and much greater starting torque is obtained than with the resistor type. In larger sizes, the capacitor is often left in circuit, and the centrifugal switch dispensed with, the motor operating practically as a two-phase unit.

Limiting Starting Current of Polyphase Induction Motor

The polyphase induction motor, when first switched on, draws a current several times full-load running value. With small units, the excessive current is of little consequence, and it lasts only a few seconds until the motor gains speed. Where larger motors are concerned, the high surge of current may seriously affect power company equipment and consumer installations in nearby establishments. Also, because initial current lags far behind voltage, starting torque is not proportional to the vast surge of current drawn from supply lines.

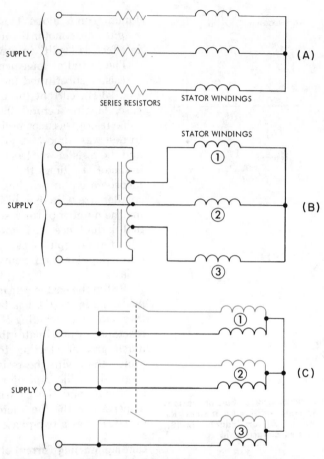

SUPPLY

SERIES RESISTORS   STATOR WINDINGS

(A)

STATOR WINDINGS

SUPPLY

(B)

SUPPLY

(C)

Fig. 24. Circuits illustrating various means of controlling starting current in polyphase induction motors. External resistors are used in circuit (A), an autotransformer in circuit (B), and divided stator windings in circuit (C).

One method for reducing this current is to insert external starting resistors in series with either two or three phase wires, Fig. 24(A). Its value is decreased in this manner without sacrificing starting torque, because current and voltage are more nearly in phase by the proportional increase in circuit resistance. Another, and quite popular method involves the use of auto-transformers, Fig. 24(B), to reduce the voltage applied to motor terminals, current taken from supply wires decreasing in direct ratio to this voltage. These, and other such devices, remain in circuit only during the starting interval.

A third scheme commonly used in the past few years, is *incremental start-*

ing, Fig. 24 (C). Stator phase windings are divided into parallel circuits so that one-half of each winding may be employed at the moment of starting, the remaining halves being connected when the motor has gained speed. The inrush of current, under this plan, is approximately one-half that for full-winding start. The term incremental means added part.

Standard single-phase induction motors are not used, today, in sizes that require limitation of starting current.

Speed Variation in Wound-Rotor Motor

The speed of a squirrel-cage motor cannot be readily altered, but with a coil-type winding on the rotor, speeds below that of normal full-load may be obtained. Fig. 25 shows how this is accomplished. Lead wires from brushes at collector rings connect the rotor winding to a set of resistance grids and a controller which permits the amount of circuit resistance to be changed at will.

The greater the amount of resistance

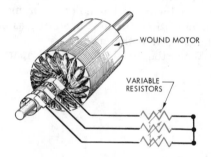

WOUND MOTOR

VARIABLE RESISTORS

Fig. 25. The drawing shows circuit arrangement of variable resistors used for speed control of wound rotor motors. The photograph shows mechanical features of the control unit.

employed, the greater the rotor slip must become in order to generate sufficient voltage to force required current through the circuit. Speed may be reduced in this way to 50 percent or less of normal full-load value. The wound-rotor unit does not require additional starting equipment, controller and resistance grids permitting a smooth, high-torque start with line current only 150 percent of normal full-load current.

The rotor winding illustrated here is of the three-phase type, but since it is altogether independent of the stator, it might be single-phase or two-phase if desired. In practice, however, the winding is invariably three-phase.

Other Methods for Providing Change of Speed

The adjustable-speed motor shown schematically in Fig. 26, has three sets of windings, two on the rotor, one on the stator. Supply conductors, connected to slip rings, furnish current to the main rotor winding, which is a standard three-phase type usually found on stators. The second rotor winding resembles that of a direct-current armature, coils terminating in commutator bars. A three-phase stator winding has six terminals, each pair connected to a set of brushes that rest on the commutator.

Brushes are attached to a mechanical device which allows those of each set to be moved toward one another, away from one another, or to be crossed over. When brushes (1) and (2) are together, (3) and (4) together, (5) and (6) together, the ends of each stator section are connected, and the commutated winding has no effect on motor operation, the unit operating as a standard induction motor.

When the brushes are moved apart,

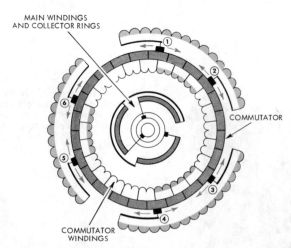

MAIN WINDINGS AND COLLECTOR RINGS

COMMUTATOR

COMMUTATOR WINDINGS

Fig. 26. Schematic of an adjustable speed motor. Speed is adjusted by moving the paired brushes closer, or further apart, to change the number of windings in the circuit. In the brush position shown, the coils shaded in color are in the circuit.

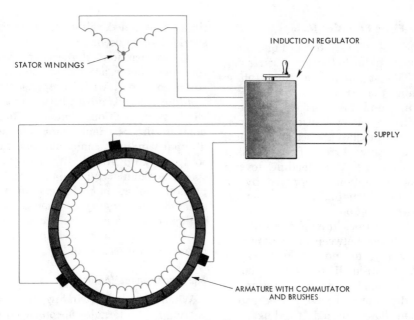

STATOR WINDINGS

INDUCTION REGULATOR

SUPPLY

ARMATURE WITH COMMUTATOR
AND BRUSHES

Fig. 27. Speed control with a polyphase induction regulator which varies both stator and armature voltages.

voltages induced in the commutated windings are placed in series with stator sections, reducing flow of current there, or increasing it as the case may be, so that motor speed for a given load may be adjusted at will. Such motors can be slowed down practically to standstill, thus suiting them to applications where "creeping" speeds are needed during "make-ready" operations such as multi-color press work or in knitting mills.

The motor of Fig. 27 has a three-phase stator winding, and an armature with a commutator. An induction regulator, connected to both stator and armature may be adjusted to furnish a variable voltage. Speeds below, and somewhat above synchronism are obtainable through this arrangement.

Multispeed Motors

Certain apparatus must be driven at different set speeds in the course of operation. The *multispeed induction motor* is employed for the purpose. It has two or more separate stator windings, each with a different number of poles. One winding may provide constant four-pole speed, a second, six-pole; or the first may provide six-pole, the second eight-pole, any desired combination being possible. Supply wires are connected, at will, to one or the other winding by means of controllers or contactors.

Single-Phase Repulsion Motor

The motor illustrated in Fig. 28 has features similar to those of a series motor. The stator has a single-phase winding, the armature a commutated winding, and the brushes are short-circuited by a low-resistance jumper. When the brushes are set in a vertical line, a heavy current flows through them, but the force tending toward clockwise rotation is opposed by an equal force tending toward counterclockwise rotation.

If the brushes are in a horizontal line, voltages between brushes nullify one another, and no current flows. At points between these two extremes, however, tendencies for rotation in both directions are no longer equal. Current flows through windings and brushes, and a good torque is developed. Different speeds for a particular load result from shifting the brushes in one direction or another within the

limits mentioned, although excessive sparking often occurs.

A few variations of the elementary repulsion motor are in common use, some equipped with *compensating windings,* placed midway between main windings in positions similar to interpoles in direct-current motors, the additional windings combating tendency to spark, and stabilizing armature speed. A popular type starts as a repulsion motor, but changes to the straight induction type when a centrifugal device short-circuits commutator bars as the armature approaches normal operating speed.

The Synchronous Motor

When the stator winding of a revolving-field alternator is connected to three-phase supply, the field structure comes up to speed like the rotor of an induction motor. If its field poles are then excited by direct current, mag-

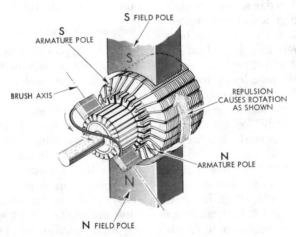

Fig. 28. The single-phase repulsion motor with short-circuited brushes

netism of the poles "locks in" with that of shifting stator poles, and the unit revolves at precisely synchronous speed. Thus, the revolving-field alternator becomes a synchronous motor that will drive mechanical load.

The motor has low starting torque, limiting its use to such applications as air compressors which can start unloaded. A light, pole-face, squirrel-cage winding is often used to provide initial torque, this winding becoming inactive at synchronous speed. Flux from the rotor poles generates a counter-emf in the stator winding. At no-load, when centers of stator poles and those of shifting flux poles are directly in line with each other, the counter-emf is directly opposed to the supply voltage.

As load comes on, rotor speed cannot change, but the rotor poles slide back a small amount with respect to stator poles so that a lagging phase

difference exists between counter-emf and line voltage, permitting more current to flow. If the field poles are overexcited, they slide forward with respect to positions of shifting flux poles, generating a counter-emf which leads supply voltage, causing the motor to draw a leading current from the line. Because of this feature, synchronous motors are often employed to raise the power factor of long feeders, and called *synchronous capacitors.*

Wound-Rotor Frequency Changer

The maximum synchronous speed of a 60-hertz induction motor is 3600 rpm, obtainable with a two-pole unit. Because of slip, as noted earlier, the speed is somewhat lower than this amount, perhaps 3400 rpm, or even less, depending upon the design. Higher speeds are desirable in some industrial processes, such as woodworking, where re-

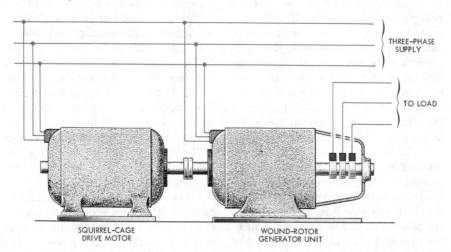

THREE-PHASE SUPPLY

TO LOAD

SQUIRREL-CAGE DRIVE MOTOR

WOUND-ROTOR GENERATOR UNIT

Fig. 29. The frequency doubler. The squirrel-cage motor on the left drives the wound-rotor unit on the right from which the output is taken.

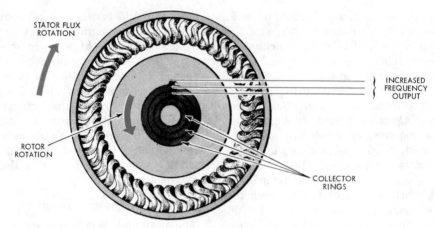

STATOR FLUX
ROTATION

INCREASED
FREQUENCY
OUTPUT

ROTOR
ROTATION

COLLECTOR
RINGS

Fig. 30. The rotor of the frequency changer is driven in the opposite direction to the stator flux rotation.

volving cutters operate more efficiently at 6000 rpm than at 3600.

For such applications, a *frequency doubler,* Fig. 29, is often employed. A drive motor, usually of the *squirrel-cage* induction type, is directly connected to a wound-rotor induction unit. Both stators are connected to the three-phase supply circuit, and output current is taken from the collector rings as shown. Rotation of the driving motor is chosen so that it turns the wound rotor in a direction opposite to stator flux rotation, as indicated in Fig. 30.

If the rotor, Fig. 30, is at rest, its conductors generate 60-hertz voltage. If made to turn in the opposite direction at a speed equivalent to that of the shifting stator flux, 120-hertz voltage is generated. The four-pole unit of the illustration is driven at a speed of 1800 rpm to accomplish this result. Other frequencies are possible from the

same arrangement if the speed of the drive motor is different from that of the driven unit. Thus, a two-pole drive motor in the present instance would produce an output frequency of 180 hertz.

The usual practice is to employ two units having like nameplate speeds so that 120-hertz current is obtained. Unless a synchronous drive motor is used, the frequency is somewhat less, because of slip, and is often given at the nominal value of 100 hertz. Voltage is increased in the same ratio as frequency so that a wound rotor whose normal voltage is the same as that of the stator, say 220 volts, produces a voltage in the neighborhood of 400. Power output of the combination is greater than input power to the drive motor, because energy is transferred from stator to rotor of the generating unit by transformer action, which will be studied in the next chapter.

1. What purpose do interpoles serve in a motor?

2. State the rule for current flow in regard to brushes of direct-current generators and motors.

3. Why is more power required to drive a generator armature as load comes on?

4. What is counter-emf?

5. What is the relative polarity of motor interpoles with respect to main poles?

6. Describe the connection of motor interpoles in circuit.

7. The armature of a 115-volt armature has a resistance of .1-ohm, and draws a current of 20 amps. What is its counter-emf?

8. If the full-load speed of the above armature is 1500 rpm, and the no-load current is 4 amps, what is the no-load speed?

9. How does the speed of a compound motor compare with that of a similar shunt motor under full-load conditions?

10. Describe the behavior of a series direct-current motor under varying load.

11. Describe two-phase stator flux rotation.

12. Does three-phase flux rotation travel faster than two-phase?

13. What is the synchronous speed for a 50-hertz, 6-pole induction motor?

14. A 4-pole, 60-hertz, induction motor has 3 percent slip at full-load. What is its full-load speed?

15. Describe the modern split-phase resistance type motor.

16. Describe the resistance method of starting a polyphase induction motor.

17. How is a wound-rotor motor started?

18. Outline briefly the construction and operation of an adjustable-speed polyphase motor.

19. Describe a single-phase, repulsion-induction motor.

20. Describe a three-phase synchronous motor.

Chapter 10

Transformers, Voltage Regulators and Power Rectifiers

TRANSFORMERS

General Nature

Transformers are devices which change voltage or current of a supply circuit into some other desired value by means of electromagnetic induction. They are made in a large variety of sizes and shapes, including huge power transformers with outputs running to thousands of kilowatts, along with miniature space application units whose output is only a fraction of a milliwatt.

Low-frequency transformers like those which furnish power to home and industrial locations have iron cores, but those for high-frequency applications such as dielectric heating, electron tube, or solid state circuits, usually have air cores.

In this chapter we will be concerned only with power transformers as shown in Fig. 1, although the basic principles of operation are the same for all types.

Fig. 1. Power transformer.
Courtesy General Electric Co.

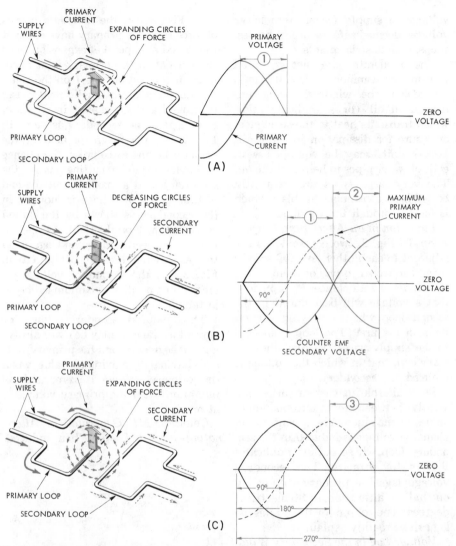

Fig. 2. Induction between adjacent loops. The drawings show in sequence the changing current conditions in the primary loop and the current which is induced in the secondary.

The Power Transformer

The power transformer must be a rugged piece of electrical equipment because it is most often mounted in outside locations where it is exposed to the elements. Its function, in most cases, is to transform dangerously high voltages of supply feeders into lower voltages deemed safe for use in homes, shops, or industrial plants.

The moderate size pole-mounted transformer common to city streets is oil cooled, the windings being immersed in oil (or a similar liquid) which transmits heat to the sheet steel enclosure for dissipation into the air. Larger units may be equipped with coils of water pipes to help cool the oil. The very large types are frequently cooled by a continuous air blast which is forced through the windings.

Induction in an Alternating-Current Loop. In Fig. 2, two copper loops are adjacent to each other, an A-C supply being connected to one of them. From the study of inductance, it is obvious that a voltage will be induced in the second loop when a varying current flows in the first. The loop connected to the supply wires is called the *primary* coil, that in which the voltage is induced, the *secondary*.

Since alternating-current varies constantly between zero and maximum values of the sine curve, a continually changing voltage is induced in the secondary loop whose outline duplicates that of the primary. The secondary voltage lags the primary voltage by one half an alteration, or 90 electrical degrees. The reason for the 90 degree lag can be readily explained.

Voltage Lag in the Secondary Winding. The study of induction, Chapter 7, showed that maximum voltage is induced in a second loop when magnetic circles of the first loop are cutting across conductors of the second at the highest rate; that is, when their number is changing most rapidly within equal spaces of time. Conversely, the minimum induced voltage occurs when the rate of change is least.

In Fig. 2(A), the magnetic circles of force from the primary have reached their maximum point of expansion when the current in the primary is at maximum at the end of time *1* on the right. At this point the magnetic field ceases to expand and the change in circles of force across the secondary is zero. In Fig. 2(B), the magnetic field of the primary begins to collapse, the change in circles of force cutting across the secondary is at a maximum at the end of time 2, and a voltage is induced in the secondary as shown by the dotted curve in time 2 on the right. As the primary field begins to expand again on the negative half the voltage alternation, Fig. 2(C), the secondary voltage increases to a positive maximum as shown in time *3*.

Thus, induced secondary voltage (or current as the case may be) will always lag 180 degrees behind the primary voltage, having its maximum value when the primary voltage is zero, and its minimum when the primary voltage is at maximum.

Counter-EMF Generated in Alternating-Current Loop. When induced

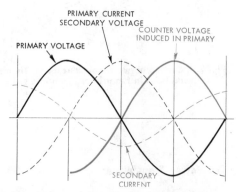

Fig. 3. Counter-voltage induced in the primary winding by current in the secondary.

current flows in the secondary, it creates circles of force which react on the primary winding to induce a voltage which lags secondary voltage by 90 degrees, as shown in Fig. 3. Using the negative alternation of primary voltage for explanation, the voltage induced in the primary winding follows exactly the same pattern, though of lesser value, in the positive direction, opposing the primary voltage. For this reason it is termed a counter voltage, or counter-emf, similar to that generated in the windings of a direct-current motor.

Elementary Transformer

Fig. 4 illustrates a simple transformer, consisting of primary and secondary loops around a laminated iron or steel core which forms a closed magnetic circuit. The core, like that in a direct-current electromagnet, increases the magnetic flux originated by the circles of force. The magnetic lines in the core of the transformer, with alternat-

ing current flowing in the primary coil, are the combined resultant of primary flux and the counter-flux of the secondary current. In effect, they constitute a single flux, and will be considered as such.

Since the iron magnetic path conducts lines of force with far greater ease than air, the primary coil may be placed on one leg of the transformer and the secondary on the other leg, without decreasing their mutual effects. Fig. 4 shows this arrangement, a primary coil of a single turn being placed at the left, and a two-turn secondary coil on the right. Both coils are exposed to the same magnetic flux, which sets up an induced voltage in the secondary winding and a counter-emf in the primary.

Step-Up Transformation. When the same number of magnetic lines cut across a coil of two turns, twice as much voltage is created as a coil of one turn. The voltage induced in the secondary of the transformer in Fig. 4, then, must be twice that generated in the primary as counter-emf.

The counter-emf of a transformer, like that in a direct-current armature, is only slightly less than the supply voltage. Suppose the supply is 6 volts and the counter-emf 5.9 volts. In this case, the secondary voltage is equal to twice 5.9, or 11.8 volts. Thus, the transformer accomplishes a transformation of voltage, taking 6 volts from the supply wires and changing the voltage to 11.8 for whatever use may be desired. This form of voltage change is termed a *step-up*, and the unit is known as a step-up transformer.

Step-Down Transformation. Suppose that a 12-volt supply line is connected to the coil of the right-hand leg of the core, so that it becomes the primary

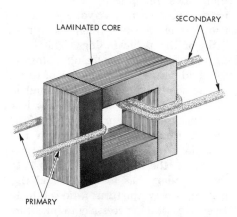

LAMINATED CORE

SECONDARY

PRIMARY

Fig. 4. An elementary transformer.

winding of the transformer. The magnetic flux cutting across the single-turn secondary winding will be the same as that cutting the two-turn primary winding. If the counter-emf of the primary winding is 11.8 volts, the voltage generated in the secondary will be only one-half this amount, or 5.9 volts. In this case, the voltage has been "stepped down," and the unit is known as a step-down transformer.

Current Available in Primary and Secondary Windings

The step-up transformer may appear to create something out of nothing, taking in 6 volts from the supply wires and providing 11.8 volts at the terminals of the secondary winding. Further investigation, however, shows otherwise.

The term *ampere-turns* was defined, in the study of electromagnetism, as the product of amperes and turns. There, it was pointed out that the strength of magnetic flux induced in an iron core is directly proportional to the number of ampere-turns in the magnetizing coil or solenoid. The same rule applies to the basic transformer, which is nothing more than a pair of solenoids.

Since the counter-emf generated in the primary by current in the secondary is approximately equal to the voltage induced in the secondary by current in the primary, their ampere-turn values must be nearly equal. When the primary winding has a single turn and the secondary two turns, twice as much current may flow in the primary winding. Where the situation is reversed, and the primary has twice as many turns as the secondary, current in the primary winding is but one-half that in the secondary.

One should note that the product of volts and amperes is practically the same for the two windings. In the step-up transformer, a primary voltage of 6 and a current of 10 amperes $(6 \times 10 = 60)$, provide a secondary voltage of 12 and a current of 5 amperes $(12 \times 5 = 60)$. The same facts, taken in reverse order, apply to the step-down transformer.

Turns-Ratio

A 6-volt, single-turn, primary winding, produced 12 volts in a secondary winding of two turns. That is, twice as many turns in the secondary winding resulted in a secondary voltage twice as great as the primary. Or, considering the reverse order, a secondary with one-half the number of turns as the primary resulted in one-half the voltage of the primary. In the same way, the voltage of the secondary may be made four times that of the primary, or ten times, or twenty, or any chosen value, simply by making the relative number of turns that have this same ratio.

This fact is expressed by saying that: The relationship between primary and secondary voltages in a transformer depends upon the turns-ratio of the two windings. The term *turns-ratio* means the result obtained by dividing the larger number of turns in one winding by the smaller number in the other. Where the secondary has two turns and the primary one turn, the turns-ratio is 2-to-1. Where the secondary has ten turns and the primary one, the turns ratio is 10-to-1. If the secondary has five turns and the primary thirty, the turns ratio is 6-to-1. In each case, it is necessary to take into consideration the matter of whether it is a step-up or a step-down transformer before applying the turns-ratio to determine the voltage of the other winding.

For purposes of illustration, transformers having but a few turns in primary and secondary windings have been considered. The ordinary commercial transformer, a cut-away view of which is shown in Fig. 5, employs primary and secondary windings that have hundreds or even thousands of turns, the actual number depending upon engineering design factors.

Transformer Ratings

Small transformers are usually rated in watts, the rating being obtained by multiplying primary current and voltage. Larger ones, such as that in Fig. 5, are rated in volt-amperes or kilovolt-amperes (thousands of volt-amperes), because actual power output depends upon power factor of its load.

This method of rating becomes necessary, since the windings are designed to carry only a certain maximum current without overheating. For example, a 500-volt, 50 kva (kilovolt-ampere)

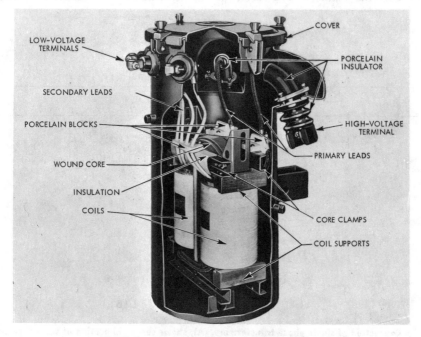

LOW-VOLTAGE TERMINALS

SECONDARY LEADS

PORCELAIN BLOCKS

WOUND CORE

INSULATION

COILS

COVER

PORCELAIN INSULATOR

HIGH-VOLTAGE TERMINAL

PRIMARY LEADS

CORE CLAMPS

COIL SUPPORTS

Fig. 5. Cutaway view of a power transformer.
Courtesy McGraw Edison

transformer could supply a load of 50 kilowatts when voltage and current were exactly in phase with one another, or at 100 percent power factor. At 80 percent power factor, however, it could supply only: 500 × 100 × .8 (power factor), or 40,000 watts, even though the current in the windings is 100 amperes in both cases.

Single-Phase Transformer Connections

The low-voltage winding of single-phase distribution transformers that supply power to homes is ordinarily made in two sections. Schematic diagrams of a single-phase transformer with the secondary leads brought out for 230 v service is shown in Fig. 6 (A). Here, the two sections are connected in series, and the voltages induced in each section are added. In Fig. 6 (B) the secondary windings are connected in parallel. This connection is used to supply a heavy load requiring 115 v service.

A common practice is to employ the three-wire type of circuit studied in Chapter 4, a third conductor being attached to the junction point of the series connection in Fig. 6 (A). Fig. 7 illustrates the arrangement, where the third wire, or *neutral,* is grounded. Grounding lessens severity of shock, since the voltage to which one is exposed when accidentally touching an outer wire is only 115 volts, whereas under certain conditions it could be 230 volts if the ground were not present.

Three-Phase Transformer Connections

Standardization of generating equipment, motors, and appliances, has brought about the acceptance of three-phase, 60 hertz current throughout the whole country. Except for network systems, which will be referred to later, a

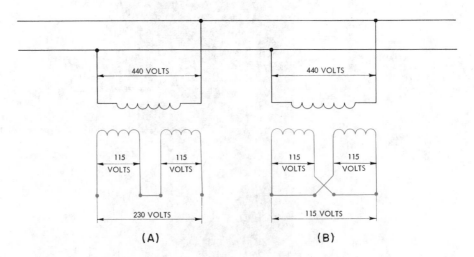

Fig. 6. Connection of single-phase transformers. (A), shows series connection of secondary coils. (B), parallel connection of secondary.

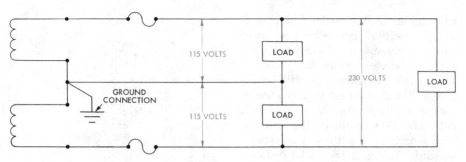

Fig. 7. Three-wire circuit with grounded neutral.

group of three single-phase transformers, known as a transformer bank, is usually employed in transforming three-phase current from one voltage to another. The three primary windings of the transformers, and the three secondary windings, are connected together in one of two ways: star or delta.

Voltage Relationships in the Star Connection. This connection, also known as the *Y* connection, is illus-

trated in Fig. 8 (A) which shows three star-connected primary windings of transformers supplied by a "2200 volt," three-phase, power line whose voltage, between each pair of line wires is actually 2197. The voltage across each winding between the line wire to which it is connected and the common, or *star* point, is shown as 1270.

Fig. 8 (B) shows how the voltage across the windings of a star-connected

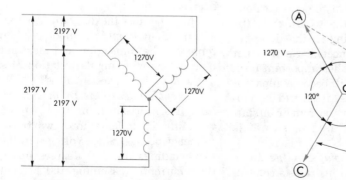

Fig. 8. *Left*, voltages in the star transformer. Diagram on the *right* illustrates the relationship of line voltages to the voltages across the windings of the star transformer.

177

transformer bank may be determined, the illustration consisting of three equal arrows spaced 120 degress apart (⅓ of a circle). From the study of three-phase generators, it will be recalled that three-phase voltages are so spaced. In the figure, the arrows A-O, B-O, and C-O represent the voltages across each of the three transformer windings in Fig. 8 (A). The distance between the ends of any two arrows, such as A and B, is a measure of the voltage between any two supply wires.

Upon measuring the distance A-B very accurately, and comparing it with the lengths of any one of the three arrows, such as A-O, it is found to be 1.73 times as long. If the length, or value, of the arrow is known, the value of the outer distance is determined by mulplying the arrow value by 1.73. If the outer value is known, the arrow value may be determined by dividing the outer value by 1.73.

In the present instance, where the line voltage is known to be 2197, the arrow or transformer voltage is equal to 2197 divided by 1.73, or 1270 volts. Expressed in the form of a rule: Line voltage in a star-connected bank is equal to 1.73 times the star voltage.

Current Value in Star Connection. One should note that the current flowing into each of the windings is the same as that flowing in the line wire to which it is connected. Such is not the case with the other important form of connection, as will become apparent shortly. Meanwhile, it should be emphasized that: Line current in a star-connected bank is the same as transformer current.

Voltage Relationships in the Delta Connection. Fig. 9 illustrates a conventional delta connection, the three windings being connected to a three-phase

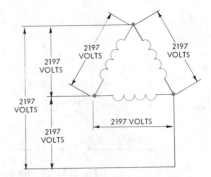

Fig. 9. Schematic of a delta connected transformer primary. In this connection line and winding voltages are the same.

supply line. It is seen that the voltage across each pair of line wires, 2197 volts, is that across the individual windings. Stated as a rule: Line voltage in a delta-connected bank is the same as transformer voltage.

Referring to Fig. 9 where each of the lines of the delta, or triangle, represent a transformer winding, it would appear by measuring the angles inside the triangle that voltages in the individual windings are only 60 electrical degrees (1/6 of a circle) apart instead of 120 degrees.

This is because the diagram is merely a convenient graphic symbol used to represent the connection and must not be confused with angular relationships of the actual voltages involved.

Current Values in the Delta Connection. The current from each supply wire, in Fig. 9 flows into two transformer windings. Since voltages in the two windings are 120 degrees apart, their currents (assuming 100 percent power factor) must be 120 degrees apart, the line current being the sum

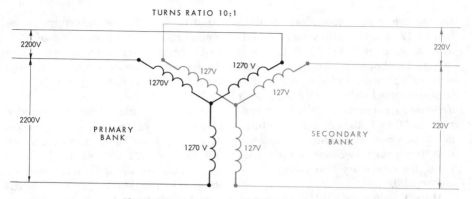

Fig. 10. A star-star connected transformer bank.

of the two. We found, in examining the star connection, that the sum of two equal voltages which are 120 degrees apart is equal to 1.73 times either one. Here, where the two currents are 120 degrees apart, the line current must be equal to 1.73 times either one. Expressed in the form of a rule: Line current in a delta-connected bank is equal to 1.73 times transformer current.

Secondary Transformer Connections. As with primary windings, the secondary windings may be connected either star or delta to suit particular requirements. Fig. 10 shows in schematic form, a transformer bank with both primary and secondary windings being connected star. This arrangement is known as *star-star*. If the turns-ratio of the step-down transformers is 10 to 1,

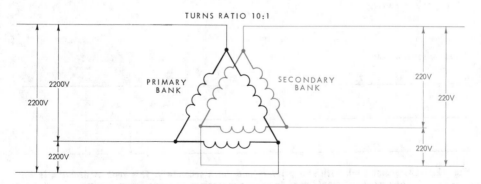

Fig. 11. A delta-delta connected transformer bank.

the voltage in each secondary winding will be 127 while that between secondary line wires is 220 (1.73 × 127).

Fig. 11 shows a transformer bank with both primary and secondary windings connected delta, and known as a *delta-delta* connection. With a turns-ratio of 10 to 1, the secondary voltage is 220 across each winding. Since each winding spans two secondary line wires, the secondary line voltage is, of course, 220.

If the transformer bank is connected *star-delta*, the primary remaining star, but the secondary being changed to delta, as in the more serviceable schematic form of Fig. 12, secondary line voltage is 127 with a turns-ratio of 10-to-1. Should the transformers be constructed with a turns-ratio of 5.77-to-1, the primary winding remaining unchanged, the secondary voltage would be 220 as before (1270/5.77 = 220).

In Fig. 13, which shows a *delta-star* connection, the secondary line voltage of a 10-to-1 set of windings becomes 398 (230 × 1.73 = 398). In this case, also, secondary voltage can be raised or lowered to any desired value by designing the windings to have a different turns-ratio.

Two Special Transformer Connections. Fig. 14 illustrates a delta-delta connection, the primary voltage 2300 and the secondary 230. Three second-

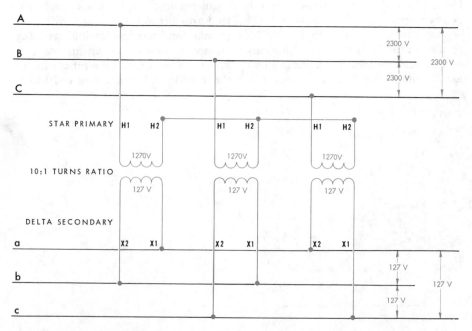

Fig. 12. Transformer bank with star primary and delta secondary. This type of drawing is the one most commonly used to show connections of transformer banks.

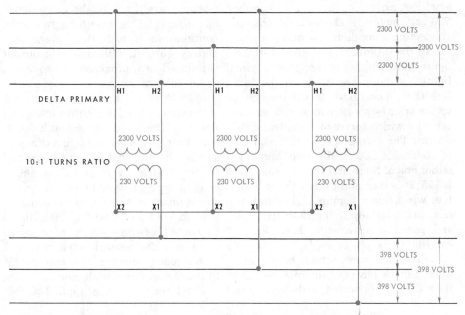

Fig. 13. Transformer bank with delta primary and star secondary.

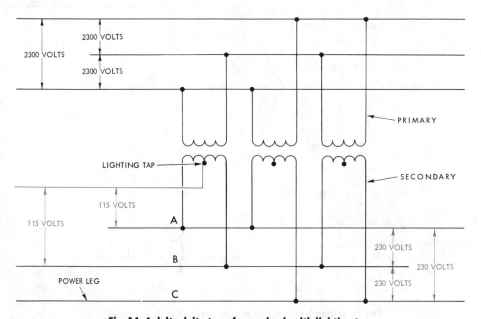

Fig. 14. A delta-delta transformer bank with lighting tap.

ary line wires: *A, B,* and *C* supply 230-volt motors and similar loads. A connection is made to the mid-point of one of the secondaries, as shown in the figure, and single - phase, three - wire lighting circuits are taken off the two line wires, *A* and *B* which are connected to the transformer secondary, and the tap wire which serves as a neutral conductor. The voltage between *A* and *B,* of course, is 230, while that between either one of them and the neutral wire is 115. It is common practice to run all four wires into a commercial or industrial plant for supplying both lighting and power requirements. Electricians call line *C* the *power leg.*

Delta-delta transformer banks are often used by power companies because they may be operated with only two

transformers if one of the three develops a fault. The remaining two units continue to furnish three-phase secondary current, although safe output is reduced to approximately two-thirds of full-load capacity.

Fig. 15 shows the popular network arrangement used for both power and lighting. The basic connection is delta-star. Voltage between any line wire and star is 120, and that across any pair of the main wires, *A, B,* and *C,* is 208. A neutral wire is brought out from the star connection to serve as the common wire for lighting circuits. Lighting is connected between all three of the main wires and the neutral conductor, and power loads between any two, or all three, of the main conductors.

The term *network* is applied to this

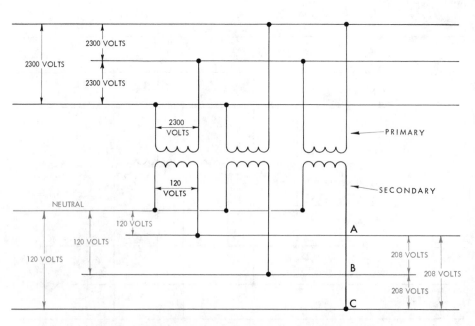

Fig. 15. Network connection of transformers with a neutral.

arrangement because it was first employed in congested downtown areas of large cities where automatically-controlled units were desired in connection with feeder networks. In this particular application, all three sets of primary and secondary windings are wound upon a single core, the unit being arranged for rapid connection or disconnection as the feeder load changes or as local trouble occurs.

Harmonics

When loads on the three phases of a network system are balanced, the neutral carries no current unless harmonics are present. *Harmonics* are currents whose frequencies are multiples of supply frequency. The third harmonic, Fig. 16 has a frequency three times that of the supply, or *fundamental* frequency,

so that if the fundamental is 60 hertz, the third harmonic is 180 hertz.

Some types of loads give rise to production of harmonics. Ballast transformers of fluorescent lamps, for example, generate such currents which circulate between phase wires and neutral, sometimes loading it to full capacity. The third harmonics of all three phases are in phase with one another, as shown in Fig. 16, and are sometimes called zero-sequence currents. This term means that no phase difference exists between them.

Parallel Operation of Transformers

Transformers, like generators, may be operated in parallel if they have similar electrical characteristics. Special care is required with transformers,

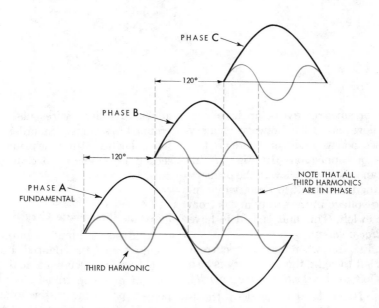

Fig. 16. Harmonics in a three-phase current. The current waves are shown separately for clarity.

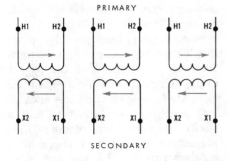

Fig. 17. Transformer connection for additive polarity. The letter designations of the leads are standard markings used on transformers and schematics.

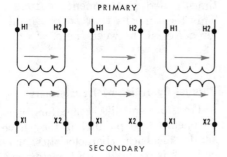

Fig. 18. Transformer connection for subtractive polarity.

however, since relative polarities of current flow may differ between primary and secondary windings. There are two types of connections. In Fig. 17, instantaneous current flow in the primary winding is from left to right, while induced voltage in the secondary is from right to left. This unit is said to have *additive polarity*.

In Fig. 18, where primary current flows left to right, the secondary is also from left to right, *subtractive polarity* exists. Nameplates of modern transformers provide this information. If transformers of like polarities are par-

alleled, the lead wires need only be grouped in symmetrical order. If of unlike polarities, either the primary or the secondary leads of one transformer must be crossed.

Power in Three-Phase Circuits

Total power input or output of an evenly loaded, or balanced three-phase transformer bank is equal to three times that of any one transformer. At 100 percent power factor, this total equals: $3 \times Voltage \times Current$. One fact to note, is that the voltage and current of the particular transformer winding must be employed in this calculation.

If the windings are connected star, we know that the transformer voltage equals line voltage divided by 1.73, but that transformer current is the same as line current. For the star connection, therefore, the calculation may be changed to read:

$$\text{Power} = 3 \times \frac{E}{1.73} \times I,$$

(NOTE: 1.73 is $\sqrt{3}$)

where E and I are line values.

If the windings are connected delta, we know that the transformer voltage is equal to line voltage, but that transformer current is equal to line current divided by 1.73. For the delta connection, the calculation reads:

$$Power = 3 \times E \times \frac{I}{1.73}$$

In either of these cases, the formula is the same as if written:

$$Power = 3 \times \frac{E \times I}{1.73}$$

Written in still another way, the formula reads:

$$Power = \frac{3}{1.73} \times E \times I$$

Now,

$$\frac{3}{1.73} = 1.73$$

In other words, 3 is the square of 1.73

To determine power taken from the line wires by the three-phase trans-former bank, then, whether it is connected star or delta, we may write:

$$Power = 1.73 \times E \times I$$

And the same calculation will apply to power delivered to the secondary line wires by the secondary windings of the transformer bank. Indeed, this formula for determining power in a balanced three-phase circuit applies not only to transformers, but to any kind of three-phase load.

If the power factor of the circuit is other than 100 percent, it is necessary to multiply the above value by that power factor. For example, if the power factor was 80 percent, or .8, the formula would read:

$$Power = 1.73 \times E \times I \times .8$$

AUTOTRANSFORMERS AND REGULATORS

Single-Phase Autotransformers

The autotransformer, shown schematically in Fig. 19 is employed for applications where an assortment of voltages is required, such as low voltage motor starting. Applications are limited to some degree as the primary and secondary are tied together resulting in low ratios. The iron core is exactly like that of the ordinary transformer, but the usual primary and secondary windings are combined into one. As indicated in the figure, a number of voltages may be obtained by making tap connections to the winding at suitable points.

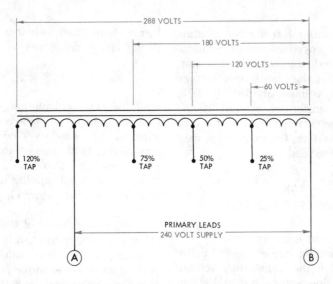

Fig. 19. A single-phase autotransformer. Voltage values available at the various taps are shown in color.

In Fig. 19, the 240-volt supply circuit is attached to input, or primary, lead wires (A) and (B). A number of taps are shown on the output, or secondary side. At the 25 percent tap, which includes one-quarter of the turns of wire between the primary leads, the voltage is 25 percent of 240 volts, or 60 volts. The 50 percent tap provides 120 volts, the 75 percent 180 volts, and the 120 percent 288 volts.

Observe that voltage at each tap point equals the counter-emf generated by the number of turns included between the tap and the (B) end of the winding. Since the number of turns at the 120 percent end is equal to 120 percent of the number in the whole primary winding, the counter-emf there is higher than the primary voltage.

The load in Fig. 20, connected between the 50 percent tap and lead wire (B), draws 10 amps. An ammeter in the supply circuit shows that 5 amps are furnished by primary wires, passing downward through one-half the whole winding to the load as indicated by the arrow. The remaining 5 amps flows upward in the tapped portion, in the direction of counter-emf, to the load. Conductor (B), like (A), carries only 5 amps.

If the upper wire for a load requiring 10 amps is placed on the 25 percent tap, Fig. 21, the ammeter shows that only 2.5 amps are supplied by the primary winding, so that 7.5 amps must have been induced in the tapped sec-

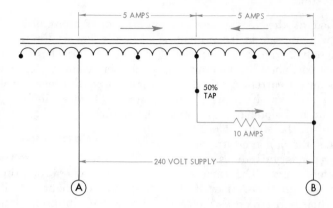

Fig. 20. An example of current relationships in an autotransformer with a load connected to the 50 percent voltage tap.

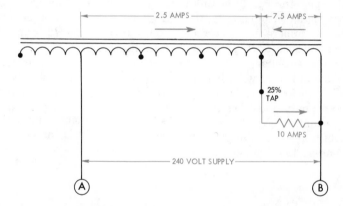

Fig. 21. Current relationships in an autotransformer with a load across the 25 percent portion of the winding.

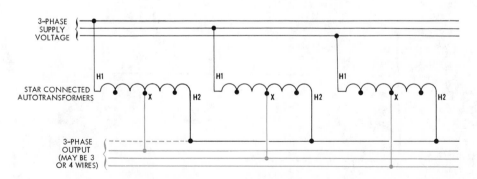

Fig. 22. Three-phase autotransformers using the star connection. Four-wire output is provided by utilizing the H2 connections.

187

tion. A load at the 75 percent tap would draw 8 amps from the primary wires, with only 2 amps being supplied by the lower portion of the winding.

The amount of current that will be delivered by primary wires can be readily determined beforehand. It was pointed out, in the study of ordinary transformers, that ampere-turns of primary and secondary windings must be approximately equal. The same rule applies to the autotransformer if the tapped portion is regarded as the secondary, and the remainder as the primary. At the 25 percent tap, primary current had to pass through three times as many turns as the secondary. For the same number of ampere-turns, current in the primary needed to be only one-third that in the secondary.

The load current here may be thought of as divided into four parts, one supplied by the primary and three by the secondary, so that, in Fig. 21, the primary contribution is one-fourth of 10 amps, or 2.5 amps, the secondary three-

fourths of 10 amps, or 7.5 amps. Thus, 7.5 amps flowing through one-quarter of the turns furnishes the same number of ampere-turns as 2.5 amps through three-quarters of them. Investigation of the example shown in Fig. 20, gives similar results.

Three-Phase Autotransformer

The three-phase autotransformer is the same in operating principle as the single-phase. Figure 22 illustrates a three-phase autotransformer hookup utilizing the star connection. This connection provides either a three-, or four-wire output. If a four-wire output is desired, the fourth wire is obtained by connecting to the H2 terminals which form the star point.

The open-delta autotransformer connection, Fig. 23 is generally employed for starting polyphase induction motors. It delivers a low voltage from the selected tap points until the motor gains speed and is then cut out of the circuit.

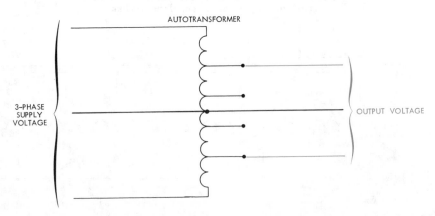

Fig. 23. Three-phase autotransformer connected in open delta. Output voltage is selected by changing the tap connections of the two outer wires.

REGULATORS

Single-Phase Induction Regulator

Voltage on long feeder lines fluctuates widely due to drops caused by resistance and inductive reactance. To overcome these losses, induction regulators may be employed to maintain a fairly constant voltage level at locations where considerable power is required, such as large industrial plants.

The induction regulator is an auto-transformer whose primary winding is connected across the supply line, and its secondary in series with the load, as shown in the simplified schematic of Fig. 24. In its construction it resembles a wound-rotor induction motor, which it is essentially, having a wound rotor (primary), and a wound stator (secondary), Fig. 25. The rotor has an additional winding, a short-circuited coil, placed at 90 degrees to the rotor winding.

In operation, the output voltage is regulated by the relative positions of the rotor and stator to each other, which changes the magnetic flux linkage between them. The movement of the rotor is generally limited to 90 degrees or

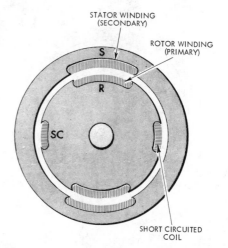

Fig. 25. Cross section of the induction regulator showing the winding arrangement. These windings are wound in slots similar to the induction motor.

less and may be controlled by hand, or by electronic motor controls.

Fig. 26 illustrates how output voltage is affected by various positions of the rotor in relation to the stator. *Position 1* shows maximum linkage of the magnetic field of the rotor with the stator. In this position, maximum voltage is induced in the secondary winding of the stator, thus making load voltage maximum. In *Position 2*, the rotor has been rotated 90 degrees from the stator; there is no flux linkage between them, and no voltage is induced in the secondary. However, since the secondary is in series with the supply, it will produce a magnetic field of its own and thus a reactive effect. In this position, the stator is adjacent to the short-circuited coil of the rotor. Since the stator is producing its own magnetic field, the short-circuited coil is being cut by a

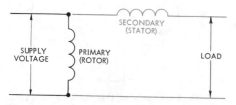

Fig. 24. Simplified schematic of the inductive voltage regulator.

189

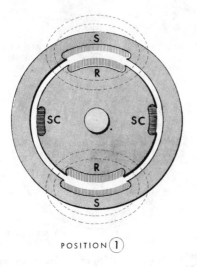

POSITION ①

maximum amount of flux and becomes, in effect, a short-circuited secondary, setting up an emf which opposes that induced in the stator and reducing the effect of the regulator to zero. In this position, value of the load voltage is near its minimum. Further rotation of the rotor beyond *Position 2* will cause the primary and secondary voltages to oppose each other and thus reduce the load voltage to its minimum value.

The inductive regulator varies the load voltage through a range of approximately 10 percent, with intermediate values between maximum and minimum depending upon the amount of flux linkage between rotor and stator. They are made both in single, and three-phase units.

Three-Phase Induction Regulator

This regulator is similar in construction and principle of operation to the single-phase, except that both rotor and stator have three windings each, placed 120 degrees apart, as shown in the schematic of Fig. 27. Each winding of the

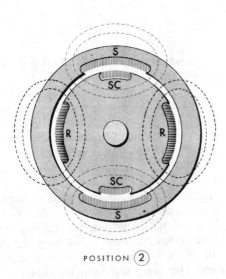

POSITION ②

Fig. 26. Operation of the induction regulator. Position 1 illustrates maximum flux linkage between rotor and stator, producing maximum load voltage. In Position 2, the rotor has turned 90 degrees, so that rotor flux of the primary has no linkage with the secondary. Flux of the stator is cancelled by the short-circuited coils so that it has no effect on the voltage.

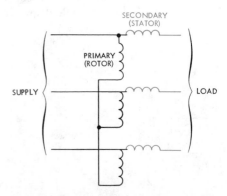

Fig. 27. Schematic of the three-phase induction regulator. Since the windings are 120 degrees apart, any movement of the rotor will cause the same voltage change in each phase.

primary is connected to a separate phase wire, as is the secondary, and any movement of the rotor subsequently affects all phases simultaneously.

For three-phase requirements, a single-phase unit may be placed in each line, or a single three-phase unit may be used.

Tap-Changing Autotransformer Regulator

This type autotransformer, shown in the schematic of Fig. 28, utilizes switching contacts to select voltages for the primaries of three transformers whose secondaries are in series with individual phase conductors, their voltages adding to or subtracting from line voltages. Tap-changing may be done under load by switches, making it unnecessary to cut off power to the unit. This is an essential feature at industrial plants where it may be necessary to adjust voltages several times a day.

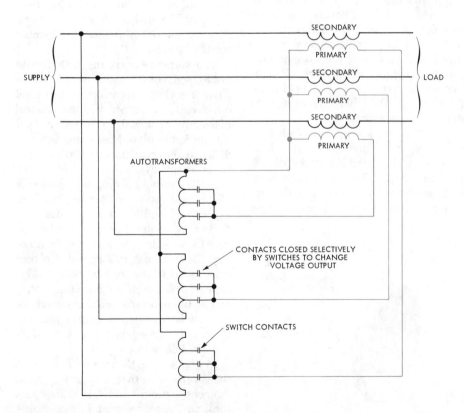

Fig. 28. Diagram of a tap-changing autotransformer regulator. Voltage of the transformers shown in color is determined by the selection of taps through switch operated contacts.

POWER RECTIFIERS

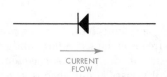

CURRENT
FLOW

Fig. 30. Schematic symbol for the rectifier. Current flow is opposite to the arrow in the symbol.

Although all but a very small percentage of electricity is generated as alternating current, it is often necessary to have a supply of direct current. It might be obtained from a small direct-current generator, but it is much easier to change the alternating current into direct current, which is termed *rectification*. This can be done by one of several available methods.

Dry-Disk Rectifier

The dry-disk rectifier, also known as a junction rectifier, makes use of the principle that certain combinations of metals permit current to flow in only one direction. Two of the more popular are the *selenium rectifier* and the *copper oxide rectifier*. The copper-oxide rectifier described here illustrates the manner of operation.

Fig. 29 shows a copper disk which is about 1¼ inches in diameter and about ⅛ inch thick. One face of the disk is polished copper. The other face of the disk has been burned so that the surface metal has combined with oxygen

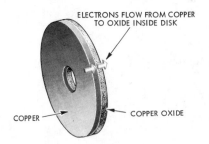

ELECTRONS FLOW FROM COPPER
TO OXIDE INSIDE DISK

COPPER

COPPER OXIDE

Fig. 29. Copper rectifying disk.

in the air to form a compound of copper which is called *copper oxide*. When an alternating voltage is applied to the opposite faces of the disk, electrons flow from the copper to the copper oxide face, but not in the reverse direction. That is, the disk rectifies alternating current. Fig. 30 illustrates the symbol for the rectifier.

The voltage across any disk should not exceed 5 or 6 v and is usually somewhat less than this value. When used on circuits of higher voltage, several disks are arranged in series, a lead washer being placed between each two disks in order to improve contact between them, Fig. 31.

In its simplest form, this device is also a half-wave rectifier as shown in Fig. 32(A) and Fig. 32(B). Four such units may be arranged in a *bridge* circuit to provide full-wave rectification. Fig. 33 illustrates this circuit. Current is supplied to the load by way of four copper oxide rectifying units: *1, 2, 3* and *4*. In each case, the copper side of the disk is represented by an arrow and the copper oxide face is represented by a short, heavy line.

In Fig. 33(A), the current flows from the supply wire that is negative at the moment to units *1* and *4*. It cannot pass through *1* because that is the copper oxide face, but it passes readily through

Fig. 31. A small and a large rectifier made up of individual disks.

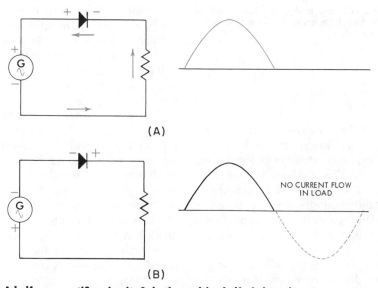

(A)

(B)

Fig. 32. A half-wave rectifier circuit. Only the positive half of the voltage wave appears across the load.

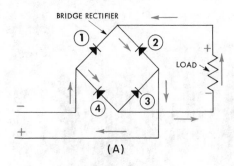

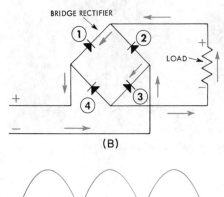

(C)

Fig. 33. A full-wave bridge rectifier circuit. In (C), the waveform across the load is shown.

unit *4* to the negative side of the load. After passing through the load, the current flows to the junction between units *1* and *2*. The current cannot flow through *1* because of the pressure of the negative supply line on the other side; therefore, the current flows through unit *2* to the positive supply wire.

When the supply voltage reverses polarity, the current flow through the bridge rectifier is shown in Fig. 33(B).

This flow is from the negative terminal of the supply, through unit *3*, through the load (this direction remains unchanged), and through unit *1* to the positive supply terminal. We will note that the current flows through the load in the same direction during each alternation of the supply voltage, resulting in a full-wave direct-current supply to the load.

Mercury-Arc Rectifier

In industrial applications, where higher amounts of current are desired than can be obtained from the selenium or copper oxide rectifiers, we find application for the glass-bulb *mercury-arc* rectifier. This rectifier employs a small pool of mercury, a plate element and a starting element, the pool serves the same purpose as the cathode of other electron tubes; that is, it furnishes the supply of electrons. This is termed a cold-cathode rectifier because the mercury pool, or cathode, is unheated.

In this tube, conduction (current flow) will not begin when a positive voltage is applied to the plate. Conduction will only begin when a positive voltage is applied to the plate and an arc is established between the mercury pool and the plate.

There are different methods for establishing the arc. One is with a starting rod, Fig. 34, which can be tilted into the mercury pool and then removed. Withdrawal of the rod causes a spark which vaporizes some of the mercury, and releases electrons at the surface of the pool. Meanwhile, one of the anodes, A for example, in Fig. 34 has been made positive by that half of the transformer secondary to which it is connected, so that it attracts the free electrons. On the way to the anode,

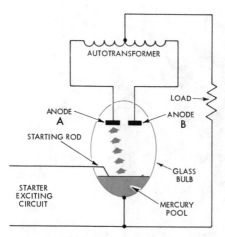

these electrons collide with molecules of vaporized mercury, releasing more electrons in a modified form of chain reaction. Molecules stripped of outer electrons become ions, that is, they take on a positive charge.

The mercury pool, at this time, is of negative polarity because it is connected to the center tap of the transformer, which is negative with respect to anode A. An arc forms between P and A, an immense volume of electrons flowing in the circuit, while heat from the arc maintains the process. As

Fig. 34. A simplified drawing of the glass-bulb rectifier. With polarity of the autotransformer as shown, current flow is to anode A. When polarity changes on the next alternation, current will flow to anode B.

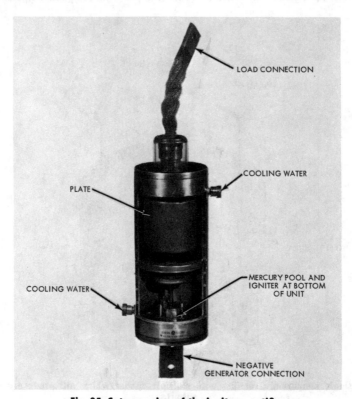

Fig. 35. Cutaway view of the ignitron rectifier.

195

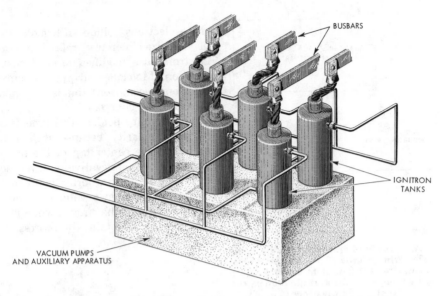

BUSBARS

IGNITRON
TANKS

VACUUM PUMPS
AND AUXILIARY APPARATUS

Fig. 36. A group installation of ignitrons to provide a polyphase output.

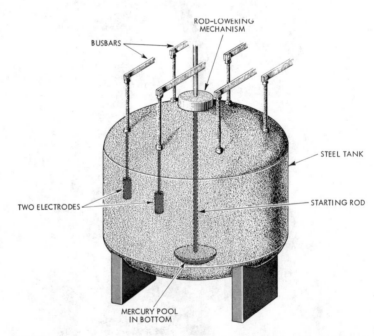

ROD-LOWERING
MECHANISM

BUSBARS

STEEL TANK

TWO ELECTRODES

STARTING ROD

MERCURY POOL
IN BOTTOM

Fig. 37. A single mercury-arc rectifier with multiple anodes which will produce a polyphase output.

A becomes negative during the next alternation, and therefore unable to attract electrons, anode B becomes positive, the arc shifting over to it until B is again negative and A positive. Current through the load is always in the same direction, alternations having been rectified.

The mercury vapor condenses on the wall of the bulb, and returns to the pool where it is again volatilized by the arc. It is worth noting that the arc is composed of electrons traveling from cathode to anode, and by positive mercury ions traveling in the opposite direction, attracted by the negative cathode. While this glass bulb device is rapidly becoming obsolete, its essential principles govern operation of the modern metal enclosed units.

Ignitron Mercury-Arc Rectifier

The *ignitron rectifier,* Fig. 35, differs from the device of Fig. 34, having a metal enclosure, and a single anode. For this reason, it delivers current only on those alternations when the anode is positive, so that two units like that of Fig. 35 would be required to supply full-wave rectification. With a single anode, the arc goes out during the negative alteration. If the same type of ignition were used as in the glass-bulb type the arc would have to be withdrawn each time, a most impractical procedure.

To overcome this difficulty, the ignitron makes use of a rod whose roughened surface material makes such poor contact with the mercury that tiny arcs are drawn so long as its circuit is excited, and free electrons are present. When the next positive alternation oc-curs, the arc is automatically reestablished. Ignitrons are usually water-cooled, the liquid circulating between inner and outer jackets. Small and medium size units are hermetically sealed at the factory to preserve the vacuum.

Polyphase Mercury-Arc Rectifiers

Ignitrons and similar units are often grouped, as in Fig. 36, for the purpose of changing polyphase current into direct. These larger assemblies are not permanently sealed at the factory, but are provided with a pumping system for maintaining the vacuum. Such installations have entirely supplanted the rotary converter formerly used to supply D.C. power to city and interurban street-railway systems. Ignitron groups are also employed for industrial processes that require large amounts of direct current, such as rolling mills and ore refining plants. A competing unit, Fig. 37, has a single large tank which utilizes multiple electrodes. After being started, this rectifier will continue to operate automatically, the arc shifting from one anode to another with the changing polarities of the supply voltage alternations.

Fig. 38 illustrates the method of changing three-phase current to six-phase for supplying either single or grouped-tank installations. The primary transformer windings are delta connected, while the secondaries are two separate, four-wire star groups with interconnected neutrals, the polarities of the phase windings in the two sections being 180 degrees apart. That is, secondary windings A and A_1 which are associated with primary winding P_1, are reversed with respect to each other.

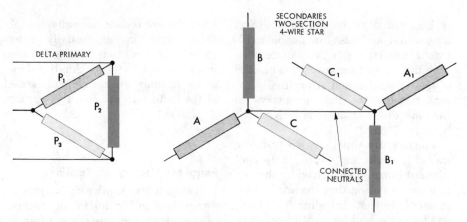

Fig. 38. Transformer connection for six-phase output. The primary is delta, the secondary is two star transformers with connected neutral. The usual schematic symbol for the windings have been simplified to better illustrate the relationship between the windings.

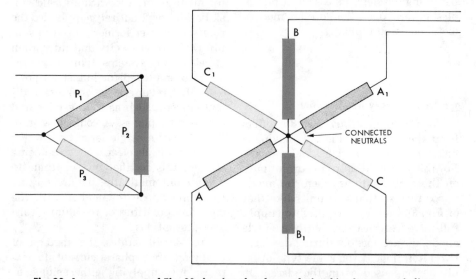

Fig. 39. A rearrangement of Fig. 38 showing the phase relationships between windings.

Windings B and B_1, associated with primary P_2, bear a like relationship, while C and C_1, associated with primary P_3, complete the pattern.

The diagram of secondary windings may be rearranged as in Fig. 39, which has exactly the same connections as Fig. 38, but which illustrates the fact that adjacent windings, such as A and C_1 are separated by only 60 electrical degrees. Thus, the secondary windings provide six-phase output.

REVIEW QUESTIONS

1. By how many degrees does induced voltage lag the inducing current?

2. State the rule for determining the value of induced voltage at any point on the sine wave of inducing current.

3. At what point on the current wave is the highest voltage induced?

4. By how many degrees does primary counter-emf lag primary voltage in a transformer?

5. A 7.5 kva 240-60-volt, step down single-phase transformer has 1000 primary turns. How many turns in the secondary winding?

6. What is the primary current when the power factor is 1?

7. What is zero-sequence current?

8. State one of the two phase rotations to be observed when changing secondary transformer connections from star to delta?

9. If the new voltage is 230 volts in Question 8, what was the original secondary voltage?

10. If the 2300-volt star-connected primaries of Question 9 remain as before, what is the ratio of transformation?

11. What are harmonics?

12. Explain the terms additive — and subtractive-polarity.

13. What are the essential differences between transformers and autotransformers?

14. If the voltage between the 40 percent tap and the near end of an autotransformer is 96 volts, what is the primary voltage?

15. If the current taken by the load at the 40 percent tap is 15 amps, what is the primary current?

16. The current delivered to the load by the 120 percent tap of an autotransformer is 20 amps. What is the primary current?

17. Describe the polyphase induction regulator.

18. Describe the operating cycle of a glass-bulb, single-phase, mercury-arc rectifier.

19. Explain how an ignitron is restarted at the beginning of each positive alternation.

20. Tell, briefly, how six-phase current is obtained.

Chapter 11

Electrical Measuring Instruments

Use of Meters

Electrical meters are used to measure volts, amperes, ohms, watts, and numerous other factors associated with electrical circuits. The operation of electrical instruments depends basically upon electron flow through a moving coil, the amount of current determining the amount that the coil will move. A pointer attached to the moving coil indicates on a calibrated scale the appropriate quantity that the meter was designed to measure. Direct-current meters almost universally use a coil arrangement known as the D'Arsonval movement. In essence, a meter is an electro-mechanical Ohm's law computer which makes use of electron flow, calibrated resistances and voltages (in an ohmmeter), to indicate the desired electrical quantities.

The D'Arsonval Movement

The D'Arsonval movement, shown in Fig. 1, consists of a permanent magnet and a moving coil with a pointer. The magnet is constructed of a ferromagnetic material which has a high degree of magnetic retentivity. The moving coil has a large number of turns of small diameter wire wound on an aluminum frame, which is mounted on jeweled bearings to reduce friction of movement. The number of turns and the diameter of the wire used in the coil is determined by the intended sensitivity and application for which the meter is designed. Fig. 2, and Fig. 3 illustrate the moving coil and its jewelled bearing.

Certain details are worth noting. The magnetic circuit is made up of the ferromagnetic unit, soft-iron pole pieces whose faces are accurately machined to the proper radius, and a soft-iron core. The coil frame is mounted on the soft-iron core and is mounted between the pole faces of the magnet. The soft-iron elements insure an even distribution of magnetic flux in the airgap between the pole faces and the core. If the whole circuit were of the ferromagnetic material, pole faces could be distorted in the heat treating process, rendering the instrument inaccurate.

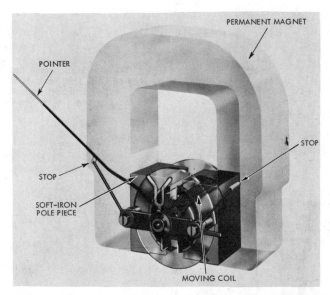

Fig. 1. The D'Arsonval meter movement showing the essential parts.

Courtesy Daystrom Inc., Weston Instrument Div.

The D'Arsonval movement operates on the principle of repulsion between like magnetic poles; the magnetic poles of the permanent magnet reacting against the poles established by current flow through the coil. Reference to Fig. 4 will help in understanding the operation of a direct-current meter. The coil is held in place by disks that are rigidly fastened to a shaft. The shaft and coil are centrally placed between the poles of the permanent magnet, and turn against the tension of the spring which tries to hold them in the position shown.

When the meter is not in operation, and the coil is in the position shown in Fig. 4, the pointer indicates zero on the meter scale. Now we will assume that the meter is arranged in a circuit to read the desired electrical quantity—in this case voltage or current. When the meter is connected, current will flow into the meter and through the coil. The current flow establishes an N pole on one side of the coil and an S pole on the other. The repulsion between the N pole of the magnet and the N pole of the coil, and also the S pole of the magnet and the S pole of the coil, causes the coil to turn against the tension of the spring, moving the pointer across the scale.

The amount the coil rotates depends on the strength of the repulsion between the two sets of like magnetic poles.

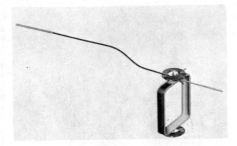

Fig. 2. The aluminum frame and winding of the moving coil. The frame is mounted on a cylindrical soft iron core.
Courtesy Daystrom Inc., Weston Instrument Div.

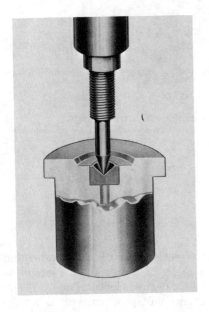

Fig. 3. Enlarged view of the jeweled bearing of the moving coil.
Courtesy Daystrom Inc., Weston Instrument Div.

Since the strength of the magnetic pole set up in the coil is dependent on current flow through it, rotation is in proportion to the amount of current through the coil.

The aluminum frame upon which the coil is wound, serves to stabilize, or dampen, the whole moving element.

The metal frame acts like a short-circuited loop, a voltage being induced in it as it moves. The resulting current reacts with the magnetic flux to oppose any tendency to oscillate before the pointer comes to rest. This action does not affect accuracy of the instrument, because the induced current becomes

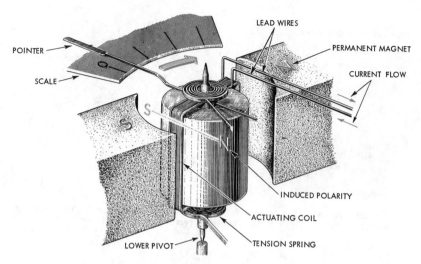

POINTER

SCALE

LEAD WIRES

PERMANENT MAGNET

CURRENT FLOW

INDUCED POLARITY

ACTUATING COIL

LOWER PIVOT

TENSION SPRING

Fig. 4. Simplified drawing illustrating the operation of the D'Arsonval movement. When current flows through the moving coil, poles are set up as indicated. The force of repulsion between like poles moves the coil in the direction shown in the large arrow.

zero when motion ceases. Errors caused by external magnetic fields are avoided by enclosing the mechanism in a soft-iron case which furnishes an easy path around the mechanism for any stray flux.

Direct-Current Ammeters

Figures 5 and 6 illustrate a typical ammeter and the method for connecting it into the circuit to measure current. It should be noted that an ammeter is *always* connected in series with the current it is desired to measure, and that all the current must go through the meter. This makes necessary a word of caution. If the ammeter is built to measure 5 amps, we must be certain that less than 5 amps are flowing, for if more current goes through

the meter than the maximum amount shown on the scale, the fine wire of the moving coil may be burned out.

Another point to note is that in connecting a direct-current ammeter polarity must be observed. That is, the leads are attached so that the current flow is into the terminal that has a negative (−) marking. Connecting the meter with reversed polarity can damage the movement by causing deflection in the wrong direction.

The movements used in ammeters are designated by the amount of current necessary to cause full scale deflection: full scale deflection being the amount of coil movement that will indicate maximum reading on the meter scale. Since most movements require such minute quantities of current for full scale de-

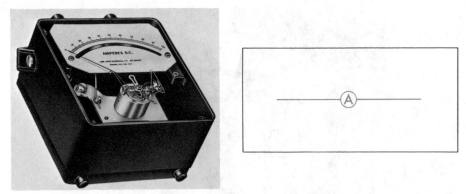

Fig. 5. A d-c ammeter with the front removed to show the movement, which has only one range of measurement. At right is the ammeter symbol used in schematic drawings.
Courtesy Daystrom Inc., Weston Instrument Div.

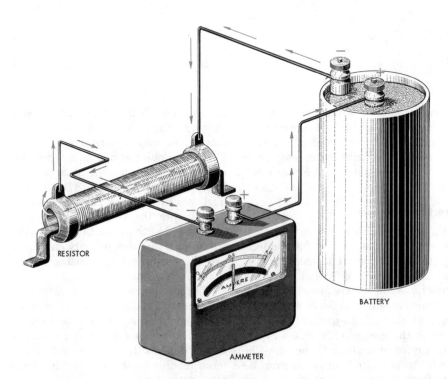

RESISTOR

AMMETER

BATTERY

Fig. 6. The d-c ammeter connected into a circuit. It is always connected in series with the current to be measured and correct polarities must be observed.

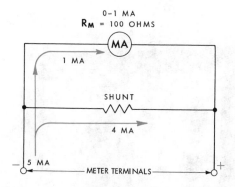

0-1 MA
R_M = 100 OHMS

MA

1 MA

SHUNT

4 MA

5 MA

METER TERMINALS

Fig. 7. Shunt used to extend the range of an ammeter. Higher currents can be measured because the shunt bypasses the major portion of the current around the meter movement.

$$E = .001 \text{ amp} \times 100 \text{ ohms}$$
$$E = .100 \text{ volts}$$

Since the voltage across the movement will be the voltage across the shunt, the shunt resistance is calculated as follows:

$$R_{shunt} = \frac{E_{shunt}}{I_{shunt}}$$
$$R_{shunt} = \frac{.100 \text{ volts}}{.004 \text{ amps}}$$
$$R_{shunt} = 25 \text{ ohms}$$

flection, they are obviously inadequate for a great many practical applications requiring a larger range of current measurement. This difficulty is overcome by the use of resistors connected in parallel across the movement which extend the range of the meter. In practice these resistors are referred to as *shunts*.

To illustrate the method of extending the range of the basic meter movement by the use of shunts refer to Fig. 7 which has a 0-1 ma movement with a resistance of 100 ohms. Suppose, for example, that we wish to increase the range of measurement to 5 ma. Since no more than 1 milliamp must go through the movement, then the shunt must bypass 4 milliamps of current.

Our first step in solving for the shunt is to determine the voltage across the movement at full scale using Ohm's law. We write the formula:

$$E = I \times R$$

Substituting known values, we have:

By the use of a shunt to divert the major portion of the current through the meter, the range can be extended to measure far greater values of current than that for which the movement is designed.

Meters may be designed as single range units such as that illustrated in Fig. 5, or as multiple range units, Fig. 8. The use of external shunts with the single range meter allows the extension of their range to almost any desired value. In practice it is rarely necessary to calculate the value of an ammeter shunt as was done in connection with Fig. 7. Many ammeters are furnished with accurately calibrated shunts which are marked with their ampere ratings instead of their resistance. Thus, where the ammeter has a full-scale rating of 5 amperes, shunts may be furnished for 15 amps, 30 amps, or whatever range is desired.

In the multiple range ammeter, the shunts are incorporated into the circuitry of the meter and are selected by a range knob for the desired range of the current reading. Fig. 9 illustrates a typical circuit schematic for an ammeter with three current ranges.

Fig. 8. A multiple range microammeter. The range to be measured is selected by the use of a range selector knob on the left.

Courtesy Keithley Instruments

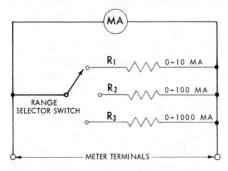

Fig. 9. Circuit for a three range ammeter. The maximum current range is determined by the selector switch which connects the shunt resistors across the meter movement.

The Direct-Current Voltmeter

Since the voltmeter is designed to measure voltage drop it must be connected *across* the load that causes the voltage drop as shown in Fig. 10. The same caution must be exercised with the voltmeter as with the ammeter—proper polarity must be observed and

the voltage being read must not be greater than the meter range.

The same meter movement used in the ammeter can be used in the voltmeter, but the voltmeter requires a current limiting resistor in *series* with the meter movement instead of a shunt. This series resistor is called a *multiplier,* and its value determines the maximum range of the meter.

Fig. 11 illustrates a typical voltmeter circuit with its multipliers, using a 1-milliampere movement with a resistance of 100 ohms. Let us suppose that we desire full-scale deflection for 1.5 volts. The resistance of R_l needed in series with the meter can be determined by Ohm's law, for we know the voltage (1.5) and the current (.001) to produce full-scale.

$$R = \frac{E}{I}$$

$$R = \frac{1.5 \text{ volt}}{.001 \text{ amp}}$$

$$R = 1500 \text{ ohms}$$

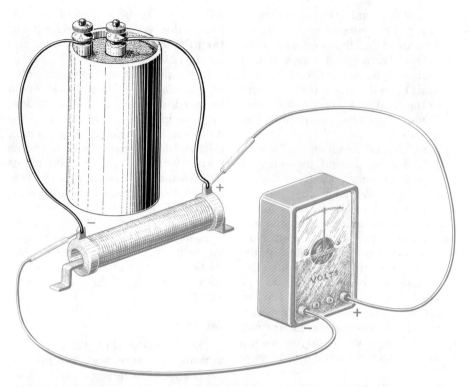

Fig. 10. A d-c voltmeter connected in a circuit for measurement. Voltmeters are connected across the circuit whose voltage is to be measured. Correct polarity must be observed.

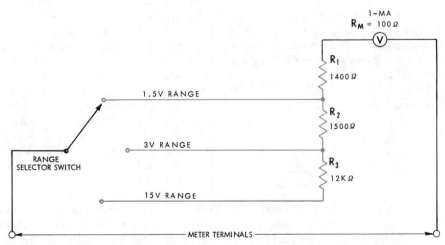

Fig. 11. Circuit for a three range voltmeter.

This is the total value of resistance needed to limit current to 1 milliamp, and we know that the resistance of the meter is 100 ohms. Therefore, to obtain the value of the multiplier, we simply subtract the resistance of the movement from the total resistance, (1500 ohms −100 ohms = 1400 ohms, multiplier resistance).

With the additional multipliers R_2 and R_3 we can extend the range of measurement from 1.5 volts to 3 volts and 15 volts. We know that the meter will give a full-scale deflection of 1.5 volts when a multiplier resistance of 1400 ohms is used in the circuit. Now we must determine the value of the multiplier for a range of 3 volts. The same approach is used as that to calculate the multiplier for the 1.5 volt range. We know that the applied voltage will be 3 volts and the current will be 1 milliamp (.001 amp), so we use Ohm's law to find the total resistance.

$$R = \frac{E}{I}$$
$$R = \frac{3 \text{ volts}}{.001 \text{ amp}}$$
$$R = 3000 \text{ ohms}$$

Since we already have 1500 ohms in the circuit, it is necessary to add only 1500 ohms in series with the movement to obtain the resistance needed for the 3 volt range.

To increase the range to 15 volts, the process is repeated.

$$R = \frac{E}{I}$$
$$R = \frac{15 \text{ volts}}{.001 \text{ amp}}$$
$$R = 15,000 \text{ ohms}$$

The resistance of the movement, and the multipliers for the 1.5 and 3 volt range already in the circuit add up to 3,000 ohms, so for the 15 volt multiplier value we subtract this amount from the total required resistance and obtain 12,000 ohms.

In the previous voltmeter examples, all multipliers were in series. It should be noted, however, that some designs may use individual resistances for the multiplier in each range. In such a case, each resistance would be of the value needed for the range of reading minus the resistance of the movement. Taking the 15 volt range of the voltmeter in Fig. 11, for example, the total resistance required was 15,000 ohms. In a meter using individual resistances for each range, the multiplier value would be 15,000 ohms minus the 100 ohms resistance of the movement = 14,900 ohms.

The Ohmmeter

The ohmmeter, Fig. 12, is used to measure resistance, and, like the ammeter and voltmeter, uses a D'Arsonval movement. The movement is usually more sensitive, however, requiring less current for full-scale deflection.

Some important differences exist between the ohmmeter and the meter circuits studied so far. In the previous meters, currents and voltages were present for normal operation of the circuit whose values were being measured. The ohmmeter must be used *only in an inoperative circuit with the power turned off.* The movement is so sensitive that normal circuit voltages would burn it out. An ohmmeter does not require power from the circuit being measured because it uses a battery to supply its own current and voltage.

Fig. 13 illustrates the basic circuit of the ohmmeter, consisting of the meter movement, a current limiting resistor and a battery. Full-scale deflection is

Fig. 12. A typical ohmmeter with five resistance ranges. Note that the scale reads from right to left.

Courtesy Triplett Electric Instrument Co.

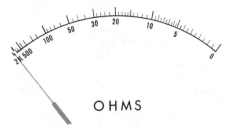

Fig. 14. The scale of an ohmmeter. Because it reads from right to left it is known as a *backup* scale.

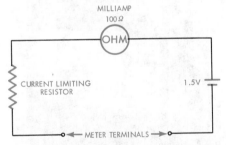

Fig. 13. Schematic diagram of the basic ohmmeter circuit.

obtained with no external resistance being measured and with the meter leads shorted together. Thus, the current limiting resistor of the meter circuit must limit the current to the value for which the movement is rated. Let

us see how this resistance is determined. Referring to Fig. 13, we know that the battery produces 1.5 volts, and that current must be limited to .001 amps, so we use Ohm's law.

$$R = \frac{E}{I}$$
$$R = \frac{1.5 \text{ volts}}{.001 \text{ amps}}$$
$$R = 1500 \text{ ohms}$$

This represents the total resistance required for full-scale deflection. Since there are already 100 ohms resistance in the movement itself, it is necessary to subtract this amount from the total in order to find the value of the current limiting resistor.

1500 ohms − 100 ohms = 1400 ohms

The scale of the ohmmeter, Fig. 14, known as a *backup scale* has its zero position on the right which corresponds to full-scale deflection. That is, maximum deflection of the meter indicates zero ohms of resistance. The scale is calibrated in relation to battery current and total resistance — total resistance being the resistance of the meter plus the resistance being measured.

For example, in Fig. 15, suppose we take the basic ohmmeter circuit, add a

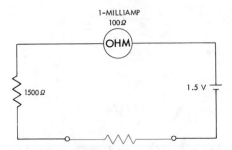

Fig. 15. An ohmmeter circuit with a resistance equal to the meter across the terminals will cause half-scale deflection.

The Multimeter

The multimeter, Fig. 16, incorporates the functions of the voltmeter, ohmmeter, and milliammeter into one unit which gives a versatility and economy that could not be obtained by the use of individual instruments. The multimeter is also called a volt-ohm-milliammeter, or, in more common usage is simply called a VOM.

Primarily the multimeter is an ammeter with a D'Arsonval movement. The design for the various meter circuits is in accordance with the meter circuits already studied; their combination into one unit being achieved by the

resistance equal to that of the meter across the terminals and determine where the meter will indicate on the scale. To do this we must first calculate the current through the meter. We know that battery voltage is 1.5 and that the resistance is 3000 ohms, so,

$$I = \frac{E}{R}$$
$$I = \frac{1.5 \text{ volts}}{3000 \text{ ohms}}$$
$$I = .0005 \text{ amps}$$

This is 500 micro-amps which is half the full-scale current, so the meter will indicate at ½ scale. This point is marked on the scale as 3,000 ohms rather than as amperes. The other points on the scale are established in a similar manner.

It should be noted that the ohmmeter scale, Fig. 14, is non-linear, and that pointer deflection becomes less as the resistance increases, because the greater the resistance, the smaller the value of electron flow for any given voltage.

Fig. 16. A multimeter which can be used to measure voltage, current, and resistance, with several ranges for each function.
Courtesy Simpson Electric Co.

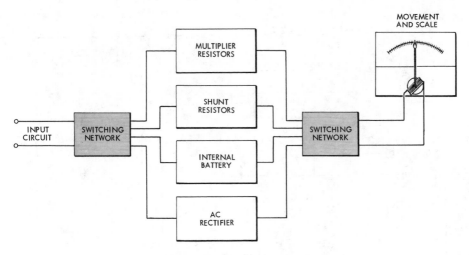

Fig. 17. Block diagram of a typical multimeter.

use of multiple selector switches which connect them into the circuit with the meter movement. The block diagram of Fig. 17 illustrates the basic circuit arrangement of a multimeter.

Since we are already familiar with the basic circuitry, let us examine the switching arrangements of the multimeter and see how the various functions are combined. Fig. 18 is the schematic for the multimeter shown in the block diagram.

Beginning first with the voltmeter circuit, and tracing from the left-hand meter terminal of Fig. 18, we come to switch S_1. Here we have two voltmeter ranges: switch *Position 1*, and *Position 2*. Following the circuit through *Position 1*, we find multiplier resistor R_1 in series with the meter movement, through switch S_3 to the other meter terminal, which completes the circuit through the meter. We have not ac-

counted for switch S_2, so we return to S_1 and trace to S_2, where we find that there is no connection in *Position 1* and *Position 2*. Thus switch S_2 is not utilized in the voltmeter circuit.

Next, we trace the milliammeter portion of the circuit which has two ranges, switch *Position 3* and switch *Position 4*. Beginning at the left hand meter terminal of Fig. 18, through S_1, we go through the meter movement to S_3, through S_3 to the right-hand meter terminal. Returning to S_1, the circuit then goes through S_2, to the shunt resistors R_3 and R_4 which are across the meter movement, and then to the right-hand meter terminal which completes the circuit.

The ohmmeter section of the multimeter, switch *Positions 5* and *6*, is traced in the same manner as the above circuits. Note, however, that *Positions 5* and *6* on switch S_3 are shorted to-

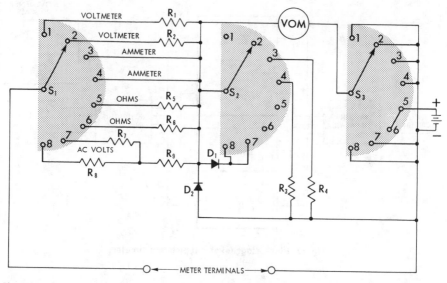

Fig. 18. Schematic of a multimeter circuit having two ranges for each function. The areas shaded in color are rotary selector switches which give the unit its versatility.

gether so that the battery remains in the circuit for either range.

The multimeter illustrated here has one meter circuit that has not been studied: the a-c voltmeter. While individual a-c instruments will be dealt with at length in a later section, the inclusion of this function in many multimeters justifies a preliminary discussion of the circuit.

The D'Arsonval movement is basically a direct-current instrument. If alternating-current is applied to the coils of the movement, the meter would deflect in the proper direction during one alternation of the sine wave. When the other alternation began the current through the coil would reverse and the pointer would deflect in the wrong direction to the left — banging the pointer against the stop. This would

continue with the pointer swinging back and forth, and would give no reading of any value whatsoever.

This difficulty is overcome by the use of rectifiers in switch *Positions* 7 and 8 of the multimeter. Let us trace the circuit for the a-c voltmeter and see how the rectifiers are employed. We begin, as in the other circuits, with the left-hand meter terminal, which brings us to switch S_1. Tracing through for *Position* 7, from S_1 we encounter R_7, which is a multiplier, to R_9, and diode rectifier D_1, through the meter movement to S_3 and through S_3 to the other meter terminal. This does not complete the circuit, because passing through R_9 there was a circuit branching off containing rectifier D_2. Returning to the circuit at the point where D_2 branches off from R_9 we trace through D_2 to the

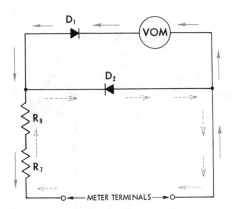

Fig. 19. A simplified schematic of the rectifier used in the a-c circuit of the multimeter. Arrows show current flow for successive alternations of the sine wave.

right-hand meter terminal. Thus D_2 is across the meter movement. The combination of R_9, D_1 and D_2, make up an a-c rectifier circuit. Referring to Fig. 19, which is a simplified schematic of the voltmeter circuit, will help to clarify its operation.

Current during the first alternation is through the meter, D_1, R_9 and R_7, as shown by the solid arrows on the schematic. As D_2 is connected to oppose current flow on this alternation, no current goes through it. During the next alternation, when polarity changes, electron flow is indicated by the dashed arrows on the schematic. No current can go through D_1, so it bypasses the meter through D_2. Thus the meter will operate only on ½ hertz of the a-c sine wave.

Meter Sensitivity

The sensitivity of meter movements are specified in two different ways. One system rates the sensitivity in terms of the voltage drop necessary across the moving coil to cause full-scale deflection. Thus, if a meter required 100 millivolts for full-scale deflection, it would be referred to as a "100 millivolt movement." For example, if the moving coil of this meter has a resistance of 100 ohms, then 1 milliamp would be required for full scale deflection,

$$E = IR$$
$$E = .001 \text{ amp} \times 100 \text{ ohms}$$
$$E = 100 \text{ millivolts}$$

The second system of rating meter sensitivity rates the meter in *ohms per volt.* Using the specifications of the previous meter, let us suppose that we desire a meter with a full-scale reading of 1 volt. Since 100 millivolts will cause full-scale deflection, a multiplier resistance will be required to limit the current through the movement. We can obtain the value of the multiplier using Ohms law, using the 1 volt required for full-scale, and the 1 milliamp of required current.

$$R = \frac{1 \text{ volt}}{.001 \text{ amp}}$$
$$R = 1000 \text{ ohms}$$

The sensitivity of the meter is then said to be "1000 ohms per volt." If the movement required 500 microamps (which is half of 1 milliamp), the meter sensitivity would be 2000 ohms per volt.

The Megohmmeter

The megohmmeter, more often called a *megger,* is an instrument used to measure very high resistances, and derives its name from the fact that these resistances are in the megohm range. (The term *mega,* means million. In this

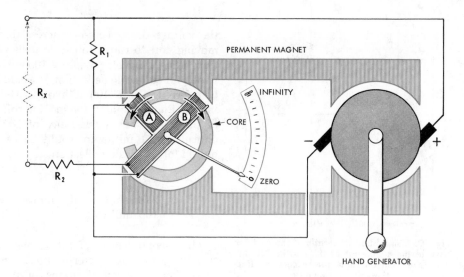

Fig. 20. The megohmmeter and a diagram of its internal construction. The permanent magnet supplies the magnetic field for both the movement and the generator.

case it has been shortened to *meg.*) The megger consists of a hand-driven direct-current generator and a meter movement; the magnetic field for both being supplied by a common permanent magnet, Fig. 20.

The meter used is an opposed coil type, having two coils, *A* and *B*, mounted over a gapped core. The coils are wound on a light frame, similar to that of the D'Arsonval movement, and rotate around the core which remains stationary. The current for the coils is supplied by the hand-driven generator.

To explain the operation of the unit, it is necessary to examine the action of the coils with open terminals; with the terminals shorted; and with a resistance across the terminals.

When the terminals are open, current flow is from the generator, through *A* and R_1 which are in series with the generator. Since the terminals are open, no current flows through coil *B* to oppose movement and coil *A* will swing *counter-clockwise* to a position over the gap in the core. In this position the pointer indicates infinity.

With the terminals shorted together,

a larger current flows through coil B than through coil A and the greater force in coil B moves the pointer *clockwise* to the zero position on the scale. Resistor R_2 is a current limiting resistor which prevents damage to the meter in this situation.

If a resistance is connected across the terminals, current flows through coil B, R_2, and the unknown resistance R_x. This current attempts to move coil B clockwise, but the opposing force created by current through coil A tries to move it counter-clockwise. The final position of the coils is determined by the magnitude of the current through R_x, and the coils will stop at a point where the forces tending to move them are at a balance. The pointer then indicates the value of the unknown resistance on the scale. No springs are used in the movement since the opposing forces in coils A and B balance the pointer when a reading is being taken. Having no springs to hinder its movement, the pointer floats freely back and forth across the scale when the megger is not in operation.

Meggers may be obtained with different voltage ranges; the more common being 500 and 1000 volts. The higher the resistance range to be measured, the higher the voltage required to actuate the movement for a reading. Friction clutches are used to hold the generator to its rated voltage output. In operation, these clutches are designed to slip if cranked over a certain rate of speed, thus dropping the output to a safe value.

It should be noted that in the operation of the megger a very high voltage exists at the terminals which can be dangerous. Proper procedure in operation is to make sure that all power to the circuit to be tested is turned off: the test terminals are then attached and the reading taken.

The Electrodynamometer Movement

In the electrodynamometer, Fig. 21, the permanent magnet common to the D'Arsonval type meter has been replaced by an electromagnet. With the addition of a rectifier, it is suitable for use in both a-c and d-c circuits. The

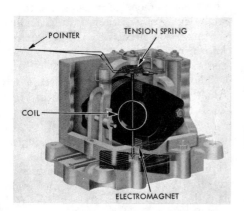

Fig. 21. A cutaway view of the electrodynamometer movement. This movement is adaptable to both a-c and d-c measurements.
Courtesy Weston Electrical Instrument Corp.

D'Arsonval meter, because of its higher efficiency, predominates over this type in the direct-current field, but it is used in numerous a-c instrument applications such as wattmeters, frequency meters, and power-factor meters.

The D-C Wattmeter

The d-c wattmeter, Fig. 22, uses the electrodynamometer movement described in the previous paragraph. It incorporates features of both the ammeter and voltmeter and is somewhat similar in construction to the d-c voltmeter.

The field coils are connected in series with one of the circuit wires as an am-

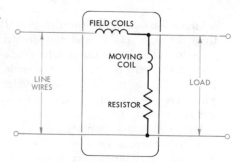

Fig. 23. Circuit connection of the a-c or d-c wattmeter.

Fig. 22. The d-c wattmeter. This model is calibrated to indicate kilowatts and has two ranges.

Courtesy Weston Electrical Instrument Corp.

meter. The moving coil is connected across the circuit as a voltmeter, as illustrated in Fig. 23. Torque developed by the moving coil is dependent upon the voltage across the moving coil and the current through the field coils. Since movement of the coil is dependent on both current and voltage, of the circuit, the scale may be calibrated in watts, because *volts × amps = watts.*

The Wheatstone Bridge

The Wheatstone bridge, Fig. 24, is a precision laboratory device used for the measurement of resistances to a high degree of accuracy. It is basically a comparison circuit: measuring an unknown resistance against precision resistors of known value.

The Wheatstone bridge circuit is usually shown schematically as in Fig. 25. It derives the name bridge from the fact that a sensitive D'Arsonval movement, known as a galvanometer, *bridges* the two sides of the circuit. This is typically a zero-center meter having a sensitivity of 20 to 50 microamps and is called the *detector.* The resistors R_1 and R_2 are termed the *ratio arms* and form one side of the bridge circuit.

Fig. 24. A wheatstone bridge. The meter on the left in the photo is called the detector. The knobs on the right are used to adjust for zero current through the circuit.
Courtesy Leeds & Northrup Co.

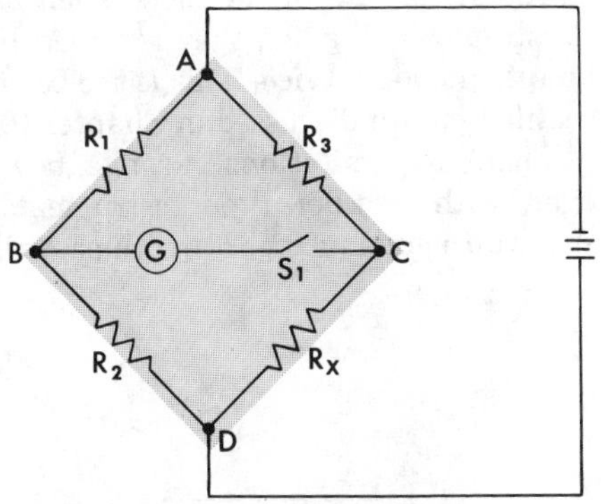

the *multiplier*. The resistor R_x is the unknown resistance whose value is to be determined. The other side of the bridge is made up of R_3 and R_x.

In operation, the known resistances, R_1, R_2, and R_3 are adjusted so that the detector indicates zero on the scale, which means that the voltage and current of the two sides of the circuit are equal. The circuit is then said to be in balance. The switch, S_1, is used to cut the meter out of the circuit in cases of severe imbalance. It is cut in momentarily while the adjustable resistors are used to achieve a near balance and is then left "on." When the detector reads zero, the unknown resistance may then be read from calibrated dials or calculated from the values of known resistance. Remember that the detector is used only to indicate circuit balance and *does not* indicate the value of the known resistance.

Fig. 26 illustrates a balanced bridge circuit. In order to have a balanced cir-

Fig. 25. Schematic of the wheatstone bridge circuit. It is traditionally drawn in the diamond shape illustrated here.

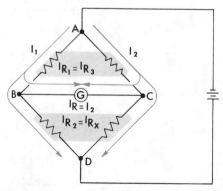

Fig. 26. The conditions in a wheatstone bridge circuit at balance. Current and voltage are equal through both sides of the circuit, and no current flows through the detector.

They are adjustable precision resistors with values in steps of 10 or 100 which simplifies calculation of the resistance values. R_3 is also an adjustable precision resistor, sometimes referred to as

cuit, the following conditions are necessary. Voltages are balanced, so that the voltage from point A to point B equals the voltage from point A to point C. Likewise, the voltage from B to D must equal voltage from C to D. The relationship may be indicated in terms of the IR drop (voltage drop) as follows:

$$\frac{IR_1}{IR_2} = \frac{IR_3}{IR_x}$$

However, this is usually simplified,

$$\frac{R_1}{R_2} = \frac{R_3}{R_x}$$

In order to find the value of R_x we must transpose,

$$R_x = \frac{R_2 R_3}{R_1}$$

$$\text{or } R_1 R_x = R_2 R_3$$

On the type of bridge shown in Fig. 24, the values of R_1, R_2, and R_3, are indicated on calibrated dials, the resistance of R_x then being determined by the use of the above formula.

In an unbalanced bridge, Fig, 27, current will flow through the detector, either to the right or left, until R_1, R_2, and R_3 are adjusted so that voltages at points X and Y are equal.

ALTERNATING-CURRENT METERS

Introduction

Most instruments employed with d-c are of the permanent magnet D'Arsonval type, and would not be suitable without change for use with a-c. Through the use of rectifiers, such as illustrated in Fig. 28, they are readily adapted to a-c service. This is the basic rectifier circuit discussed in Chapter 10.

There are limitations in range, however, with rectifier type instruments. A-c voltmeters using a rectifier will

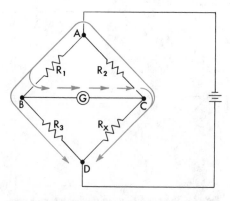

Fig. 27. An unbalanced wheatstone bridge circuit. Voltage at points X and Y are not equal and current flows through the detector.

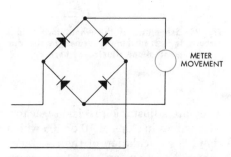

Fig. 28. Schematic of the typical rectifier circuit used in a-c meters.

provide accurate measurements up to a frequency of 10,000 hertz, and a-c ammeters up to 15 milliamperes. They are not suitable for measuring larger currents because the temperature co-efficient of the rectifier makes it inadvisable to use larger shunts, since higher currents create a great amount of heat that could damage the rectifier. Also, because of the inductive effects found in a-c circuits, current through the shunts is affected and a proportionate distribution of current is not obtained and would give an inaccurate reading.

While the D'Arsonval movement is unrivaled in its particular field, another movement, known as the *moving iron* type, is more common for industrial uses in the a-c field. The term "moving iron" covers several types of instruments, such as the *solenoid, repulsion-vane,* and *inclined coil* meters, all of which use a soft iron core as the moving element. These instruments are suited to a wide variety of applications, each of which will be discussed in turn.

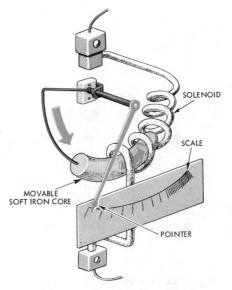

Fig. 29. The solenoid meter mechanism.

Solenoid Meters

The solenoid movement illustrated in Fig. 29 may be employed in ammeters and voltmeters for both a-c and d-c circuits. Its operation is very simple: current through the solenoid pulls the soft iron core into the coils, moving the pointer across the dial scale. The degree of deflection is determined by the amount of current through the coil and the strength of the magnetic field. The scale is calibrated in volts or amperes depending on the type of coil used. If the meter is to be used as an ammeter, the coil is made up of a relatively small number of turns of heavy wire. For use as a voltmeter, the coil requires a large number of turns of fine diameter wire.

Inclined Coil Meter

Fig. 30 shows the mechanism of the inclined coil meter. Here, both the electromagnetic coil and the moving iron vane are placed at inclined angles to the shaft on which the pointer rotates. When current flows through the coil, a magnetic field is set up which induces magnetic poles in the iron vane. The vane turns, attempting to align its magnetic poles with those of unlike polarity in the coil. The forces of attraction and repulsion between magnetic poles in the vane and those of the coil results in a torque which works against the tension of the pointer spring. Thus, the movement of the pointer is proportional to the strength of the current through the coil.

219

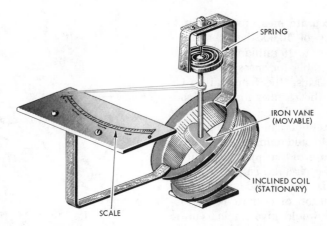

Fig. 30. The inclined coil meter. Both moving vane and the coil are at an angle to the shaft.

Repulsion Vane Meter

Another type of moving vane instrument is illustrated in Fig. 31. There are two iron vanes, X and Y. When current flows in the coil, both vanes are magnetized so that they have N poles at one end and S poles at the other. Vane X is attached to the side of the coil and is stationary although it may be adjusted. Vane Y is attached to the shaft and moves against the pressure of a spring (not shown in the drawing). Repulsion between like poles of X and Y causes Y to move; the degree of movement being determined by the

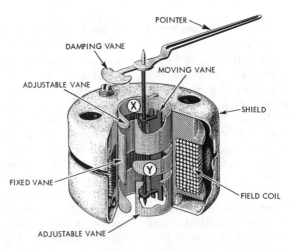

Fig. 31. Cutaway view of the repulsion-vane meter movement.

Courtesy General Electric Co.

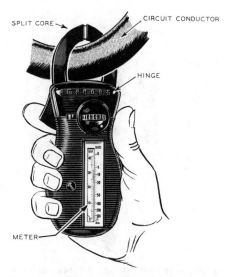

Fig. 32. A clip-on meter showing the method of use. The core is hinged so that it may be closed around the conductor whose current is to be measured.

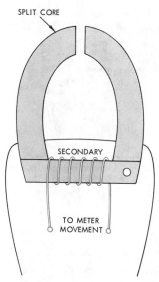

Fig. 33. Diagram of the core and secondary of the clip-on meter.

strength of the magnetic poles. The strength of the magnetic poles is determined by the amount of current flowing in the coil, and the scale is calibrated accordingly in either volts or amperes.

Clip-On Volt-Ammeter

The clip-on volt-ammeter, Fig. 32, is a popular testing device for a-c circuits. The core is split and hinged so that it may be clipped around the circuit wire whose voltage or current is to be measured. The core is made of soft steel and forms the primary of a transformer, Fig. 33. The secondary winding is wound on the portion of the core inside the instrument and is connected to the meter. Current through the circuit wire sets up a field in the soft steel core which is transferred through the secondary to the meter movement.

The type shown has several ranges which may be selected. Should the current be so low that it gives less than half-scale deflection, the circuit conductor may be looped through the core two or more times, thereby extending the range as illustrated in Fig. 34. For example, if the lowest scale is 5 amps, and the pointer shows a reading somewhere around 1 amp, the wire can be looped through the core twice as in Fig. 34, so that the increased current in the core will give a larger scale reading.

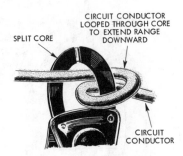

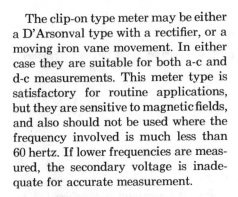

Fig. 34. The range of the clip-on meter may be extended by winding additional loops of the conductor around the core.

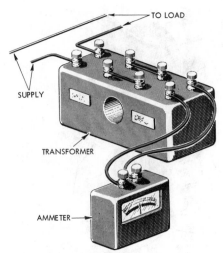

Fig. 35. A current transformer connected in circuit with an ammeter.

The clip-on type meter may be either a D'Arsonval type with a rectifier, or a moving iron vane movement. In either case they are suitable for both a-c and d-c measurements. This meter type is satisfactory for routine applications, but they are sensitive to magnetic fields, and also should not be used where the frequency involved is much less than 60 hertz. If lower frequencies are measured, the secondary voltage is inadequate for accurate measurement.

Current Transformers

Since shunts cannot be used to extend the ranges of a-c ammeters, due to various factors mentioned earlier, the range can be extended by the use of current transformers, Fig. 35. The primary winding is connected in series with the load; the secondary terminals to the meter. These transformers are generally used with a 5-ampere meter, and the ratio between primary and secondary must be used as a multiplying factor to obtain the correct reading. If the meter is so designed, the value may be read directly from the scale. Terminals are provided for changing the transformer ratio, so that full-scale de-

flection may be utilized in all cases. The ranges and connections are usually marked on the nameplate. Where a heavy current (over 100 amperes) is to be read, the circuit wire is passed through the central hole; this wire acting as the transformer primary winding.

Potential Transformers

Potential transformers are basically the same as current transformers, except that they are used to extend the range of a-c voltmeters. In this case the primary is connected *across* the load, with the secondary connected to the meter, Fig. 36. These transformers are usually used in conjunction with voltmeters having a range of 100 to 150 volts, and are designed to extend the range beyond 600 volts, which is the maximum range for the majority of a-c voltmeters. A selection of turns ratios is provided to allow for the selection of

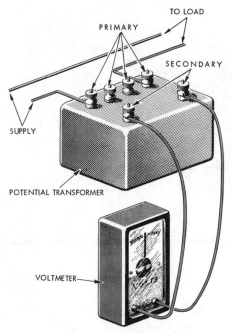

Fig. 37. The kilowatt hour meter.
Courtesy Duncan Electric Mfg. Co.

Fig. 36. A potential transformer connected in circuit with a voltmeter.

the proper range. Thus, if a 150 volt meter is to be used to measure a voltage of approximately 1500, a ratio of 10:1 would be required in the potential transformer.

Kilowatt Hour Meters

When the electric power company supplies electricity to homes and businesses, the amount of power used and charged for is based on the kilowatt hour. The kilowatt hour meter, is illustrated in Fig. 37, and can be seen in the basement, or on the side of any home.

The kilowatthour meter measures the amount of power being used and the length of time that it is used. It consists of a motor element, a set of gears, and the indicating dials. The motor drives the gears which are attached to the pointers, the dials being calibrated on the basis of power and time. Thus, a hundred watts consumed steadily for a period of one hour will cause the gear train to indicate the same number of revolutions as three-hundred watts for a space of twenty minutes.

Fig. 38 illustrates the motor whose laminated core has three poles, one on the top side and two below. An aluminum disk, mounted on the vertical shaft projects between the pole pieces. Current coils on the lower poles are in series with the load, while the potential coil on the upper one is across the load. The potential coil has more inductance than the current coils, so that a lag exists between the two currents. A rotating flux is created, inducing currents in the aluminum disk, causing it to rotate in a way similar to the rotor of an induction motor.

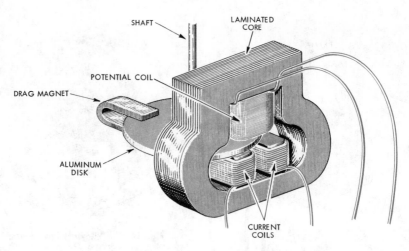

SHAFT

LAMINATED CORE

POTENTIAL COIL

DRAG MAGNET

ALUMINUM DISK

CURRENT COILS

Fig. 38. An internal view of the kilowatt hour meter showing current and potential coils.

A drag magnet at the outer edge of the disk establishes a permanent flux which is cut by the rotating disk, setting up a motor force that opposes rotation. The combination of motor effort and magnetic drag causes speed of rotation to depend on the rate at which power is consumed. The higher the current through the series coils, the faster the disk tries to move; and the faster the disk turns, the greater the opposing force of the drag magnet. A balance is established for each change in current so that speed of revolution indicates the rate at which power is consumed.

The polyphase kilowatt hour meter has two units arranged one above the other in a frame, their disks fastened to a common shaft. Current coils of one motor are in series with phase conductor A, Fig. 39, the potential coil across wires A and B. Current coils of the sec-

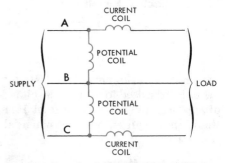

SUPPLY

LOAD

A

CURRENT COIL

POTENTIAL COIL

B

POTENTIAL COIL

C

CURRENT COIL

Fig. 39. The circuit connection of kilowatt hour meters for measuring polyphase power in a three-wire circuit.

ond motor are in series with phase conductor C, its potential coil across wires C and B. The meter then will indicate the total power consumed through all three wires.

ELECTRONIC TEST EQUIPMENT

Introduction

The meters described up to this point have been those used mainly for power electricity measurements. Frequency, almost universally 60 hertz, is a matter of little concern and hardly enters into calculations. Years ago electronics, which employs high frequencies over a tremendous range, was associated almost entirely with communications: first wireless telegraphy, then radio, and later television and radar.

Today the industrial uses of electronics in scores of applications outnumber those of communications. The maintenance of electronic equipment used in industry has become increasingly important, and the technician or engineer knowledgeable in at least some field of electronics as well as power electricity is in an enviable position.

For these reasons the more basic test equipment used for electronic operational checkouts and troubleshooting is now considered fundamental. Major instruments in this category are the oscilloscope and the signal generator. Both of these, including their uses, are described in the following paragraphs.

The Oscilloscope

The oscilloscope is a test instrument used to display voltage waveshapes on the screen of a cathode ray tube. Fig. 40 is a typical oscilloscope. The reader will note that the oscilloscope consists of a cathode ray tube (CRT) and various knobs or controls. The CRT is described in detail under the next heading, and controls and their functions are described in Table I.

Cathode Ray Tube (CRT). The cathode ray tube, hereafter called a CRT, is the basic component around which the oscilloscope is constructed. In Fig. 41 a typical cross section of a CRT is illustrated.

The elements of the tube are tied to power and controls by the pins at the base of the tube. The filament provides heat to the cathode from a-c power input. The cathode is a cylinder made of nickel. At the forward side of the cathode a material that emits electrons easily has coated the outer area of the cathode cylinder. This material is usually in the form of an oxide such as strontium or barium. The control grid acts much the same as the control grid in the standard vacuum tube. It provides control of the space current from the

Fig. 40. General purpose oscilloscope.

Courtesy EICO Electronic Instrument Co.

Fundamentals of Electricity

TABLE I

CONTROLS	FUNCTION
INTENS (Intensity)	Controls the brightness of the trace
FOCUS	Controls the sharpness or definition of the trace on the scope screen
ASTIG (Astigmatism)	Affects spot shape and is used to obtain a trace of uniform thickness
PHASE	Effective only in 60 Hz position of SYNC SELECTOR. Permits phase adjustment of the 60 Hz sync voltage to shift the starting point of the waveform display to any desired point on the wave form.
AC POWER SWITCH	Located on the PHASE switch and is turned on by CW movement from OFF position. Applies power to all oscilloscope circuits.
VERTICAL GAIN	Allows continuous adjustment of the vertical amplifier gain
VERTICAL POS (Position)	Adjusts the vertical location of the trace on the screen
VERT ATTEN V/CM (Vertical Attenuator in Volts per Centimeter)	Provides a choice of no attenuation or three decade steps of frequency compensated attenuation of the input voltage fed to the vertical amplifier
VERT AC/DC (Vertical Capacitive or Direct Coupling)	Provides either direct coupling (DC) or capacitive coupling (AC) to the vertical amplifier
SWEEP SELECTOR	Selects frequency band over which the SWEEP VERNIER can be varied
SWEEP VERNIER	Varies frequency adjustment of the internal linear sweep, and serves as the horizontal input selector
HORIZONTAL GAIN	Allows continuous adjustment of the horizontal amplifier gain
HORIZONTAL POS (Position)	Adjusts the horizontal location of the trace on the screen
SYNC SELECTOR	Permits selection of four positions to the sweep oscillator
	+ internal sweep starts during positive excursion of sine wave
	− internal sweep starts during negative excursion of sine wave
60 Hz	60 Hz power line frequency applied to sweep oscillator to sync it at line frequency
	EXT for external synchronizing voltage
SAW TOOTH jack	Provides saw tooth signal from jack to ground from the output of the sweep circuit oscillator
SYNC/HORIZ jack	Provides input to horizontal amplifier
VERT INPUT jacks	Provides input to vertical amplifier

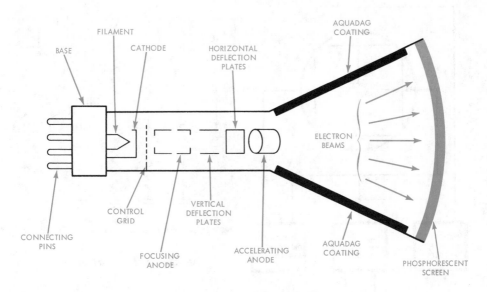

Fig. 41. The cathode ray tube (CRT).

cathode to the other elements of the CRT. The focusing anode and the accelerating anode provide focusing action to concentrate the electron beam and control intensity on the screen.

The accelerating anode is the most positive point in the CRT. While most of the electrons are attracted to the accelerating anode, some pass through a diaphragm inside the accelerating cylinder. An electrostatic field is built up between the focusing and accelerating anodes to control electron stream to the screen.

When the electron beam strikes the screen (made of phosphor), the electrons are shown up as a trace on the screen. The electrons must have a path back to the cathode. This path is made through an aquadag coating made of graphite which covers the CRT (except for the screen) and is joined to the cathode for a negative return path. The rest

of the CRT consists of deflection plates. These plates are connected in parallel horizontally and vertically, and they deflect the electron beam according to the strength and direction of the input signals.

Oscilloscope Operation. Fig. 42 is a block diagram of the basic oscilloscope. Signals are placed into the oscilloscope by means of leads to vertical and horizontal input jacks. The vertical input feeds the vertical deflection amplifier, which provides vertical deflection of the trace along a Y axis. The horizontal input feeds the horizontal deflection amplifier, which provides horizontal deflection of the trace along an X axis. The horizontal deflection amplifier is also controlled by a time base generator which generator controls speed that the trace may travel across the screen horizontally. The generator also controls how many cycles may be visually re-

227

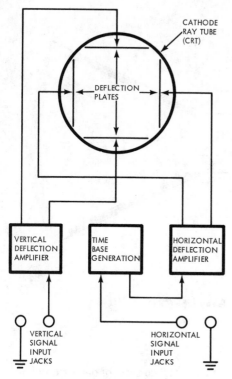

Fig. 42. The basic oscilloscope.

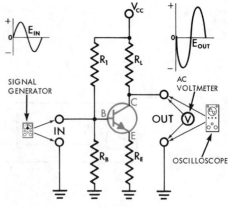

NOTE: Signal in at <u>base</u> provides amplified signal at <u>collector</u>

Fig. 43. Use of the oscilloscope.

flected on the screen. The rest of the controls on the oscilloscope bias the oscilloscope circuits to provide pictures of good scope. They also serve to set up the range or amplitude of the signal.

Use of the Oscilloscope. The oscilloscope may be used for nearly any function that a voltmeter is used. The scope patterns may be provided to show voltage patterns in relation to time, frequency, or in comparison to other signals such as output-to-input voltages.

A typical example of the use of the oscilloscope is in the test circuit in Fig. 43. Note that in the figure the oscilloscope is monitoring the output a-c sine wave of an amplifier. The a-c meter is measuring these same points.

The Signal Generator

The signal generator is a test instrument that develops sine waves or square waves at specified frequencies and amplitudes. There are many generator types that range in frequency from very low to very high and that range in power (amplitude) from low to high power and provide waveshapes such as square waves, sawtooth waves, and sine waves. Whatever the shape and size, the signal generator is a useful piece of test equip-

Fig. 44. Typical signal generator.
Reprinted with Permission of Heath Company

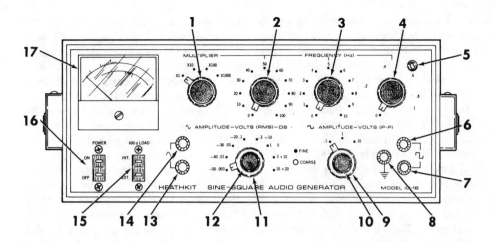

1 MULTIPLIER SWITCH

Multiplies the frequency set on the FREQUENCY (Hz) switches and control by 1, 10, 100, or 1000.

2 TENS FREQUENCY SWITCH

Selects first figure of frequency.

3 UNITS FREQUENCY SWITCH

Selects second figure of frequency.

4 FREQUENCY CONTROL

Indicates third figure of frequency.

5 PILOT LAMP

Indicates when power is on.

6 SQUARE WAVE OUTPUT

7 CIRCUIT GROUND

8 CHASSIS GROUND

Directly connected to chassis and to line cord earth ground.

9 SQUARE WAVE AMPLITUDE SWITCH

Coarse amplitude adjustment. Three ranges: .1, 1, and 10 volts peak-to-peak.

10 SQUARE WAVE AMPLITUDE CONTROL

Fine amplitude adjustment. Adjusts square wave output from zero to maximum of Square Wave Amplitude switch setting.

11 SINE WAVE AMPLITUDE SWITCH

Coarse amplitude adjustment. Eight ranges, from .003 volts to 10 volts rms (-50 to +20 dB).

12 SINE WAVE AMPLITUDE CONTROL

Fine amplitude adjustment. Adjusts sine wave output from zero maximum of Sine Wave Amplitude switch setting.

13 CIRCUIT GROUND

14 SINE WAVE OUTPUT

15 600Ω LOAD SWITCH

Functional on six lowest sine wave output ranges. Non-functional on sine wave 3 volt and 10 volt ranges.

Internal (INT): Connects internal 600Ω load across sine wave output.

External (EXT): Disconnects internal load from output.

16 POWER ON-OFF SWITCH

17 PANEL METER

Monitors the sine wave output. Has two voltage scales and one dB scale.

Fig. 45. Sine-square wave generator controls and their functions.

Reprinted with Permission of Heath Company

ment because it is designed to do the job with great consistency and accuracy.

In Fig. 44 a typical signal generator is pictured. This particular type is a sine-square-wave audio generator. It provides sine waves at amplitudes of .003 volt to 10 volts at frequencies of 1 Hz to 100 kHz. It also provides square waves at amplitudes up to 10 volts at frequencies of 5 Hz to 100 kHz. Fig. 45 lists the generator's controls and their functions. The basic signal generator consists primarily of an oscillator, a tuner, an output attenuator, and an output meter. Fig. 46 is a block diagram of the basic signal generator.

The Signal Generator Oscillator. The basic part of a signal generator is the oscillator. The many types of oscillators such as the Colpitts, Clapp, Hartley, and Wien-Bridge are explained in some detail in Chapter 15, so we shall not repeat this explanation. However, the requirements of a good oscillator should be understood at this point.

In order for an oscillator to function, it must have a power supply, an amplifier, regenerative feedback to feed the amplifier, and a frequency determination network. Once these things are available, the oscillator must produce an output wave form that does not vary in frequency or magnitude. The signal generator is probably the one piece of test equipment that utilizes these characteristics to the fullest.

The oscillator in a signal generator must be stable. Therefore any tuner, amplifier or power supply used in the generator must be of high quality. The tuner must have a stable resonant circuit. The amplifier must be biased so that thermal problems are not present. This can be done with thermistors, temperature-sensitive capacitors, and other such devices. The power supply must be regulated so that the d-c level is not affected by the load requirements. In the final assessment, we wish to tune the oscillator to a desired frequency and know that this frequency will not be distorted in any manner.

The Attenuator. The attenuator is basically a device with which to extract the oscillator signal to output jacks. This output attenuator is made adjustable to any desirable level. A potentiometer may be used as an output attenuator. In this event, the output is not calibrated and r-f (radio frequency) leakage is encountered.

Some of the leakage may be avoided by placing the attenuator in a shielded compartment and by shielding power leads. Other methods are by shielding the case or adding r-f inductors (chokes) into power leads. To provide a calibrated output is more expensive because of the addition of many parts. Some attenuators use a bridge network, some use step switching. In any event, r-f leakage must be held to a minimum when transferring the oscillator output to the load.

The Meter. The meter of a signal generator is usually an rms voltmeter and

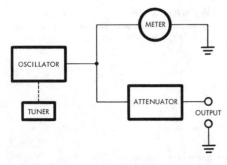

Fig. 46. Block diagram of basic signal generator.

decibel meter combination. The voltmeter is usually graduated in step levels matching amplitude levels on the generator's controls. More elaborate generators may have frequency meters added.

Use of the Signal Generator. The signal generator is used primarily to check gain and distortion in an amplifier. The signal generator output is placed on the input of the amplifier and an oscilloscope or volt meter is placed on the output. Fig. 44 shows this simple test setup on a single stage amplifier.

A generated frequency is placed into the amplifier at a known level. This is monitored by the oscilloscope, thus gain calculations can take place. The oscilloscope at the same time can readily see sine-wave distortion. The generator can also be placed on the input jacks of a multistage amplifier. This may serve as an immediate check of amplifier operation.

It may be necessary to place a capacitor in series with the input to an amplifier to pass the a-c signal from the generator and block d-c voltage. It may also be necessary to match the generator's output impedance with the amplifier's input impedance. In this event, a voltage divider is necessary.

One other use of the signal generator is for frequency measurement. This is accomplished by placing the unknown frequency into an oscilloscope's horizontal jacks and the signal generator output into its vertical jacks. The signal generator is then adjusted until a circle is shown on the oscilloscope. The frequency set in by the signal generator is now the same as the unknown frequency. The test pattern on the oscilloscope is called a Lissajous pattern and is only one of many types.

The signal generator and the oscilloscope may be used together to perform many types of electronic tests. It would behoove the reader to become more familiar with this valuable piece of test equipment. Prior to use of the signal generator, it is most advisable to thoroughly read the operation and application instructions that come with the signal generator.

1. Describe the D'Arsonval movement.

2. A 1500-volt oltmeter draws .025 amp for full-scale reading. How much external resistance is needed to adapt the instrument for use on a 450-volt circuit?

3. What is the purpose of the shunt resistor in a d-c ammeter?

4. A 10-amp d-c ammeter has a resistance of .2 ohm. What value of shunt is required for use on a 100-amp circuit?

5. Explain use of a millivoltmeter to measure current flow.

6. Define the term calibration.

7. What is a back-up scale?

8. How does the term ohms-per-volt indicate sensitivity?

9. Describe the operation of a megger.

10. How does a simple wattmeter movement differ from a D'Arsonal?

11. State the principle of operation applying to the Wheatstone Bridge.

12. How does frequency affect a clip-on instrument?

13. Explain briefly how a repulsion-vane instrument works.

14. What is a potential transformer?

15. How may the range of an inclined-coil ammeter be extended?

16. Describe the operation of a hinged-core clip-on ammeter.

17. How is output voltage of the megger controlled?

18. What is the principal feature of an electrodynamometer instrument?

19. Describe the motor element of a single-phase watthour meter.

20. Why is it impractical to use shunts for a-c ammeters?

21. What is the purpose of an oscilloscope?

22. Why is a time generator used in an oscilloscope?

23. Does a signal generator produce any waveforms besides sine and square waves?

24. What are the basic elements of a signal generator?

25. Can the signal generator and the oscilloscope be combined to perform test functions?

Chapter 12

Electron Tubes

INTRODUCTION

"Electron tube" is the modern replacement for the older term "vacuum tube." Both terms apply to the same thing and both are accurate since in most tubes electrons move through a vacuum. (Some heavy duty tubes for industrial equipment are filled with inert gases.) Electron tubes have applications in the fields of radio, television, medical instruments, and radar. Fig. 1 illustrates some of the representavite types.

Although most new equipment designs use solid state devices for amplifiers and rectifiers, an understanding of electron tube operation is still valuable. There are still many devices which use tube type amplifiers and many others using special purpose tubes such as television picture tubes, magnetrons and klystrons which generate and amplify microwaves, image intensifiers, and vidicon tubes for TV cameras. Although the operation of the special purpose tubes is beyond the scope of this chapter, the information contained here is necessary for understanding their operation also.

This chapter will deal with basic theory and circuits common to electrical and electronic applications.

Though electron tubes seem to have an endless list of various types, they are all basically similar. This may be clarified if we begin with what an electron tube does; its basic purpose is to *amplify*. This is the key word in dealing with most electron tubes. However sophisticated it may become, the tube's basic purpose remains the same. All its design features have the objective of achieving amplification more efficiently. In its simplest form it will only rectify, but principles remain the same in all cases.

The logical question that arises at this point is: If the function of an electron tube is to amplify, why so many variations? Variations exist to handle different operating conditions with regard to voltage quantities, temperature, and mechanical requirements. It is a case of fitting a basic design to operate within a set of specified conditions and requirements. No matter how the tube design varies in size or shape or construction, its similarities to other tubes remains the same.

BASIC PRINCIPLES

The electron tube has as its basic principle the fact that electrons will move through a vacuum. This phenomenon, discovered by Edison dur-

Fig. 1. Electron tubes are manufactured in a variety of sizes, shapes, and internal construction to serve a large number of applications.

Courtesy RCA

ing experiments with the incandescent lamp, was subsequently called the *Edison effect*. This effect, very simply stated, is that a free electron in a vacuum is attracted by a positively charged electrode (called the plate), and is repelled by a negative one.

To obtain a source of free electrons, *thermionic emission* is utilized. Thermionic emission means that electrons are given off when a material is heated. If a wire filament is heated by an electric current, electrons inside the metal

are speeded up enough so they can fly off the surface of the metal and form a cloud of electrons around the filament.

This brings us to the reasons for the vacuum in the electron tube. If oxygen were present in the tube, the filament would burn up almost immediately. Within a vacuum, the filament can emit electrons for long periods and remain intact. The vacuum also serves another purpose. If some of the gases found in the normal atmosphere were not removed they would interfere with the passage of electrons, increasing transit time and decreasing the efficiency of the tube.

However, small amounts of an inert gas are sometimes introduced into the vacuum to increase the output efficiency of tubes in many applications. In a vacuum, electrons may be emitted from the filament faster than the plate electrode, or anode, can absorb them and form a negative cloud, or space charge, that repels some of the electrons back into the filament.

When an inert gas is present, collisions between electrons and gas molecules create positive *ions* (electrified molecules) and neutralize the space charge, assisting the flow of electrons. Whether or not gas is to be used is dependent on the design of the tube and its application. In general, gas is used in tubes where the application calls for a large flow of electrons.

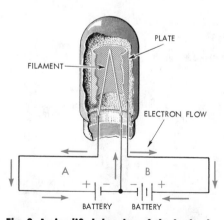

Fig. 2. A simplified drawing of the basic electron tube circuit. The A battery provides current to the filament, while the B battery provides a positive charge to the plate. Electron flow through the circuit is indicated by the arrows.

The Diode

Fig. 2 is a simplified drawing of a basic electron tube and its circuit.

A hot filament is used as the cathode and must be connected to a battery or transformer to provide the power required to heat it. Batteries used to heat filaments are generally called *A* batteries, while batteries (or power supplies) used to supply the anode circuit are called *B* batteries (or plate power supplies).

The plate is the anode of the tube because it is the electrode by which the main stream of electrons leave. It is connected to the battery *B*, which in turn is connected to the cathode.

The electrons emitted from the cathode will be attracted to the plate if it has a positive charge, and repelled if it is negative. Therefore, electrons will flow through the tube and complete the circuit only if the plate is positive.

This tube is called a *diode* because it has only two electrodes. It is the most elementary electron tube and cannot amplify; it can only rectify. Tubes are now seldom used for rectifiers. They have been largely replaced by semiconductor (solid state) rectifiers, which require no heater power.

CONSTRUCTION

Figure 3 illustrates the internal construction of a typical electron tube. This particular tube is a miniature vacuum tube. The parts are similar, however, to all electron tubes.

The elements of the tube are enclosed within an envelope which may be either metal or glass. During manufacture, the

235

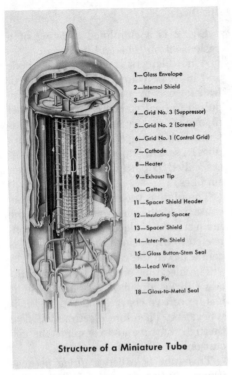

1—Glass Envelope
2—Internal Shield
3—Plate
4—Grid No. 3 (Suppressor)
5—Grid No. 2 (Screen)
6—Grid No. 1 (Control Grid)
7—Cathode
8—Heater
9—Exhaust Tip
10—Getter
11—Spacer Shield Header
12—Insulating Spacer
13—Spacer Shield
14—Inter-Pin Shield
15—Glass Button-Stem Seal
16—Lead Wire
17—Base Pin
18—Glass-to-Metal Seal

Structure of a Miniature Tube

Fig. 3. A cutaway view of a miniature electron (vacuum) tube.
Courtesy RCA

The silvery appearance of many tubes is due to the coating left after the burning of the magnesium.

The Cathode. Figure 4 illustrates the construction of a cathode of the *indirectly* heated type. It is called indirectly heated because a filament is used to heat the cathode which emits the electrons. In a tube that is *directly* heated only a filament is used, but it is usually referred to as a cathode.

The need for two types of emitters is due to different operating requirements and conditions. The filament, or directly heated emitter, does not emit a constant flow of electrons when heated by alternating current. The cathode of an indirectly heated tube is somewhat massive in comparison to the filament and, when heated, will maintain a constant stream of electrons.

electrodes and the envelope are assembled and as much air as possible is pumped out. This process does not create a complete vacuum, however, and the remaining gases must be removed. This is usually accomplished by putting a small amount of magnesium on a part of the tube referred to as the "getter" during the assembly. When as much air as possible has been evacuated by pumps, the tube is sealed and then placed inside a series of coils, and a high frequency current is applied. This generates heat that ignites the magnesium, which then combines with the residual gasses, rendering these gases inactive.

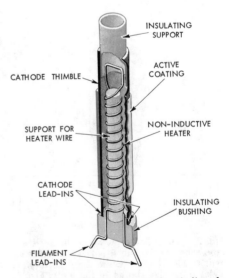

INSULATING SUPPORT

ACTIVE COATING

CATHODE THIMBLE

NON-INDUCTIVE HEATER

SUPPORT FOR HEATER WIRE

CATHODE LEAD-INS

INSULATING BUSHING

FILAMENT LEAD-INS

Fig. 4. The construction of the indirectly heated cathode.

Indirectly heated cathodes also allow the use of a one-filament power source even though the cathodes of different tubes are operated at different voltages.

Tungsten, thoriated tungsten, and alloys coated with various oxides such as barium and strontium are some of the materials used for cathodes. The particular material used is determined by the application of the tube and the required efficiency of emission. Tungsten, for example, is used in tubes with directly heated cathodes which are designed to handle large amounts of current and power. Its emission efficiency is low, compared to other materials, but it will withstand the high currents and heat involved in large power applications.

Emission efficiency is determined by the number of electrons that will be emitted for a given temperature. The low efficiency referred to in the last paragraph means that high temperatures are necessary to obtain electrons. High efficiency emissions would give off more electrons with less heat.

Thoriated tungsten and oxide coated alloys are used in cathodes of tubes designed for lower power applications. The oxides have higher emission efficiency, giving a greater output of electrons at lower temperatures, but will stand high currents.

The Plate. As illustrated in Fig. 3, the plate usually surrounds the cathode and must be relatively massive to withstand the bombardment of electrons which reach extremely high speeds in the transition from cathode to the plate. It must also be able to handle the large amounts of heat generated by high currents.

Some plate materials in common use are nickel, graphite, tungsten, tantalum, and molybdenum. As in the case of the cathode, the use of a particular metal is determined by the intended applications and the power handling requirements.

Tube Pins and Sockets

The elements inside the tube are connected through airtight seals to the outside by means of *pins* which are inserted into sockets to which the circuit is wired. Fig. 5 illustrates typical pin arrangements for various tube types. To assure that the tube will be replaced properly each time it is removed for inspection or replacement, a keyed post in the center of each tube base fits into a corresponding hole in the socket. Other types, such as miniature tubes, do not have a center post but have a spaced gap which allows the tube to fit into the socket only in the proper way.

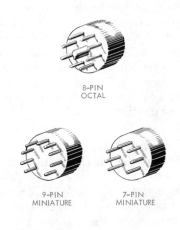

8-PIN
OCTAL

9-PIN
MINIATURE

7-PIN
MINIATURE

Fig. 5. Typical pin arrangements for electron tubes. In the loctal arrangement at the upper right, a clip snaps over the indentation in the post.

Fig. 6. The numbering sequence for tube pins begins with the first pin to the left of the keyed post. This is the view from the *bottom* of the tube. The sequence for the tube socket is the same when viewed from the wiring side.

The pins have a definite numbering sequence. That is, looking at the tube pins from the bottom of the tube, the sequence is clockwise, beginning with the first pin clockwise of the key or space, whichever is used as an indicator. The numbering sequence for the socket is the same as that of the tube when looking at the socket from the wiring side. Fig. 6 illustrates how tube pins and sockets are numbered. On electrical wiring diagrams these numbers are used as a guide in wiring the socket for the particular tube involved.

Special Pin Connectors

Some of the larger tube types now have a nine-pin *novar base* or a twelve-pin *duodecar base*. Each of these devices has several size bulbs. Modern tubes no longer have guide pins, notches, or lead keyways. Pins are spaced so that the tube can only be placed in the socket in one position.

Electron Tube Types

The number of electrodes contained in each tube determines its type. The simplest type (the diode) contains only two: an electron emitting cathode and a plate to receive the electrons after they have passed through the tube. Whatever the number of electrodes, these two—the cathode and plate—are common to all. Additional elements, called *grids*,

may be added to control electron flow through the tube.

Each tube type has a name relating to the number of electrodes it contains, as shown in the list below.

Diode   = 2 electrodes

Triode   = 3 electrodes

Tetrode = 4 electrodes

Pentode = 5 electrodes

Special Purpose Tubes

The *compactron tube* is one of the newer tube devices. The compactron is a multipurpose tube that provides the user with more than one tube function in a single envelope. For instance, a compactron may have a triode and a pentode in one envelope. Again, it may have a double diode (duo-diode) and a duo-triode in one envelope.

The *cathode ray tube* is a tube that has a large envelope housing an electron gun, a fluorescent screen, and deflection plates. The tube is described in detail as part of the oscilloscope in Chapter 11.

The *beam power tube* is a pentode tube which uses beam power plates instead of a suppressor grid to provide low screen current. This allows the screen grid to be at plate potential.

The *phototube* is sensitive to light. The cathode of the tube is the sensitive element and emits electrons when exposed to light. The anode of the tube collects the electrons. In a television camera, light is changed into video signals.

The *thyratron tube* is a gas-filled tube which handles high current. The gas allows fast passage of electrons. In a thyratron, a signal is supplied to the grid. After current starts flowing between the cathode and the (graphite) plate the grid voltage may be taken away and current will continue to flow. The hot-cathode mercury vapor tube of the same type, called a *phanostron,* can handle currents up to 50 amperes. The more recent solid state device called a *silicon*

controlled rectifier (SCR) provides the same characteristics as the thyratron and is a suitable and reliable substitute for the high current tube.

The *black and white television picture tube* consists of an electron gun, several control electrodes, a glass envelope with a silicate or sulfide screen, and a deflecting yoke to deflect the electron beam. Figure 7 shows an electron gun for television picture tube. The envelope has been removed to show the electron gun details.

The *color television picture tube* (see Fig. 8) is basically the same as the black and white tube but is different in three ways. Each *dot trio* on the silicate screen is capable of emitting light in one of three colors (red, green, and blue). A

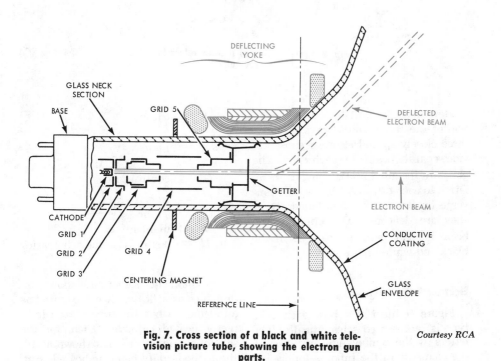

Fig. 7. Cross section of a black and white television picture tube, showing the electron gun parts.

Courtesy RCA

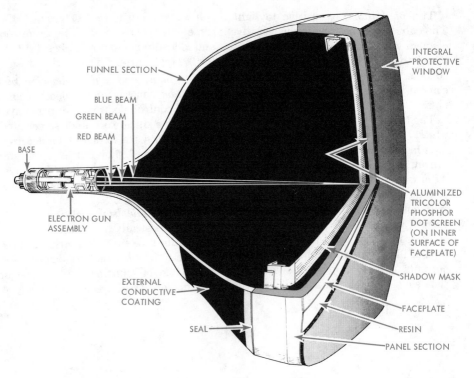

Fig. 8. Cross section of a color television picture tube.

Courtesy RCA

shadow mask is used for color separation of the three color arrays. Finally, three closely spaced electron guns provide separate beams for each color. The guns are installed at 120° angles in relation to each other. All three guns converge on the screen simultaneously as they are deflected. Deflection of the beam is made with a yoke, as in the black and white picture tube.

Electron Tube Diagrams

Figure 9 illustrates how each tube type is represented schematically. The letters by the symbols identify each of the elements in the tube as follows:

$K =$ cathode

$P =$ plate

$G_1 =$ control grid
$G_2 =$ screen grid
$G_3 =$ suppressor grid
$F =$ filament
or, $H =$ heater on some schematics

The filament, sometimes shown on drawings as a heater, it not counted as an element when the tube has an indirectly heated cathode. Often, for the sake of simplicity it is not shown at all, the cathode *only* being indicated. But,

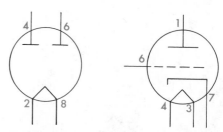

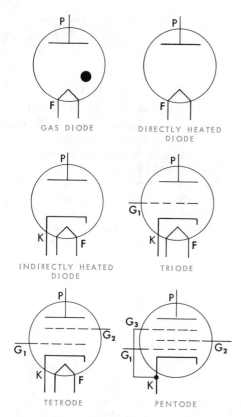

Fig. 9. The schematic symbols for electron tubes illustrating the electrodes and their arrangements in the various types.

Fig. 10. These symbols illustrate the method of identifying pin numbers on most schematic diagrams.

even though it is not counted in the number of elements, two pins are always required for its wiring. Thus, a diode having a plate, cathode and heater would have four pins. This point should be remembered to avoid possible confusion between the number of pins and the number of elements.

Note that in the schematic symbols the pin numbers are not indicated. This is because we are dealing with basic

tube types rather than specific tubes. In the field the technician will be working with schematics and wiring diagrams published by the manufacturers for their electrical and electronic equipment. In these the pin numbers are usually given and the identifying letters or names of the elements are omitted as shown in Fig. 10.

In the "Tube Manuals" published by the tube manufacturers, *base diagrams,* Fig. 11, indicate both pin numbers and electrode names. The grids in these diagrams are usually designated as G_1, G_2, etc. rather than control grid, screen grid and so on. In any case, while the method of identifying pins and electrodes may vary according to how they are presented, the schematic symbols for them are standard.

Figure 12 illustrates some of the shapes and sizes in which a vacuum tube may be purchased. Tubes may come in metal or glass envelopes and in many shapes and sizes. Their shape, size and material is dependent on the job that the tube is used for, along with the type of equipment in which the tube is installed..

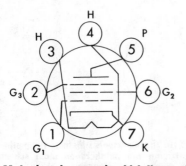

Fig. 11. In the tube manuals which list specifications for individual tube types, the pins and electrodes are identified as shown above. It should be noted that the connections do not come straight out of the symbol as in the typical schematic of Fig. 9, but are shown connected according to the actual physical arrangement.

Courtesy RCA

AMPLIFIER TUBES

The Triode

The triode, Figs. 13 and 14, with three elements, is the most basic amplifier tube. Its ability to amplify is due to the addition of the grid to the basic diode.

The grid is physically located between the cathode and the plate. It is formed of wire which may be wound spirally or in flat form on supports around the cathode as shown in the illustration. Because the grid is directly in the path of elec-

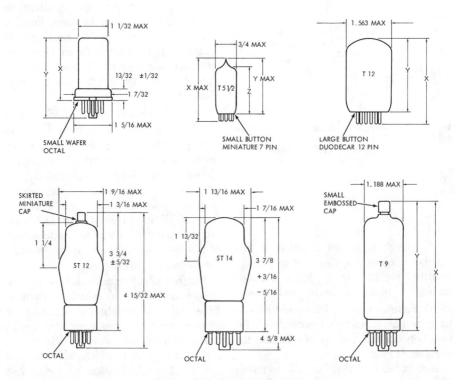

Fig. 12. Typical tube outlines.

Courtesy RCA

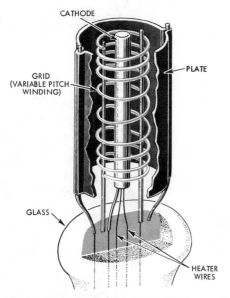

CATHODE

GRID
(VARIABLE PITCH
WINDING)

PLATE

GLASS

HEATER
WIRES

Fig. 13. The construction of a triode tube. The design illustrated here uses a spiral winding with an indirectly heated cathode.

trons flowing from the cathode to the plate, a large negative voltage applied to it will repel the electrons and prevent them from passing through and reaching the plate. By varying the voltage on the grid, the flow of electrons to the plate can be varied.

The grid requires almost no current as the electrons are repelled by its negative charge.

All the current that the external circuit has to supply to the grid is just enough to charge it to the desired voltage. This current is so small that it is usually neglected except in high power tubes like those that radio transmitters use in the output stages.

The voltage on the grid controls the current flowing through the tube so the plate signal duplicates the grid voltage waveform on a much larger scale, as shown in Fig. 15. Note that the ampli-

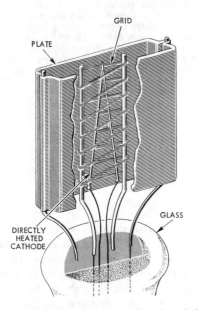

GRID

PLATE

GLASS

DIRECTLY
HEATED
CATHODE

Fig. 14. This form of construction for triodes uses the flat form and is the directly heated type.

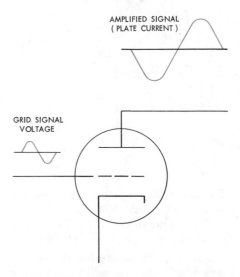

AMPLIFIED SIGNAL
(PLATE CURRENT)

GRID SIGNAL
VOLTAGE

Fig. 15. Comparison of the grid and plate waveforms. The signal is inverted in the plate circuit.

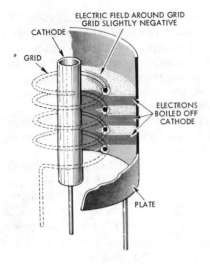

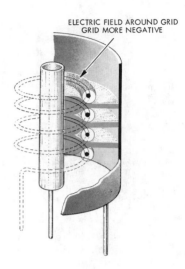

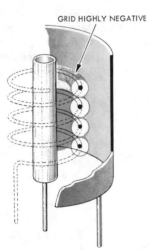

Fig. 16. These drawings illustrate how the grid controls the flow of electrons from the cathode to the plate. In the top drawing, the grid is slightly negative, and a large flow of electrons passes between the fields of the grid wires. As the grid voltage increases, the field around the grid wires gets stronger, gradually cutting off the flow of electrons to the plate and repelling the electrons back into a cluster around the cathode as shown in the right and bottom drawings.

fied signal on the plate is inverted 180 degrees out of phase with the grid signal.

The manner in which the grid controls plate current is shown in the three drawings of Fig. 16. In the top drawing, a very small negative voltage is applied to the grid. This voltage causes a negative field around the grid wires, its strength dependent upon the voltage. This negative field will repel the electrons emitted from the cathode. In this case it is very small, thus allowing the electrons to pass on to the plate.

In the second drawing, the grid voltage has been made more negative, effectively decreasing the effect of the plate voltage on the electrons near the cathode. Some of the electrons from the cathode are repelled and only a small portion of them move through the space between the grid. This reduces the number of electrons reaching the plate and plate current is reduced in proportion to the amount of voltage on the grid.

In the bottom drawing, the grid is highly negative. The field around the grid wires completely shields the electrons around the cathode from the plate wires. The negative field holds the electrons clustered around the cathode and none of them reach the plate. There is no current flow and the tube is said to be *cut off*.

If the average grid voltage is between zero and cutoff and a varying voltage—a sine wave, for example—is applied to the grid, the field around the grid wires gets stronger or weaker following the applied voltage of the waveform. The electrons passing through the grid to the plate are controlled accordingly, and the waveform of the voltage on the grid is reproduced on the plate.

Tube Voltages

The voltage applied to each element of the tube must be of the proper polarity with reference to each other if the circuit is to operate properly. Fig. 17 is a triode circuit showing the various voltages and their polarities. The cathode is usually the reference point for all voltage tests that might be made on the circuit. The grid is negative, and the plate positive with reference to the cathode. Should the grid become positive, electrons would be drawn to it rather than the plate, which would result in a distorted signal and a loss of current in the plate circuit.

The filament of an indirectly heated cathode does not have to be referenced to the cathode because it does not affect the operation of the circuit. For this

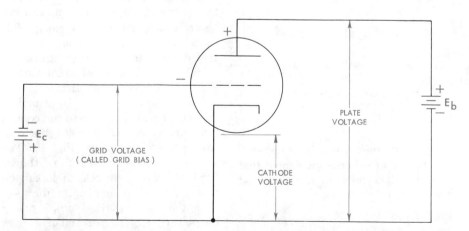

Fig. 17. The operating voltages and their polarities for an electron tube circuit. Polarity of the cathode voltage is not indicated because it is the reference point for the other voltages.

reason it is frequently not indicated on the original flow portion of schematics.

There are two sets of voltages required for the tube. One is the d-c voltages required to determine the operating characteristics of the tube, called *operating voltages*. The full names of the voltages are frequently not used in technical literature, but are referred to by the symbols given below.

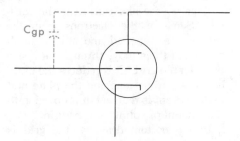

Fig. 18. The equivalent electrical circuit for interelectrode capacitance. With a large capacitance, the signal on the grid is coupled directly to the plate.

$$E_b = \text{d-c plate voltage}$$

$$E_c = \text{d-c grid voltage}$$

$$E_k = \text{d-c cathode voltage}$$

The other voltages are the a-c signal voltages, the voltages which are amplified by the tube. To distinguish between these two voltages which appear in the tube circuit simultaneously, a different set of symbols using lower case letters are used in referring to the a-c signals.

$$e_b = \text{a-c signal plate voltage}$$

$$e_c = \text{a-c signal grid voltage}$$

$$e_k = \text{a-c signal cathode voltage}$$

To avoid confusion between operating and signal voltages, remember that capital letters refer to d-c and that lower case refers to a-c.

The Tetrode

It should be pointed out again that tubes other than the triode were devel-

oped to handle applications for which the triode was not suitable. Triodes are still used for many purposes.

The tetrode is similar in basic construction to a triode with another element, called a *screen grid*, added. This screen grid was added because *interelectrode capacitance* was a problem in some triode applications. Interelectrode capacitance means that, because of their nearness to each other, the control grid and plate have a capacitance between them. At high frequencies, this capacitance can couple a signal on the grid directly to the plate, without the signal being amplified. Fig. 18 illustrates the equivalent electrical circuit.

The Screen Grid. In construction, the screen grid is similar to the control grid and is located between the control grid and the plate, Fig. 19. The screen grid accomplishes three things. First, it reduces interelectrode capacitance. Second, it accelerates the electrons toward the plate. Third, it makes the plate current independent of changes in plate voltage over most of the operating range of the tube.

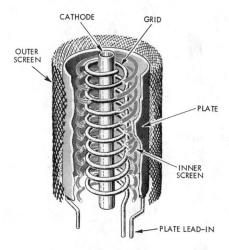

CATHODE GRID

OUTER SCREEN

PLATE

INNER SCREEN

PLATE LEAD-IN

Fig. 19. A cutaway view showing the construction of the tetrode tube. The screen grid is shown in color.

Capacitance exists in the tetrode between the plate and the screen grid and between the screen grid and the control grid. But the screen grid is usually bypassed to the cathode with a large capacitor so that capacitance is very small and the signal is not coupled to the screen grid. Instead, the electrons are accelerated toward the plate.

The tetrode is always operated with a positive voltage on the screen grid. Electrons are accelerated to the plate because the positive voltage of the screen grid aids that of the plate, creating a larger positive attraction for the electrons from the cathode. However, the plate voltage must always be more positive than the voltage on the screen grid, or electrons would be attracted to the screen rather than the plate, especially at low plate voltages.

Except for very low plate voltages, plate current of a tetrode is independent of plate voltage because the

steady, positive d-c voltage on the screen grid accelerates the electrons toward the plate at all times. Higher amplification can be obtained with a tetrode than with a triode for this reason. Tetrodes are still used for special purposes, but for most TV or radio receivers they have been replaced by pentode tubes.

The Pentode

The pentode, Fig. 20, has five electrodes, with an additional grid, called the *suppressor grid,* added to those already discussed. It was developed to solve certain problems created in using tetrodes.

The Suppressor Grid. The screen grid accelerates electrons to the plate with such a high velocity that they knock other electrons out of the plate. These electrons from the plate are known as *secondary electrons.* They can be attracted by the positive voltage on the screen grid, causing an increase in screen current and a decrease in plate current. This is usually not desired, although special control circuits have been designed which use this *negative resistance* effect.

The suppressor grid reduces this *secondary emission* effect. Normally connected to the cathode, the suppressor grid creates a negative charge near the plate. Secondary electrons knocked out of the plate are repelled back into it by the negative voltage of the suppressor grid. Connection to the cathode is external in some tubes and internal in others. Fig. 21 illustrates a typical pentode circuit.

Tube Characteristics

The relationship of the d-c operating

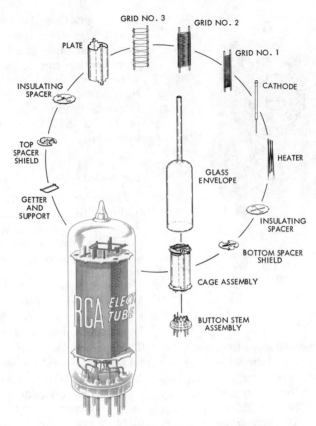

Fig. 20. An exploded view of the pentode tube.
Courtesy RCA

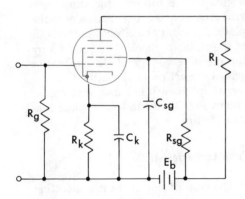

Fig. 21. A typical pentode amplifier circuit.

voltages on the various elements of the tube determine how the tube will operate in a particular circuit. The operation of an electron tube is usually described in terms of the effect a change in grid or plate voltage will have on the plate current and are shown by graphs called *characteristics curves*. The two most commonly used are the *transfer* characteristics and the *plate* characteristics curves. Both of these are referred to as *static* characteristics because they indicate how the tube will operate with only the d-c operating voltages present.

Fig. 22 is a family of plate characteristics curves. These curves are plotted using a circuit like that of Fig. 23. The grid voltage is set at the desired value and the plate current is measured as the voltage on the plate is varied over a specified range. The grid voltage is then changed and the whole procedure repeated for each curve.

Fig. 24 is a family of transfer characteristics which show plate current as a function of grid voltage with a constant value of voltage on the plate. The circuit shown in Fig. 23 is used in determining the curves. In the circuit illustrated, a particular value of plate voltage is selected, the grid voltage is varied over a specified range and the resultant changes in plate current are plotted to give a curve for the selected value of plate voltage. The value of the plate voltage is then changed and the test repeated, giving a curve for each of the desired plate voltages.

The characteristics curves illustrated here are merely representative examples of typical curves. Each particular tube type has its own set of curves which show its operation over a wide variety of conditions. There are, of course, variations between individual tubes, but these are usually of no con-

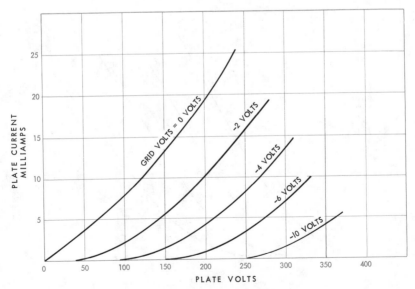

Fig. 22. A family of plate characteristics curves.

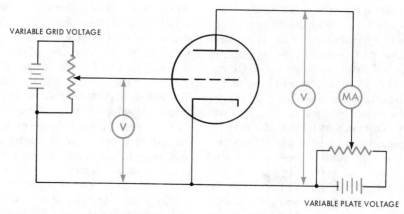

Fig. 23. The test circuit used in plotting static characteristics curves. Both the grid and plate voltages may be varied over the operating range of the tube.

sequence. In the production process, the manufacturer tests a large number of samples, for example, the 6SN7 tube. The results are averaged and a charac- teristics curve typical of all 6SN7 tubes is obtained.

The *dynamic* characteristics are obtained when an a-c voltage is put on

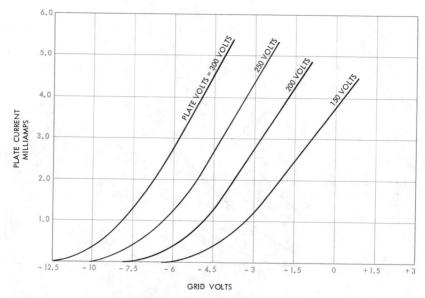

Fig. 24. A family of plate characteristic curves.

the control grid and d-c voltages are on the other electrodes. The dynamic characteristics are important because they show how the tube will behave in actual operation. Although the dynamic characteristics may be shown as curves, this is not usually done. Dynamic characteristics include the *amplification factor, dynamic plate resistance*, and the *transconductance*.

Information on tubes is published by the various manufacturers in booklets called "Tube Manuals." These manuals contain characteristics curves, tube pin arrangements, plate resistances, and are a necessity for the technician who works with tubes.

Amplification Factor

One of the dynamic characteristics of the tube is the amplification factor, represented by the Greek letter Mu (μ). The amplification factor indicates the tubes ability to amplify the input signal on the grid and depends only on the characteristics of the tube and not on the circuit in which it is used. It is the ratio between a change in plate voltage to a change in grid voltage (in the opposite direction) with the plate current remaining constant.

In a particular tube, a positive change of 40 volts on the plate requires a negative change of 2 volts on the grid to keep the plate current constant. The amplification factor is found by dividing the *change in plate volts* by the *change in grid volts* as follows:

$$\mu = \frac{\Delta e_b}{\Delta e_g}$$

$$\mu = \frac{40\ volts}{-2\ volts}$$

$$\mu = 20$$

In the formula, the Greek letter Delta (Δ) signifies "a small change." The amplification factor for triodes may vary from 5 to 100, the tubes having the higher numbers being referred to as high mu triodes. Lower case letters are used in the formula because the amplification factor applies only to the a-c signal that is to be amplified.

Plate Resistance

The plate resistance of a tube is the resistance to the flow of electrons through the tube. This resistance is not constant but varies over the operating range of the tube when either plate voltage or plate current changes, because, as we know from Ohm's law, resistance is determined by the relationship of current and voltage. Plate resistance then can only be specified for a particular value of voltage on the plate with a constant value of voltage on the grid.

Dynamic Plate Resistance. The dynamic resistance is the resistance of the internal path between cathode and plate to the a-c signal. Because the a-c signal is always changing, the change must be considered in our calculations. Here we use a specified small change in plate voltage that causes a particular small change in plate current.

For example, in a particular tube, a change of 50 volts on the plate causes a 2 milliamp change in plate current with a particular bias on the grid. The dynamic plate resistance is calculated by Ohm's law using the values of *change*, as follows:

$$r_p = \frac{\Delta e_b}{\Delta i_b}$$

$$r_p = \frac{50\ volts}{.002\ amp}$$

$$r_p = 25,000\ ohms$$

Again, the Greek letter Delta (Δ) is used to indicate that these are values of change in voltage and current rather than constant values. As in the case of d-c plate resistance, the grid voltage must be kept constant. In most cases, the dynamic plate resistance, r_p, is the important resistance.

Transconductance

Transconductance (g_m) is the ease with which an electron tube conducts electrons. It is also referred to as *mutual conductance*. Transconductance in electron tube may be more technically defined as the ratio of a change in plate current to the change in grid voltage that caused it, with the plate voltage kept constant. It is determined by dividing the change in plate current by the change in grid voltage, and is expressed mathematically:

$$g_m = \frac{\Delta i_b}{\Delta e_g}$$

The unit of measurement for transconductance is the *micromho*. This formula expresses g_m in terms of i_b and e_g, and is used here to illustrate that relationship. However, the most common method of determining g_m is by using the amplification factor (μ) and the plate resistance (r_p).

$$g_m = \frac{\mu}{r_p}$$

This is possible because both μ and r_p have been determined on the basis of ratios of changes in voltage and current. However, because the value of both of these factors are dependent upon specific conditions of voltage,

these must be indicated in specifying the transconductance of the tube.

Ways of Obtaining Grid Bias

There are several ways of obtaining the voltage used for biasing the grid. A small battery may be used, or the voltage can be tapped from the power supply. These methods provide what is called *fixed* bias. The most common method uses the voltage developed across a cathode resistor and is called *self* bias, see Fig. 25. In the illustration the cathode resistor causes a drop of 2 volts. The grid is therefore 2 volts negative with reference to the cathode.

When a cathode resistor is used to obtain bias, the voltage across it is dependent upon plate current and will change with the signal. This is not usually desirable; a constant bias is needed on the grid in almost all cases. The most common method of obtaining

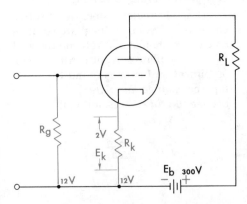

Fig. 25. Self bias is developed by utilizing the voltage drop across the cathode resistor.

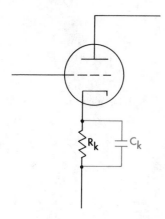

a constant grid bias is by the use of a *bypass* capacitor across the cathode resistor, Fig. 26. As we know from earlier chapters, a capacitor offers a low reactance to the passage of a-c. The a-c component of the plate current is bypassed through the capacitor, resulting in a relatively stable voltage across the cathode resistor and a stable bias voltage applied to the grid.

Use of Characteristics Curve in Amplifier Design

The manner in which the characteristics curves may be utilized in amplifier design is illustrated in Fig. 27. The characteristics curve aids in determining the operating point of the

Fig. 26. The cathode bypass capacitor stabilizes the voltage across the cathode resistor and the cathode. It prevents the a-c signal from being dropped by R_k.

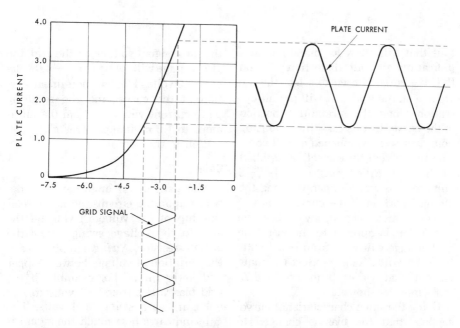

Fig. 27. The characteristics curve used to determine the operating point of an electron tube.

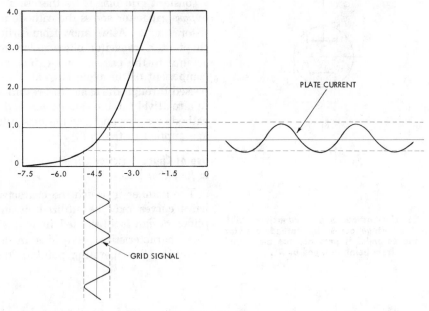

Fig. 28. Distortion occurs in the output when the tube operates at a non-linear portion of the characteristics curve.

tube that will allow for maximum amplification without distortion. To do this, the curve is used to select the d-c voltage on the grid that will permit operation along the straightest portion of the curve. In this case, the operating point has been established at −3 volts. This means that the d-c voltage applied to the grid, called bias, must be −3 volts for the tube to operate along the straight portion of the curve.

Now, if an a-c signal, say a sine wave of 0.5 volt is applied to the grid, the grid voltage will vary from −2.5 volts to −3.5 volts. This produces a plate current that varies from 1.5 to 3.75 milliamps, as shown.

Using the same characteristics curve, suppose that the bias is changed to −4.5 volts as shown in Fig. 28. With the same signal voltage on the grid, the grid voltage will swing from −4.0 volts to −5.0 volts. This is a non-linear portion of the curve, and the resulting plate current will vary from 0.4 milliamps to 1.2 milliamps. This gives distortion in the output.

Gain

The gain of an amplifier tube depends upon the circuit and is the relationship of the voltage swing on the grid to the voltage swing across the output resistor. *Swing* simply means the variation of voltage between upper and lower limits. For example, if the grid bias varies from −2 volts to −6 volts, it has a swing of 4 volts. The gain indicates how much the signal is amplified and can be found by the following formula:

$$Gain = \frac{\mu \times R_L}{r_p + R_L}$$

where R_L is the load resistance in the

amplifier circuit. There are no "units" of measurement involved. If for example, an answer of 20 was obtained, the gain would simply be stated, "the gain is 20."

1. What is the basic purpose of the electron tube?
2. What is thermionic emission?
3. Why is a vacuum required in an electron tube?
4. In what applications is the directly heated cathode most often used?
5. How is emission efficiency determined?
6. Name some of the materials used for the plate in an electron tube.
7. What element in an electron tube gives it the ability to amplify?
8. How does an electron tube amplify?
9. What is the relationship of the grid voltage to the plate voltage in an amplifier tube?
10. What are operating voltages?
11. What is the function of the screen grid in a tetrode tube?
12. What is secondary emission?
13. Explain the function of the suppressor grid.
14. How is interelectrode capacitance reduced in an electron tube?
15. What does the amplification factor indicate?
16. How is dynamic plate resistance determined?
17. Name three methods of obtaining grid bias.
18. How are characteristics curves used in amplifier design?
19. What is self bias?
20. What voltage is used in determining static characteristic curves?

Chapter 13

Electron Tube Circuits

DIODE CIRCUITS

In explaining the basic principle of the electron tube it was stated that the diode could not amplify. However, even with this seeming limitation, the diode has a multitude of uses and applications. Since electrons move through the diode only when the plate is positive, the diode will conduct in only one direction, and it is in effect an electronic switch. Every diode circuit that will be encountered, including rectifier circuits, waveshaping, and control circuits, operates in accordance with these factors. Now let us examine the operation of the diode in typical circuits.

Half-wave Rectifiers

We are already familiar with the pur-

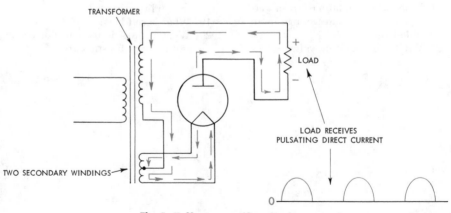

TRANSFORMER

LOAD

+

−

LOAD RECEIVES
PULSATING DIRECT CURRENT

TWO SECONDARY WINDINGS

0

Fig. 1. Half-wave rectifier circuit.

pose of a rectifier, which is to convert an alternating current and voltage to direct current. There are two primary types of diode rectifiers; the half-wave and the full-wave. Fig. 1 illustrates the half-wave circuit. Although rectifier tubes with separate heaters and cathodes are manufactured, a majority use the heated filament as a cathode and will be the type used here for the sake of simplicity.

Fig. 1 shows a diode connected to the secondary of a transformer. The secondary has two windings; one connected to the plate, the other to the cathode to supply heater current. During the positive half of the alternating current wave, the top of the transformer is positive, producing a positive voltage on the plate of the diode. Current then flows through the diode, through the load and back to the diode completing the circuit. During the next alternation of the a-c wave, the top of the transfor-

mer is negative, the diode will not conduct and no current flows through the load. On the next positive alternation current will again flow through the load. Thus, only half-waves of current flow through the load. This type of current is known as *pulsating d-c*. It is satisfactory for some applications but not for most because current flows only half the time. A more constant flow is needed for most applications. This constant current flow can be provided by full-wave rectifiers.

Full-wave Rectifiers

Fig. 2 illustrates a full-wave rectifier circuit. Two diodes are used, both enclosed in a single envelope, this single electron tube generally being referred to as a *duodiode*. A single transformer is used, as with the half-wave, but the ends are connected to the plates and a center-tap is connected to the load which in turn is connected to the tube

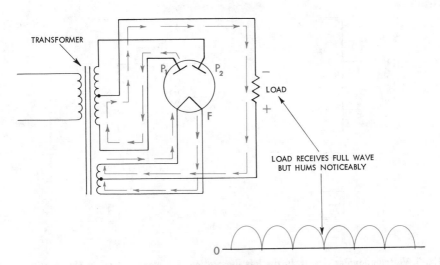

Fig. 2. Full-wave rectifier circuit.

cathode. Operation is as illustrated in Fig. 2.

When the end of the secondary winding connected to plate *1* is positive, the other end is negative. Current flows upward through the transformer to the center-tap, then downward through the load to the cathode. On the next alternation, plate *1* becomes negative and plate *2* becomes positive. Electron flow is then from plate *2* through the upper portion of the transformer to the center-tap, and again downward through the load to the cathode. Thus, both half-waves of current pass through the load in the same direction. Although full-wave rectification is preferable in most cases to half-wave rectification, pulsations in the rectified output can still cause difficulties in operation. These pulsations may be smoothed out to give a more constant current by a process known as filtering, which can be used with both half-wave and full-wave rectification.

Voltage-Doubling Circuit

Before considering methods for smoothing out pulsations in the full-wave rectifier current noted above, it is worthwhile investigating a rectifier circuit in which double-voltage plate transformers are not required. There are two diodes, V_1 and V_2, in Fig. 3, along with two capacitors, C_1 and C_2. The tubes are shown as being indirectly heated to simplify explanation, but they may be of either type.

The cathode of V_1 is connected to the plate of V_2 and to one end of the transformer secondary, S. C_1 and C_2 are in series across the plate of V_1 and the

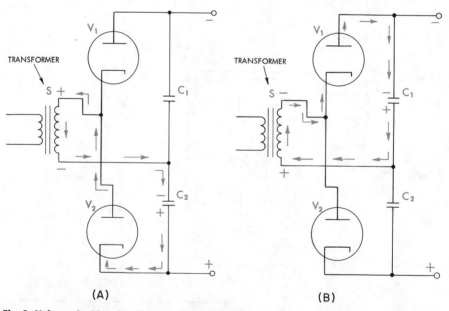

(A)   (B)

Fig. 3. Voltage doubler circuit. In the left illustration, V_2 is conducting. In the right drawing, transformer polarities have changed and V_1 is conducting.

cathode of V_2. The remaining end of secondary S is connected to the junction point between the two capacitors. When the upper end of S is positive, View (A), electrons pass from the lower end to the upper plate of C_2, flow continuing from the lower plate of C_2 to the cathode of V_2, then to the plate of V_2 and the positive terminal of S. The charge on C_2, it is apparent, gives the capacitor a negative polarity on top, a positive one on the bottom.

As the lower end of S becomes positive, Fig. 3 (B), electrons pass from the upper end of S to the cathode of V_1, through the tube to the plate, and down to capacitor C_1, from the bottom plate of which the circuit is completed to the positive terminal of S. The charge on C_1 makes the upper plate negative, the lower one positive. The two capacitors are therefore in series across the output terminals, the polarity of the upper one negative, the lower positive, and the voltage of the output circuit is twice that of either capacitor.

Filter Circuits

Filter circuits are necessary with both half-wave and full-wave rectifiers because the unfiltered pulsations will create hum in audio circuits and very poor voltage regulation in power supplies. Pure d-c with no variations or pulsations would, of course, be an ideal condition. However, whether this ideal condition is to be aimed for in circuit design is usually determined by a compromise between cost and the desired system application.

The frequency of the pulsations, called the *ripple* frequency, is equal to the supply frequency for half-wave rectifiers. For full-wave rectifiers, the ripple frequency is double the supply frequency since the rectifier conducts on each alternation of the sine wave. How much filtering is required to smooth out the pulsations is dependent on the ripple frequency; less filtering is required for full-wave than for half-wave rectifiers as less of the ripple must be filled in.

Basically, filter circuits make use of capacitors, inductors, or both, in their operation. Those using both are combination circuits, having different operational characteristics and subsequently, advantages and disadvantages for a particular application. Each type of combination circuit will be studied in turn after examining the basic filter-

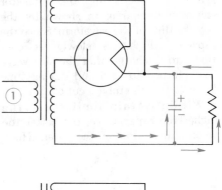

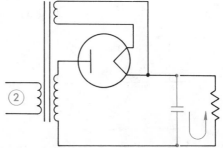

Fig. 4. The top drawing shows the charge path of current to the capacitor while the tube is conducting. The bottom drawing shows the capacitor discharge path after the tube is cut off.

ing action of individual capacitors and inductors.

Capacitor Input Filters. The simplest and most basic filter type is a single capacitor connected in parallel across the load resistor. As previously discussed, the capacitor resists a change in voltage. Fig. 4 illustrates its action in the circuit. When the tube is conducting, the capacitor charges until the voltage reaches its peak and begins decreasing to zero amplitude. At this point the capacitor begins discharging through the load in the same direction as the tube current. Fig. 5 gives a more detailed view of the capacitor's effect on the rectifier output wave.

In Fig. 5 the capacitor charges during times T_0 to T_1. At T_1, when the voltage begins decreasing, the capacitor starts to discharge, as shown by the color outline of the waveform. Since the capacitor discharges slowly, it continues to discharge after the sine wave has returned to zero and during time T_2 to T_3 that the tube is cut off.

Within certain limitations, the higher the capacitance, the better the smoothing action of the filter. How-

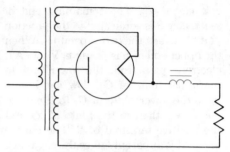

Fig. 6. Rectifier circuit with simple choke filter.

ever, the addition of the capacitor in the filter circuit increases the average current through the tube, and the current rating of the tube generally determines the highest capacitance that can be used.

Simple Inductor Filter. Fig. 6 illustrates a simple inductor filter circuit. The inductor in this circuit is usually called a *choke.* As we know from previous chapters, an inductor resists any change in current in the circuit. This characteristic provides the basic filtering action for this filter. As the current through the load begins to decrease toward zero, the inductance acts to keep the current flowing and fills in the portion between the waves in much the same manner as the capacitor filter. However, note that the inductor also resists the current change as it begins to build up from zero to maximum. The resultant waveform is similar to that of Fig. 5, the overall effect being to give a more constant average of current and voltage across the load. In practice, the capacitor and choke are seldom used alone to provide filtering. They are used in combination as *choke input* or *capacitor input* depending on the arrange-

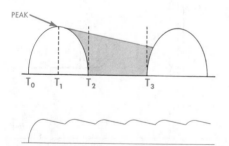

Fig. 5. The shaded portion of the voltage wave shown in color is the area filled in by the discharge of the capacitor. The capacitor begins discharge at time T_1, continues during the time the tube is cut off to time T_3.

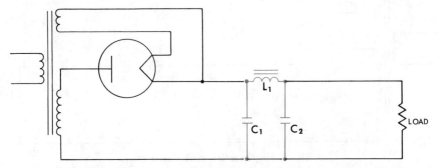

Fig. 7. Rectifier circuit with capacitor input filter. This filter is sometimes referred to as a "pi" filter.

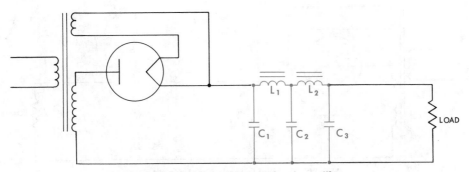

Fig. 8. The double-section capacitor-input filter.

ments of the components in the circuit.

Capacitor Input Filters. Fig. 7 illustrates a combination filter circuit termed a capacitor input filter, which uses capacitors and a choke. This particular circuit is a *single-section* filter, sometimes referred to as a *pi* filter because of its similarity to the Greek letter *pi* (π). Fig. 8 has an additional capacitor and choke and is a *double-section* capacitor input filter. In operation, these two circuits combine the effects of the individual capacitor and

choke components explained in the previous paragraphs.

Advantage of the capacitor input filter is in its high output voltage in relation to its transformer input voltage. This however, is more of an advantage when used with silicon or selenium (solid state) rectifiers, than with electron tube rectifiers. With electron tubes, the regulation is poor compared to the choke input filter, and applications are predominantly for applications where regulation is less important than high output voltage.

261

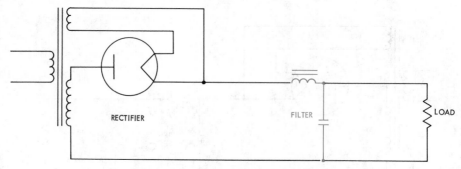

Fig. 9. Single-section choke-input filter.

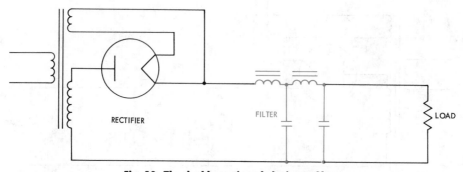

Fig. 10. The double-section choke-input filter.

Choke Input Filters. A single-section choke input filter, Fig. 9, utilizes a single choke and capacitor. A double-section choke input filter, Fig. 10, with two capacitors and two chokes, provides a more complete smoothing action of the wave than the single-section.

The choke input filter has better regulation at high load currents. Applications are primarily in power supplies handling heavy or varying loads. Because of its better regulation of current it

is well adapted for use with electron tubes.

Complete Power Supply Circuit

The circuit of Fig. 11 consists of transformer, rectifier tube, filter, and *voltage divider.* Rectification and filtering have already been explained. A voltage divider is simply a resistor which is tapped at selected points to obtain desired voltages. (It may also be a series of resistors, with taps at the

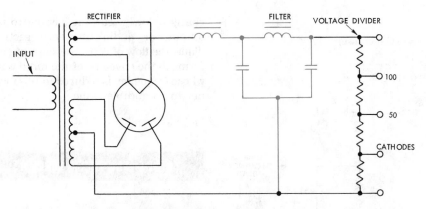

Fig. 11. Schematic of a complete power supply circuit with filter and voltage divider.

points where each resistor is connected to the next one in the series). Voltage dividers are used because different tubes require different d-c operating voltages. For example, one tube may require 140 volts on the plate, and 100 volts on the screen grid, but another in the circuit may require 250 volts on the plate. Observe that a *cathode bus* connection is made a short distance from the lower end, and that a lead from the negative end is a tap for grid voltages. The small capacitors, of course, *bypass* any alternating voltage component which might get into the various circuits.

The Electron Tube as Detector

A third important application of the electron tube, particularly in the area of radio reception, is that of detector. The high-frequency wave from the broadcasting station, Fig. 12 (*top*) has been modulated, its *envelope,* or outline, following a pattern of voice frequencies. Radio waves cannot be heard by human ears, of course, because their

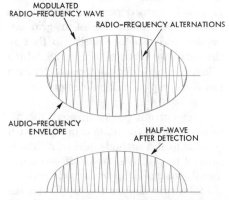

Fig. 12. Detection of a radio frequency wave using a diode electron tube.

frequency is far too high. The detector, in effect, reduces frequency to audible levels.

The first detecting operation is to remove the lower, negative half, Fig. 12 (*bottom*), by means of rectification. A number of methods have been employed, among them Class C amplification studied in connection with Fig. 21,

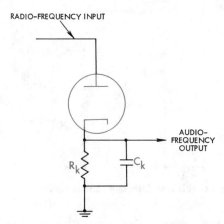

RADIO-FREQUENCY INPUT

AUDIO-
FREQUENCY
OUTPUT

R_k C_k

Fig. 13. The diode circuit for radio frequency detection.

through biasing the tube to cut-off. This scheme is termed *power detection*. An earlier plan made use of a grid capacitor and resistor similar to that in the oscillator circuits of Fig. 23, which is known as *grid-leak* detection. Modern receivers employ diode detectors similar to Fig. 13.

Since current passes through the diode only when the plate is positive with respect to the cathode, electron flow is limited to positive halves of the input signal, voltage drop in cathode resistor R_k following their outlines. Capacitor C_k is larger than the usual bypass unit, its charge and discharge rate much too slow to conform with rise and fall of individual waves. Instead of discharging when a particular half-wave falls to zero, it retains the charge and adds that from other half-waves as they arrive, so long as the envelope becomes increasingly positive.

When the outline pattern begins to decrease, however, the capacitor dis-

charges slowly, current passing to the output, or audio frequency section. Thus, the flow of capacitor current conforms to the envelope of the radio wave whose frequency is reduced in this manner to an audible value.

TRIODE AMPLIFIER CIRCUITS

Before proceeding to discuss the triode as an amplifier, it may be well to consider what is meant by the term *amplification* which is sometimes misunderstood as obtaining something for nothing. This is never the case, Nature's laws requiring that energy be transformed only through expenditure of energy. When a radio amplifier takes a minute input signal and raises its strength to a high level sufficient to drive a loudspeaker consuming several watts of power, it does so only by extracting energy from the power supply system.

Coupling

The triode, Fig. 14, has an iron core transformer, T_g, in series with grid battery C, and a similar one, T_p, in series with plate battery B. The coupling method here is known as *transformer coupling*. When sine-wave signal current flows in the primary of T_g, its secondary voltage is added to or subtracted from that of C, which is known as the *bias battery*. If the induced secondary voltage varies between maximum positive and negative voltages of $+2$ and -2 respectively, the maximum voltage applied to the grid of the tube fluctuates between -3 and -7 volts.

Current in the plate circuit follows

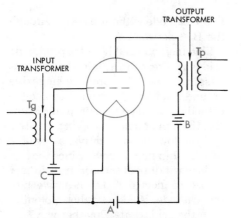

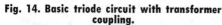

Fig. 14. Basic triode circuit with transformer coupling.

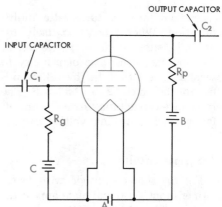

Fig. 15. Triode circuit with resistance-capacitance coupling.

the pattern of grid voltage, magnetic flux in the primary of T_p rising and falling in step with it, thus inducing a voltage in the secondary winding. If 10 volts appear at the secondary terminals of T_p, the ratio of input to output voltage is equal to: 10 volts/2 volts = 5. That is, the grid signal voltage has been amplified, or multiplied, by five.

The circuit of Fig. 15 is similar to Fig. 14, except that capacitors and resistors have been substituted for transformers in grid and plate circuits. The coupling method used here is known as *resistance - capacitance coupling*. The term *coupling* simply refers to the method for transferring energy. The input signal is imposed on the outer plate of capacitor C_1, causing the inner plate to draw electrons up through resistor R_g, and thus giving a positive charge to the grid, or repelling them down through R_g to give a negative charge to the grid, according to the outline of the input voltage wave. Resulting increase and de-

crease of plate current is accompanied by changing voltage drop across resistor R_p, thus influencing the charge on capacitor C_2. The varying charge across C_2 creates a signal which is passed on to the input circuit of the following stage in the amplifier.

Observe particularly the relationship between electron flow to or from capacitor C_1 and the resistor polarity. When electrons are drawn toward the capacitor, a + polarity is created at the top of the resistor; when they are repelled down through the resistor, a − polarity results at the top. This knowledge will prove beneficial in the analysis of complicated circuits, and the correct polarities are easy to recall in view of the fact that electrons always travel through a circuit from negative to positive.

One important difference between the two circuits should be noted. The *output transformer* in Fig. 14 furnishes voltage increase between primary and secondary through turns-ratio multiplication. Where the tube itself, for example, supplies an amplification of 2.5, and the turns-ratio of primary to secondary is 1 to 2, the overall amplification becomes: $2 \times 2.5 = 5$. The resistance-capacitance arrangement offers no such additional voltage gain.

The Triode Circuit

Fig. 16 shows a triode with transformer input. Grid and plate appear the same as before, but the cathode element has been added.

The secondary of the input transformer may be adjusted, or *tuned,* to resonance at a particular frequency in some amplifiers by means of a fixed capacitor C_1. In certain cases, the capacitor may be a variable type for obtaining a resonant condition at any number of selected frequencies. The small *bypass* capacitor C_2 between B+ and ground, offers a low impedance path for high-frequency plate current ripples which thus find their way back

to the cathode without passing through the B+ network.

The method used here for producing grid bias is readily explained. Electrons passing upward from the negative ground conductor, or ground bus, to the cathode cause a voltage drop in resistor R_k, the upper end becoming positive and the lower end negative, as per the relationship noted above. Since the secondary winding of the grid transformer is also connected to the negative bus, the cathode is at a higher potential than the grid. Stated another way, the grid is negative with respect to the cathode, just as though a battery were in circuit. The purpose of bypass capacitor C_k is similar to that of C_2, providing a means for keeping alternating current ripples out of the cathode resistor. Fig. 17 shows the identical arrangement as applied to a resistance-capacitance circuit.

The manner in which bypass capacitors work is shown in Fig. 18. In most amplifiers, the negative grid bias is insufficient to reduce plate current to

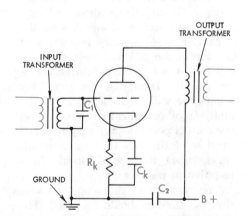

Fig. 16. Triode circuit with tuned input transformer.

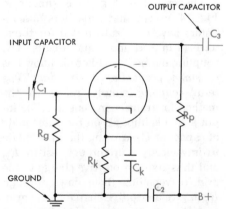

Fig. 17. Resistance-capacitance coupled triode amplifier.

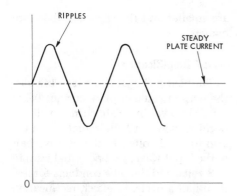

Fig. 18. Operation of the bypass capacitor.

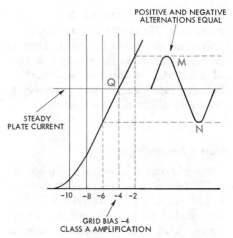

Fig. 19. The operating point bias for Class A amplification.

zero, and a steady current flows when there is no signal on the grid. As a sine wave voltage appears at the grid, however, plate current rises to a maximum value and drops to a minimum as grid voltage rises and falls. These variations are known as *ripples*. If they were to get into the plate supply network, they would affect all other tubes connected to the B+ system. The capacitor absorbs these deviations in current flow, passing them on to the cathode as alternating waves, while current taken from the B+ terminal remains practically constant. The same effect is produced by the bypass capacitor, ripples in the cathode circuit being kept out of R_k so that a steady current flows through the series resistor, and a constant bias is furnished to the grid of the tube.

Class A Amplification

Fig. 19 is a grid-voltage-plate-current curve constructed by varying grid voltage while plate voltage is maintained at a constant value. For example, this

curve might be drawn by connecting a 120-volt battery to cathode and plate, and then taking plate current readings as grid voltage changes in progressively increasing steps between two extremes such as −10 volts and 0 volts, or −10 volts and −2 volts.

Fig. 19 shows that plate current becomes zero when the negative bias is 10 volts or more. From zero, the outline curves somewhat until the bias falls to −6 volts. It straightens here, and rises steadily, moving vertically the same distance between −6 and −4 as between −4 and −2. In other words, the plate current increases at precisely the same rate throughout this region.

If normal grid bias is set at −4 volts, and the input sine wave of voltage lies between +2 and −2 volts, grid potential will rise to −2 volts and fall to −6 volts. The *no-signal grid bias* of −4 volts permits a steady plate current equal to Q. As bias drops to −2 volts, the plate current rises to a maximum value of M; as bias increases to

267

—6 volts, plate current falls to the minimum N, positive and negative halves of the cycle identical. Biasing in this manner, on the straight portion of the curve, so there is no distortion between input and output wave forms, is termed *Class A amplification*.

Class B Amplification

When normal bias is established at —12 volts, a point where plate current is just cut off, plate current will flow only during the positive half of the grid signal. This type of amplification is known as *Class B*. Such amplification is acceptable under some conditions, and the arrangement is more efficient than Class A. I^2R losses in the plate circuit

are smaller and the efficiency therefore higher.

Class C Amplification

The bias in Fig. 21 is set at —14 volts, the normal value of plate current being zero. However the plate current will not begin to flow until the signal raises the grid to —12 volts. On the positive half of the input wave, grid potential rises to —8 volts, and the tube conducts for less than half a cycle. In effect, rectification has taken place, only positive pulses of current passing through the plate circuit. This form of amplification is *Class C*, the most efficient of the three. It will be seen later that the circuit is employed, also, for another purpose than amplification.

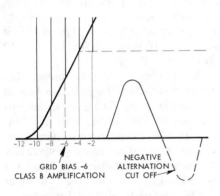

Fig. 20. The operating point bias for Class B amplification.

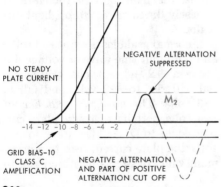

Fig. 21. The operating point bias for Class C amplification.

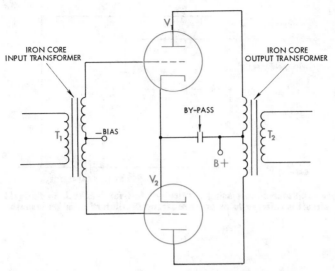

Fig. 22. Push-pull amplifier circuit used in Class B and C amplification.

Push-Pull Circuit

Fig. 22 illustrates the *push-pull circuit* often used to help eliminate distortion in the output of Class B and Class C amplifiers. This is usually the final audio stage. It consists of a center-tapped-secondary input transformer T_1, tubes V_1 and V_2, and a center-tapped-primary output transformer T_2. Both transformers are iron-core because they operate at comparatively low audio frequency.

At a given instant, the primary input wave makes the upper terminal of T_1 positive, the lower end negative. V_1 conducts for the moment, sending current through the upper half of secondary T_2. As the input wave changes polarity, the bottom terminal of secondary T_1 becomes positive, V_1 is now inoperative, and V_2 conducts.

When the input is from a Class B stage, two half-waves appear in the output, because of the so-called "flywheel effect." A positive voltage is induced in the secondary T_1 as the Type B wave rises from zero to maximum, and a negative voltage as the primary wave decreases from maximum to zero. Push-pull circuits are used often with Class A amplifiers, as well, because they eliminate even harmonics and produce more power than is obtainable from a single tube.

The Electron Tube as an Oscillator

Interelectrode capacitance in an electron tube is usually undesirable because it interferes with the amplifying duties of the tube and creates oscillations. *Oscillation* is often desirable, and special circuits are devised for the purpose of feeding energy from the plate circuit back to the grid. This energy is called *regenerative feedback*. Fig. 23

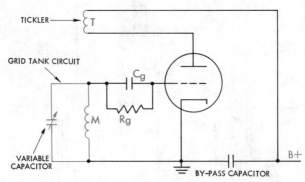

Fig. 23. Triode oscillator circuit using a "tickler" coil for feedback to the grid circuit. This circuit is referred to as an "Armstrong Oscillator" after its inventor.

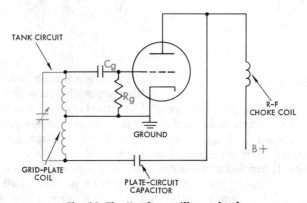

Fig. 24. The Hartley oscillator circuit.

shows the *tickler* arrangement, tickler coil T in the plate circuit inducing voltage in coil M of the tuned-grid, or *tank*, circuit. The grid becomes positive during the cycle of operation, drawing current which flows through resistor R_g to establish a bias. Capacitor C_g bypasses alternating portions of current flow.

The *Hartley circuit* of Fig. 24 is similar in principle, but the electron-coupled unit of Fig. 25 is quite different. A tetrode is employed, cathode, control grid, and screen grid forming the oscillating section of the tube, while the plate circuit is physically isolated from them. Flow of electrons between cathode and plate is altered or *modu-*

270

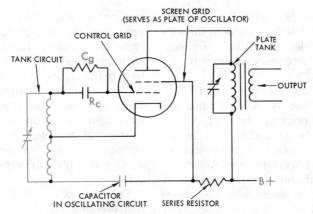

Fig. 25. The electron-coupled oscillator circuit.

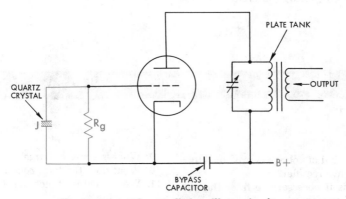

Fig. 26. A crystal controlled oscillator circuit.

lated according to demands of the oscillating circuit, increasing as flow to the screen grid decreases, decreasing as screen-grid requirements increase. Thus, the plate circuit is said to be electronically coupled to the oscillating portion of the tube. Superiority of the electron-coupled unit in certain applications rests on the fact that load variations in the plate circuit have no effect on frequency of oscillation, as sometimes occurs with other arrangements.

The electron-coupling principle is used in the *mixer* or *converter*, units of superheterodyne receivers. In these tubes, one of the two frequencies is applied to an oscillating circuit like Fig. 25. The second frequency is im-

271

posed on another grid, flow of electrons to the plate is modulated through both frequencies so that only a resultant, or *beat frequency* appears in the plate circuit.

Operation of the *crystal-controlled oscillator* of Fig. 26 is based upon the piezoelectric principle studied in Chapter 6. It will be recalled that an electric current tends to flow when a piezoelectric substance is compressed. A complementary fact, not mentioned at that time, is that the dimensions of a sample are changed when a current is forced through it.

Quartz crystal J, in the illustration, displays piezoelectric qualities. When it is tightly pressed between metallic end plates, a voltage is created in the grid circuit of the tube. This grid voltage alters the current flowing in the plate tank circuit. Interelectrode capacitance feeds some of this energy back to the crystal in the form of a grid current which alters slightly its physical dimensions. The change in the crystals dimensions, in turn, sends a voltage to the grid, the cycle continuing at a definite frequency governed by the manner in which the crystal has been cut.

The piece of quartz, in other words, is an oscillating device whose frequency is quite stable. When the tank circuit is tuned to a frequency near that of the crystal, oscillations quickly build up to maximum value.

REVIEW QUESTIONS

1. What kind of voltage is produced by a half-wave rectifier?
2. Why is it necessary to filter the output of a rectifier?
3. What is the ripple frequency of a full-wave rectifier?
4. What factor determines the highest capacity that can be used in a filter circuit?
5. What type filter is referred to as a *pi* filter?
6. What are the advantages of a capacitor-input filter?
7. For what applications is the choke-input filter best adapted?
8. The capacitor-input filter is characterized by what kind of voltage regulation?
9. What is a space charge?
10. What does the term *coupling* refer to?
11. What is the function of a bypass capacitor?
12. At what point on the operating curve is a Class A amplifier biased?
13. Describe the output of a Class C amplifier.
14. In what type of circuit is a tank circuit used?
15. What determines the frequency of a crystal controlled oscillator?
16. Which class of amplifier is the most efficient?
17. Where are push-pull amplifier circuits used?
18. How is a *beat* frequency produced?

Chapter **14**

Solid-State Devices

INTRODUCTION

Probably the most exciting branch in the electrical-electronics field is that of solid state devices. Since the birth of these miraculous electron control devices, engineers have been able to produce equipment that is lighter, more compact, more accurate, more sophisticated, and more reliable. The chapter we are about to deal with discusses solid state devices such as silicon rectifiers and controllers, zener diodes, tunnel diodes, transistors, unijunctions, field-effects transistors, light-emitting diodes, and integrated circuits. These are the special devices that in many ways have replaced the vacuum tube in much of modern day electronic hardware. Also considered in this chapter are some of the older devices such as the metallic oxide rectifier, the saturable reactor, the magnetic amplifier, the dielectric amplifier, and the transducer.

Each device covered in this chapter could justify an entire book to handle its uses and peculiarities. For the purposes of this chapter, discussion is held to general construction and operation.

SEMICONDUCTOR SCIENCE

Bell Laboratory scientists (J. Bardeen and W. H. Brattain) began experimenting with these materials which showed characteristics of both conductors and insulators, and which came to be known as *semiconductors*. Publication of their research findings in 1948 opened up a new field of solid-state physics. Two substances which received most attention were *germanium* and *silicon,* particularly the former. Both have four electrons in the incomplete outer shell, their atoms forming crystalline structures similar to that of common salt, studied in Chapter 6.

In the pure state, germanium is practically an insulator because it is so hard to produce free electrons. Groups of atoms form closely-knit crystals in which orbits are so interlaced that outer electrons of one atom are shared with others. As a result, the outer shells of

14

all contributors are effectively complete, like those of true insulators.

Covalent Bonds

Because the interwoven patterns of electron orbits are complicated, scientists have adopted a method used in chemistry to express the interaction of one atom with another. This system uses the idea of covalent bonds. The atom is represented as a circle, Fig. 1 (*top*), its outer electrons as four arms or bonds which reach out to contact arms of adjacent atoms. A simplified crystal is illustrated in Fig. 1 (*bottom*), eight atoms combining to form a cube, and eight arms projecting from the

crystal to contact similar cubes. And, so long as the endless chains of *bonded atoms* exist, no free electrons are available to form an electric current.

Doping

During the course of experiments seeking to improve its conductance, pure germanium was mixed, or *doped,* with substances of different atomic structure, especially ones having an uneven number of electrons in the outer shells. Two of the doping materials, arsenic and boron, produced striking results. *Arsenic* has five electrons in the outer shell, *boron* three. When germanium is doped with the former substance, arsenic atoms enter into crystalline formations with the result illustrated in Fig. 2 (*top*).

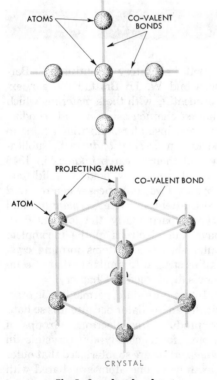

Fig. 1. Co-valent bonds.

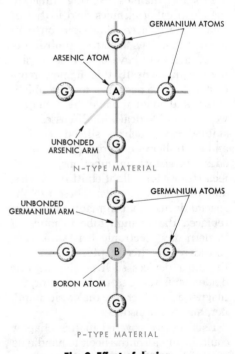

Fig. 2. Effect of doping.

The four *G* atoms are germanium, the fifth, *A*, arsenic. Four arms from the *A* atom contact arms of nearby atoms, but one arm is left unattached. This unbonded arm represents a less strongly-held electron which can be removed from the structure to become free. The arsenic atoms, therefore, are called *donors,* and the doped material is capable of becoming a fairly good conductor. Investigators called this material *N-type* germanium, the *N* signifying the presence of negatively-charged electrons.

It is worth noting that crystals of the *N*-type metal are electrically neutral. Atoms, as pointed out in Chapter 1, have just enough positive nuclear charge to offset negative charges of all orbiting electrons, including those in the outer shell. Before doping, both germanium and arsenic atoms are electrically neutral. Mixing did not alter this condition, even though crystals with arsenic content have an unbonded electron. If an electron is taken away, however, the crystal acquires a positive charge of one unit because the arsenic atom is no longer neutral.

Doping the germanium with boron, or some other substance like indium or gallium, which have three electrons in the outer shell, produces a different sort of crystal, Fig. 2 (bottom). Three covalent bonds are formed between the boron and surrounding atoms, leaving one of the germanium arms dangling. Investigators described this condition by saying that a *hole* existed in the crystalline structure.

This crystal, like the *N*-type, is electrically neutral despite the hole, but the unsatisfied bond of germanium atom creates a state of unbalance. This allows the crystal to act as if it had a positive charge of one unit, and it tries to capture an additional electron to

correct the imbalance. Since the three-electron impurity may thus be looked upon as willing to take another electron, it is called an *acceptor* atom. Material formed with acceptor atoms is termed *P-type* germanium because it acts as though it has a positive charge.

Current Flow in N-type and P-type Germanium

If a battery is connected to a sample of *N*-type metal, Fig. 3, electrons from donor atoms are attracted toward the positive terminal. The positively-

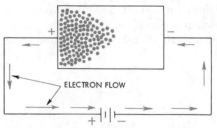

Fig. 3. Electron flow in N-type germanium.

charged atoms in the crystal then attract electrons from the negative battery terminal so that current flows as shown by the arrow. Substitution of *P*-type for the *N*-type metal produces a result shown in Fig. 4. Acceptor atoms

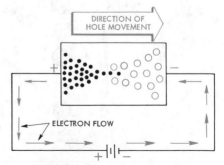

Fig. 4. Electron and hole current in P-type germanium.

draw electrons from the negative terminal, but attraction at the positive terminal takes them away, establishing a flow of electrons as indicated.

It is customary in discussing *P*-type germanium to speak as though the holes actually moved through the crystal. (This apparent hole movement is sometimes called positive current.) Only electrons move, but as shown in Fig. 5, some justification for saying that the holes move does exist.

Five acceptor atoms: *A*, *B*, *C*, *D*, and *E*, lie in a straight line, Row No. 1, all having acquired an electron except *A*, so that a hole is present at this location. An instant later, assisted by attractive force of the positive battery terminal, *A* takes an electron away

from *B*, Row No. 2, the hole appearing to move from *A* to *B*. *B* takes an electron from *C*, Row No. 3, and the hole seems to have traveled to this point. In the same way, the hole moves to *D*, and finally to *E*. This chain activity, accompanied by apparent movement of holes in a direction opposite to the flow of electrons, is termed *hole conduction*, a conventional expression adopted for the purpose of abbreviated explanation.

Potential Barrier

P and *N* types of germanium are joined to form diodes and triodes, either by the *growth* or the *fusing* method, details of which are of small concern here. Before the joining of two

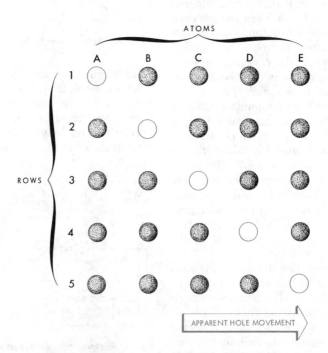

Fig. 5. Flow of current by "holes" in germanium.

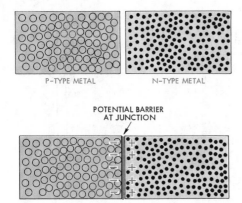

Fig. 6. Joining P- and N-types of germanium.

diode. Electrons attracted by this polarity relieve the negative concentration at the potential barrier, and others move across from the N side to fill holes on the left. This movement of electrons induces a flow from the negative battery terminal into the N-type half, current circulating as indicated by the arrows.

Change of battery polarity, Fig. 7 *(bottom)*, produces a different result. The negative terminal, which is now attached to the P half of the diode, offers no help in lowering the potential barrier, and the attractive force of the positive terminal increases barrier strength by withdrawing more electrons from the vicinity. No current flow occurs, therefore, in the external circuit.

The arrangement of Fig. 7 *(top)*, in which the P-type material connects to

samples, Fig. 6 *(top)*, the P-type at the left has a supply of holes at the near edge, the N-type a supply of electrons, but neither has an actual electrical charge.

During the fusing process, a change in relative structures takes place, atoms of the P-type material seizing electrons from the N-type in the immediate vicinity of the junction. Thus, the acceptors take on a negative charge at the left side of the junction, Fig. 6 *(bottom)*, while the donors become positively charged. The concentration of negative charges on the left gives rise to a condition similar to space charge in a vacuum-tube triode, repelling any electrons which seek to pass from the N side, and a *potential barrier* is said to exist.

SEMICONDUCTOR DEVICES

Semiconductor Diode
Biasing Technique

The battery in Fig. 7 *(top)* furnishes positive bias to the P-type half of the

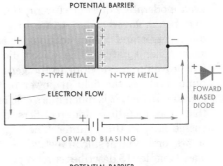

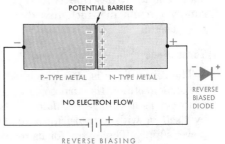

Fig. 7. Effect of battery polarity.

the positive battery terminal and the N-type to the negative, is termed *forward biasing*. The method of Fig. 7 *(bottom)*, wherein P-type is attached to the negative pole and N-type to the positive, is termed *reverse biasing*. Since electron flow results only with the first connection, a rule may be stated: *The diode conducts only when it is biased in the forward direction.*

Diode (Rectifier) Applications

In view of the fact that the diode permits a flow of electrons only when polarized in one of the two possible ways, it is obviously adapted to purposes of rectification. If an alternating current source is substituted for the battery, current will pass through the diode only during that half of the cycle which provides forward biasing.

Comparatively high resistance and sensitivity to heat render the germanium diode unsuitable for rectifying even moderate values of current, but it may be employed as a detector in a radio receiver and in applications of the same general nature. The silicon diode has proved to be superior to the germanium diode in the field of power rectification. In fact, it is much more efficient than the metallic-oxide rectifiers discussed later in this chapter. The Fig. 8 shows some typical silicon rectifier packages. Chapter 15 describes several power supply circuits in which this diode is most useful.

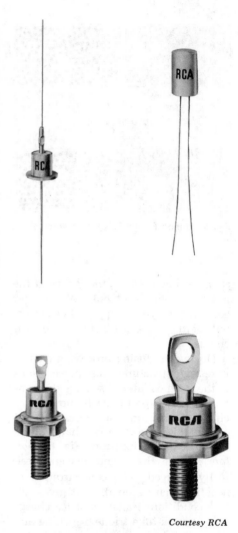

Courtesy RCA

Fig. 8. Typical silicon rectifier packages.

Zener Diode

The zener diode is closely related to the diode rectifier. Its purpose, however, is to act as a voltage regulator, either by itself or with a transistor-regulated power supply. The reason for its regulatory ability is its unique construction.

The zener diode is constructed to operate in reverse of the normal conditions present in rectifier diodes. This reverse condition is made possible by carefully planned doping conditions when the diode is manufactured. The zener diode is basically a breakdown diode. It is installed so its bias is the reverse of the diode rectifier.

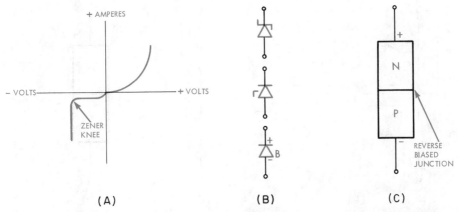

Fig. 9. Zener diode, showing: (A) current-voltage characteristic curve; (B) zener and breakdown diode schematic symbols; and (C) functional diagram of zener diode.

In the Fig. 9 (A) the current-voltage characteristic curve of a zener diode is shown. This particular curve shows normal diode operation with the zener diode forward biased. In the reverse condition the point at which the diode breaks down is called the *zener breakdown voltage* or the *avalanche breakdown*. As the voltage against the diode barrier approaches this zener point, current carrying electrons tunnel through the barrier, causing the zener effect. At the same time other electrons accelerate, knocking others loose from their valence rings in a process called *carrier multiplication*, causing the effect called *avalanche*. At that time the diode is said to be at *breakdown*. At this voltage point the current may increase greatly while the voltage remains the same. It is this reason that makes the breakdown diode a perfect voltage regulator. Refer to Chapter 15 for circuit utilization.

In actuality, the zener point and the avalanche point occur at different levels. The avalanche occurs slightly after the "knee" of the Zener. For all practical purposes, the two effects can be said to happen together. Fig. 9 (B) illustrates typical zener and breakdown diode schematic symbols. Fig. 9 (C) is a functional diagram of the zener diode. A zener diode is built into a package similar to normal rectifiers (see Figure 8).

Tunnel Diode

Because of its high frequency operation, high speed switching ability, and low power consumption, the tunnel diode has been used successfully in oscillators, micro power circuits, and computer circuits.

The tunnel diode is a two-layered solid state device which has been heavily doped to produce an effect called *tunneling* (also known as the *Esaki effect*, naming its discoverer). This effect takes advantage of the heavy doping of the device. The heavy doping (several thousand times that of a normal diode) produces a very narrow depletion region or barrier. With voltage applied, electrons can tunnel their way from the N to the P type material. This causes an

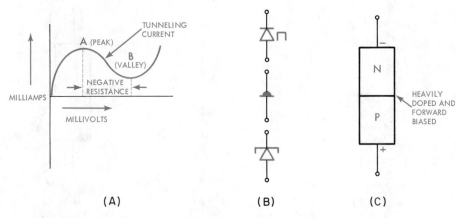

Fig. 10. Tunnel diode, showing: (A) current-voltage characteristic curve; (B) tunnel diode schematic symbols; and (C) functional diagram of the tunnel diode.

increase in current with little bias voltage, see Fig. 10 (A).

As the voltage is increased, the current suddenly disappears. Electrons cross the barrier at tremendously high speed. The effect is that the tunnel diode can be operated at much higher frequencies than other devices. The other advantage is that the tunnel diode requires little bias, less than 100 millivolts.

The valley point as shown in the Fig. 10 has little or no resistance. This particular property of the tunnel diode allows it to convert d-c power supply energy into a-c circuit energy, thus permitting it to be used as an amplifier or an oscillator. Other uses of the tunnel diode are switching circuits, rectification, and precision measurement.

The current range of the many tunnel diodes is from several microamperes to several hundred amperes. As the bias voltage is increased, the tunnel diode tends to have the same characteristics as the normal diode. The Fig. 10 (B) illustrates typical tunnel diode schematic

symbols and Fig. 10 (C) is the biasing technique.

Varactor Diode

The varactor diode is used for tuning antennas, frequency multiplication, and oscillation functions.

Fig. 11 illustrates varactor diode symbols. Note that in some instances the diode looks like other diodes with only the identity number listed next to it

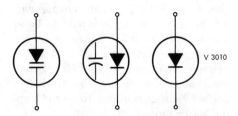

Fig. 11. Varactor diode schematic symbols. Note that the schematic at the right looks like any other diode except for the identifying number shown next to it. The other symbols contain the capacitance symbol.

on the schematic. Reference should be made to manufacturer reference books for application of the varactor.

The varactor diode is a PN junction solid-state device that has a variable reactance. This reactance is provided by the junction capacitance that varies with bias voltage. The varactor diode is doped near the barrier region so as to cause a dielectric-like area which can be controlled by bias voltage.

This capacitance is non-linear, but if the varactor diode is reverse-biased the barrier will grow and the capacitance will decrease. On the other hand, forward biasing will cause the barrier to decrease and the capacitance to increase. If forward voltage is used in great amounts, the varactor acts like a rectifier diode and there is no capacitance effect.

Light-Emitting Diode (LED)

The light-emitting diode (LED) can be used in almost any place that tungsten filament bulbs are used. The primary reason that the LED device is not used in all tungsten filament situations is that where a considerable quantity of light is required the price would be prohibitive. The LED can be used by itself as long as current flows through it. Its body lights up, thereby requiring no envelope such as a tube or a lamp case.

The LED is placed in a circuit in a forward bias direction as any forward biased junction.

In a circuit the letters LED will normally accompany a diode symbol. Light emitting diodes (LED's) are made of gallium arsenide phosphide wafers. This material is used because the recombination of electrons and holes in its lattice structure produces pockets of light called *photons*. In the lattice structure of silicon or germanium this recombina-

tion of electrons and holes produces heat.

The energy band gap at the junction barrier region in the gallium arsenide light-emitting diode determines the wave length of the photons of light. The band gap is the differential energy between the conduction band and the valence band in any material. The structure of the light-emitting diode is quite simple. The shape of the LED is determined by the needs of power and the form of the device in which the diode is installed. Problems arise because the shape may tend to cause the light to be absorbed or reflected back into the crystal.

Light-emitting diodes are used as pilot lights because they do not have to be placed in special sockets, do not require special power supplies, and do not have the failure problems of ordinary lamps. LEDs are used in applications such as punched-card readers,

Courtesy Motorola Semiconductor Products, Inc.

Fig. 12. The light-emitting diode (LED) converts electrical energy directly into light without generation of heat.

credit card and card-key verifiers, home and industrial security systems, telephone dials and bomb fuzes. Its short response time provides for uses in optical switching and film marking. The Fig. 12 illustrates one of many types of new light emitting diodes (LED's).

The Silicon Controlled Rectifier (SCR)

The silicon-controlled rectifier (SCR) is a type of diode with very low resistance when conducting and very high resistance in the "off" state. Because of these characteristics SCR's are mainly used in high-power control and switching circuits, and they may be used to control either a-c or d-c current.

In operation, the SCR is similar to the thyratron tube and therefore falls under the general solid-state category of thyristors. In construction, the SCR is a four-layer semiconductor. It is biased similar to a normal rectifier but will not conduct current until it is gated.

In Fig. 13 (A), negative voltage is applied to the cathode (N) material, positive voltage is applied to the anode (P) material and the positive gate voltage is applied to the P material in the center.

The voltage required to turn on the SCR is called *breakover voltage*. When this voltage is large enough, current breaks across the junction and the SCR will be turned on. At this time the gate voltage may be removed and current will continue to flow from cathode to anode. The SCR will remain on until power is completely removed or an open circuit occurs by switch action or some similar event.

Fig. 13 (B) provides the principal voltage-current characteristics of the SCR and Fig. 13 (C) is a schematic symbol. The SCR is useful as a diode having very low resistance when con-

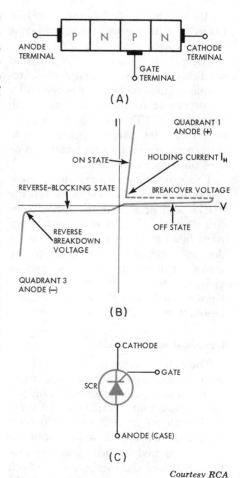

Courtesy RCA

Fig. 13. Functional block diagram of an SCR with voltage-current characteristics and schematic diagram.

Courtesy Motorola Semiconductor Products, Inc.

Fig. 14. Typical silicon-controlled rectifier (SCR) with and without external leads.

ducting and very high resistance in the *off* state.

Fig. 14 is a photograph of a typical SCR in its case.

SEMICONDUCTOR TRANSISTOR AMPLIFIERS

Although a semiconductor diode can be used as a rectifier, it offers no means of amplification. Triodes and tetrodes, called *transistors*, have been developed for the purpose. A triode consists of a *wafer* of *N*-type material sandwiched between two thicker elements of *P*-type material, Fig. 15 *(top)*, or else a *P*-type wafer between two large *N*-type pieces, Fig. 15 *(bottom)*.

The thin part of the triode is known as the *base*, the left portion as the *emitter*, and the right one as the *collector*. There are two sources of power, the emitter battery at the left, and the collector battery at the right. Observe that the emitter element, Fig. 15 *(top)*,

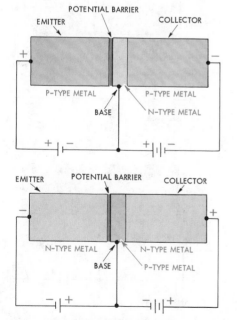

Fig. 15. Transistor biasing.

Courtesy RCA

Fig. 16. Typical transistors and symbols. The circles around the transistor are sometimes omitted in schematic diagrams.

is attached to its positive battery terminal, thus giving it forward biasing. The collector element is attached to its negative battery terminal, so that it is reverse-biased.

These biasing relationships are preserved in Fig. 15 *(bottom)*, where the left piece of N-type material is connected to the negative terminal of the emitter battery to produce forward biasing, while the piece at the right is connected to its positive battery terminal, for reverse biasing. Battery polarities will not prove confusing if it is kept in mind that P connects to *positive* in forward biasing, and N connects to *negative* in forward biasing.

A triode and the symbols applying to it are illustrated in Fig. 16. Transistors and electron tubes are similar in many respects, and it promotes understanding of transistor circuits to consider the emitter equivalent to the cathode, the base to the grid, and the collector to the plate. The triode of Fig. 16 *(left)* is termed a *P-N-P transistor*, that of Fig. 16 *(right)*, an *N-P-N transistor*. It bears repeating that the emitter, or cathode, is always forward-biased, the collector always reverse-biased.

Theory of Amplification

When the emitter is forward-biased, its resistance to passage of electrons is low as compared to that of reverse biased material in the collector end. Consider Fig. 17 *(top)*, where the emitter is in series with the secondary winding of an input transformer. If the emitter is properly biased, the potential barrier at the junction between P-type emitter and N-type base is slightly reduced.

Attraction of the positive bias terminal releases many electrons, but

some combine with holes instead of continuing on, the volume of flow through the battery being somewhat less than the number set free at the potential barrier. A few of those emerging from the negative terminal of the emitter battery are attracted by the positive terminal of the collector battery, and choose this path instead of that through the base lead to the N-type wafer. They are negligible in amount, however, because of comparatively high resistance in the collector circuit.

Arrival of a positive signal at the upper terminal of the input transformer increases bias in the emitter circuit, thereby liberating a greater mass of

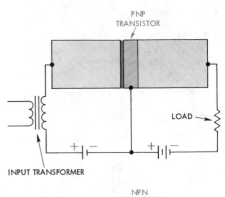

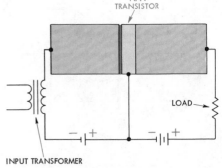

Fig. 17. Triode transistors with input transformers and loads.

barrier electrons and drawing some past the thin wafer of *N*-type material from the collector end. A proportionately larger quantity of electrons would pass normally into the collector circuit as they emerged from the negative emitter terminal, but the number is measurably increased by the attractive force of the collector element which seeks to draw electrons up through the load in order to replace those taken away from it.

Voltage drop in the load, under this condition, is much greater than voltage of the input signal, and amplification takes place. A comparable reaction occurs in the *N-P-N* transistor of Fig. 17 *(bottom)*. The potential barrier in the emitter circuit is formed with a surplus of positive charges on the left border of the juncton, negative on the right. Attraction of the positive emitter battery draws a few barrier electrons toward it, and establishes a low current. When an input signal makes the upper end of the transformer secondary negative, biasing in the emitter circuit increases, causing a surge of electrons across the barrier junction. Some of them move down the base lead to the positive terminal of the emitter battery. Others continue past the thin wafer of *P*-type metal under attraction of the positive collector bias, setting up a flow of current there. The comparatively larger voltage across the load represents amplification of the input signal, like in the *P-N-P* unit.

Comparison between Common-Emitter Transistors and the Electron Tube

Fig. 18(A) presents the schematic diagram of a common-emitter *P-N-P* transistor, Fig. 18(B) a common-emitter *N-P-N* transistor, and Fig. 18(C) a grounded cathode triode whose manner of operation is similar to that of the transistors.

Discussion to this point has been concerned mainly with the use of transistors as amplifiers, but they find wide application, too, as oscillators in radio, television, and computer circuits. The Hartley, crystal, and other common oscillators may be constructed with tran-

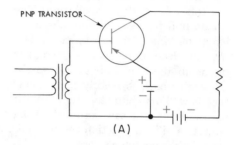

(A)

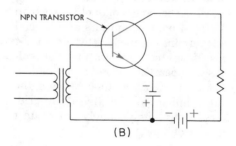

(B)

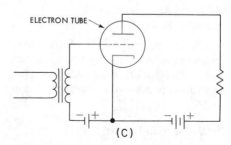

(C)

Fig. 18. Comparison of transistor and electron tube circuits.

sistors instead of electron tubes, usually with reduction in size, gain in efficiency, and overall simplification.

The *tetrode transistor*, which has a second base connection, has found limited application in the field of high-frequency operation. Many additional types have been developed to meet special requirements, such as the *unijunction* and the *unipolar transistors*, their operation founded upon principles brought out in this chapter.

Silicon diodes have largely replaced gas tubes for voltage regulation. The silicon unit possesses the unique quality of offering high resistance to passage of electrons as voltage across it builds up, then suddenly giving way at a specific point, to readily permit current flow. This characteristic makes it useful also, in the sawtooth generating circuit. A silicon junction rectifier has been developed, which normally exhibits great resistance to flow of current, but which becomes conductive when an electrical impulse appears at a special *gate* terminal in the emitter circuit. At this time, the potential barrier collapses, operation closely resembling that of a thyratron tube because the gate, similar to the thyratron grid, loses control immediately upon start of electron flow.

Additional Arrangements of Transistor Elements

Different arrangements of transistor elements known as the *common emitter, common collector,* and *common base* amplifiers are covered in detail in Chapter 15.

The Unijunction Transistor (UJT)

The unijunction transistor is a one-junction solid-state device. It is best used as an oscillator. The UJT oscillator (relaxation type) is described by its waveform, which is sawtooth in appearance.

The UJT is used in application with a capacitor. When charging, the output waveform of the UJT looks like a normal sine wave. When the capacitor is charged to maximum, current stops flowing through the UJT and the capacitor then discharges rapidly through the UJT. The UJT and the capacitor act together, producing a slow rise and a very quick descent of the waveform—hence the name *relaxation oscillator*.

The UJT can also be used to trigger silicon-controlled rectifiers (SCR's) at distinct time intervals. This allows the SCR to turn on circuit functions such as motor controls.

The UJT has three elements (see Fig. 19): an emitter lead and two base leads (1 and 2). It is formed from a large N-type silicon bar and a small P-type particle bonded to its side or top. When reverse voltage is applied across the emitter and base 1, the UJT transistor acts as a resistor. If forward voltage is

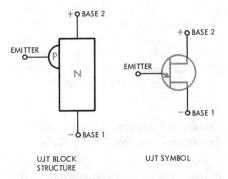

Fig. 19. Unijunction transistor (UJT), showing functional diagram of block-structure type, at left, and schematic diagram, at right.

applied to the emitter, the resistance between the emitter and base 1 decreases, therefore increasing current through the bases. More emitter voltage will cause more base current, less emitter voltage will cause less base current.

The actual construction of the UJT is in two basic forms: the bar and the cube structures. (See Fig. 20). In part *A* of the figure the PN junction is formed by alloying an aluminum emitter wire onto the N-type silicon bar nearest base 2. The transistor structure is mounted on a ceramic disc which expands with temperature at the same rate as the N-type silicon bar.

In part *B* of Fig. 20, the cube structure is illustrated. This structure is made by mounting an N-type silicon cube on a gold-plated header. Therefore base 2 is common to the transistor case and the header. Base 1 is made by alloying a wire to the top surface of the N-type silicon bar. The PN junction is formed by alloying an aluminum wire into the side of the N-type silicon bar. In both the bar and cube structures, the surface of the device is passivated and then hermetically sealed within a package.

Junction Field Effects Transistor (JFET)

The JFET can be used in small signal circuits in place of transistors. They are exceptionally useful in applications that demand a very high input resistance and can be turned on at zero voltage levels up to 0.5 volt.

The junction FET is a most useful solid state device because of its simplicity. It can handle huge input impedances of several million ohms. The JFET works on the principle that current conduction is controlled by varying an electrical field. In the Fig. 21, you will note that the JFET is made of one PN junction. A strip of N type material is sandwiched by a channel of P type material. A thin channel beneath the gate material provides a path between the material used for the source and drain connections. The PN junction is created at the interface of this thin channel.

When forward biased, current will flow readily through this thin strip because of decreased resistance. When reverse biased, current is controlled by the amount of reverse voltage as input resistance increases. Because of gate loading, a reduction in input signal may result. Another disadvantage is that there is current leakage across the PN junction as temperatures vary.

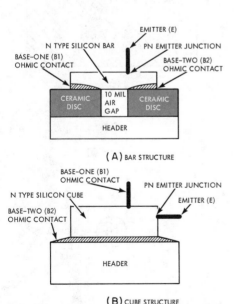

Fig. 20. Cross-sectional views of unijunction transistor (UJT) of two types: (A) the bar-structure type; and (B) the cube-structure type.
Courtesy General Electric Semiconductor Div.

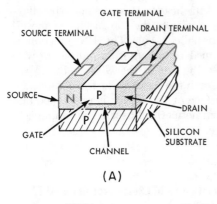

(A)

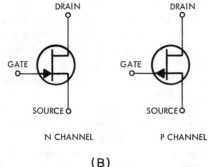

(B)

Courtesy RCA

Fig. 21. Cross section of the junction field effects transistor (JFET) in view (A) and schematic symbols for N- and P-channel devices in diagrams in (B).

Metal Oxide Field Effect Transistors (MOS/FET)

The MOS/FET can be used in place of transistors for even higher input resistances than the JFET previously discussed. However, the MOS/FET is not as sturdy as the JFET and can be destroyed by static discharge.

The MOS/FET works on the same principle as the JFET. That is, current conduction is controlled by varying an electrical field. The advantage that the MOS/FET has is that this device has its metallic gate insulated from the semiconductor surface by a layer of sili-

con dioxide. Thus we have a three-layer construction of metal, oxide, and semiconductor material. Thereby the letters naming MOS/FET are created.

By insulating the gate, an exceedingly high input resistance results. The insulation also serves as a dielectric. There are two basic types of MOS/FET: the *enhancement* type and the *depletion* type. In the enhancement MOS/FET (Fig. 22) no current flows with zero

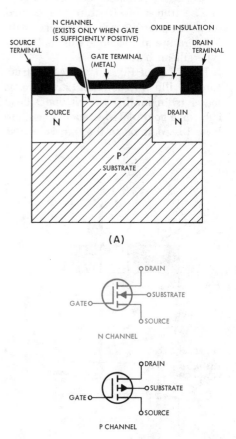

Courtesy RCA

Fig. 22. Cross section of the enhancement-type MOS/FET in view (A), and schematic symbols for N- and P-channel devices in diagrams in (B).

or reverse bias. The gate of this device must be forward biased with respect to the source in order for conduction to take place. If enough voltage is applied to the gate, the thin layer becomes saturated (changes in structure from P-type to N-type material) and conduction takes place between the source and drain. The enhancement type MOS/FET (Fig. 22), is very useful in digital and switching applications. More spe-cifically, it is used in large-scale inte-gration (LSI) devices.

The depletion type MOS/FET (Fig. 23) conducts with the gate at zero bias. That is, the thin channel underneath the gate provides a current path for conduction between the source and drain terminals. When the gate is reverse-biased to the source, the channel can be depleted of charge carriers. Therefore current conduction can be cut off if the potential on the gate is too high. Con-duction in the channel of a depletion MOS/FET can be increased by further forward bias increases. The Fig. 24 il-lustrates a typical MOS/FET in its case.

SOURCE TERMINAL

OXIDE INSULATION

DRAIN TERMINAL

N CHANNEL

GATE TERMINAL (METAL)

SOURCE N

DRAIN N

P SUBSTRATE

(A)

DRAIN

SUBSTRATE

GATE

SOURCE

N CHANNEL

DRAIN

SUBSTRATE

GATE

SOURCE

P CHANNEL

(B)

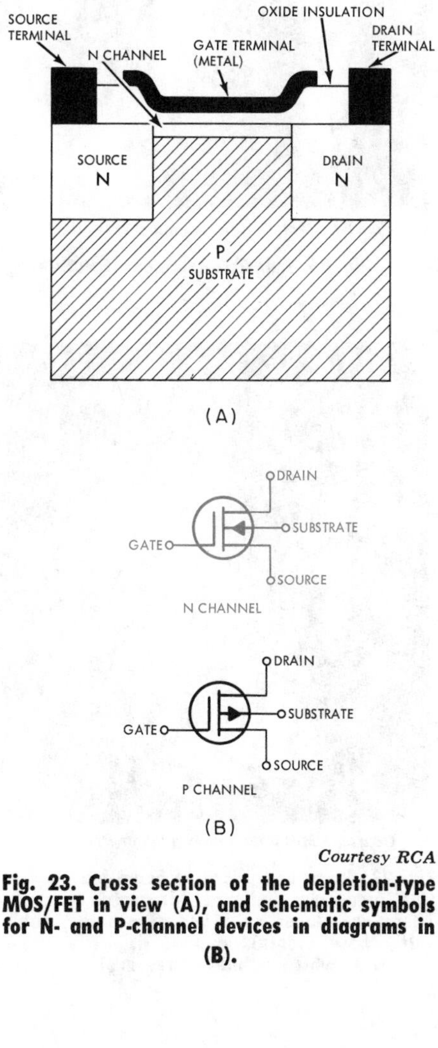

Courtesy RCA

Fig. 23. Cross section of the depletion-type MOS/FET in view (A), and schematic symbols for N- and P-channel devices in diagrams in (B).

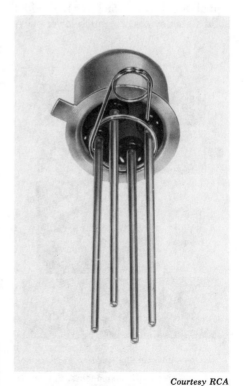

Courtesy RCA

Fig. 24. Typical MOS/FET in its case.

MICRO-MINIATURIZATION

The result of many years of research and testing has placed semiconductor manufacturers in a position to build just about what they want. They have defined the state of the art as micro-miniature and have accomplished astounding feats in the manufacture of these miraculous devices. In the Fig. 25 a photograph of some of these *chips* are shown. These are actual microcircuit components. The reader will note that the center chip is superimposed over a ten cent piece (dime), to show the actual size. Fig. 26 shows the slab construction which aids in mass production and allows the manufacturer to produce solid state devices at a much lower price.

Courtesy Motorola Semiconductor Products, Inc.

Fig. 26. Transistor slab construction.

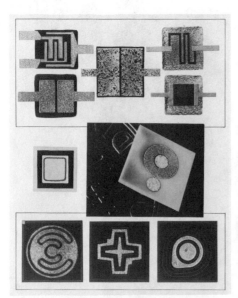

Courtesy Motorola Semiconductor Products, Inc.

Fig. 25. Microcircuit components. Entire circuits are only a fraction of the size of a dime, as can be seen in the center photograph.

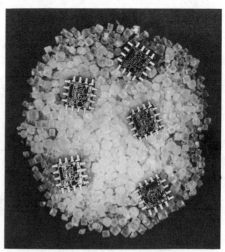

Courtesy Motorola Semiconductor Products, Inc.

Fig. 27. Magnified photo of beam lead operation amplifiers built as integrated circuits. These are placed on a bed of ordinary table salt whose crystals appear as little cubes (which actually they are).

You will note that a size comparison is made with several United States coins.

Integrated circuits (IC's) have been produced by a score of manufacturers. As an illustration of one of these IC's we have taken an operational amplifier with beam leads. The operational amplifier in Fig. 27 is sitting in table salt used in testing. (The cubic shape of the salt crystals shows the high magnification used in making the photograph.) The structures extending from the IC are the beam leads, made of gold. They are bonded to the IC internally, providing a reliable interconnection. The operational amplifier shown consists of 13 transistors, 2 diodes, and 15 resistors of various sizes. This is a typical IC. There are many types used for many purposes.

A book at least the size of the present text would be required to cover the designs, manufacturing processes, and applications of these tiny circuits which more and more are replacing solid-state circuits made of *discreet* (individual) devices discussed in this chapter. IC's appear to indicate the direction of electronics in the near and forseeable future.

OTHER SOLID-STATE DEVICES

Metallic-Oxide Rectifiers

Chapter 2 revealed that materials in general are divided into conductors and insulators. Electrons in outer shells of conductor atoms are loosely held, readily freeing themselves from attraction of the nucleus. Electrons in outer shells of insulator atoms, on the other hand,

are so tightly held that, under normal conditions, they cannot escape.

Although certain borderline materials were recognized as having both fair conducting and fair insulating properties at the same time, little practical use was made of this knowledge until the past few years. *Metallic oxides,* especially those of copper and selenium, are two such materials, behaving like conductors where electron flow is established in one direction, and behaving almost like insulators where electrons are forced in the opposite direction.

If a small copper disk, Fig. 28, is brought to a high temperature while one side is exposed to the air and the other is protected from it, a coating of *copper oxide* forms on the exposed face. Placed in circuit with a dry cell, electrons are found to pass readily from copper to copper oxide inside the disk,

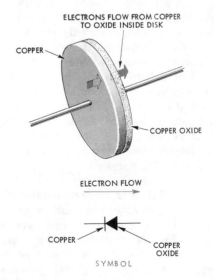

Fig. 28. Copper oxide rectifying disk and symbol.

but to move only with difficulty from *copper oxide* to copper. *Selenium oxide,* when sprayed onto a base of iron or aluminum, forms a combination with properties similar to those of the copper-oxide disk, and somewhat more efficient because its resistance in the forward, or readily conducting direction is lower.

Either unit, when connected in an alternating-current circuit, provides a means of rectification, one alternation passing freely, the other only partially. Where employed for low-voltage applications, reverse current is so small, compared to forward current, that acceptable rectification is achieved. Because some current does flow in the opposite direction, however, these units are inferior to the vacuum tube.

Photoelectric Semiconductors

Both copper oxide and selenium oxide show sensitivity to rays of light, particularly the latter. Practically all semiconductors show like properties, some to a much higher degree. Many are *photoconductive,* that is, their resistance is decreased when they are exposed to light. Others are both photoconductive and *photovoltaic,* this term indicating generation of a voltage under influence of light rays. Selenium is the best known photovoltaic substance, having been employed for years in photographic light meters.

Germanium diodes and triodes have proven efficient in photoconductive circuits. Fig. 29 *(top)* represents a germanium diode applied to this service, while Fig. 29 *(bottom)* shows a germanium triode. The rays must be focused accurately upon the potential barrier area in all cases, so they may provide energy at the precise spot

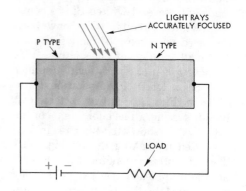

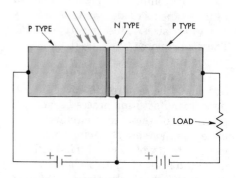

Fig. 29. Light sensitive semiconductors.

where release of electrons can start a flow of current. The light-sensitive elements are confined in glass or transparent plastic receptacles.

Saturable Reactor

Discussion of the *saturable reactor,* Fig. 30 (A), was deferred in the preceding chapter, when dealing with the subject of phase-shifting methods. It consists of a three-legged magnetic core like that of a transformer, upon which three coils are wound. Those on the outer legs are in series with load and alternating current supply wires. The one on the middle leg is in a circuit of its own which provides variable direct-current excitation.

Current through the outer coils builds up an alternating magnetic flux which saturates the core in these legs, causing domains to change alignment from one direction to the exact opposite with each alternation. Under this condition, the inductance of each coil is at the highest point, limiting the amount of current supplied to the load.

When the middle coil is excited by

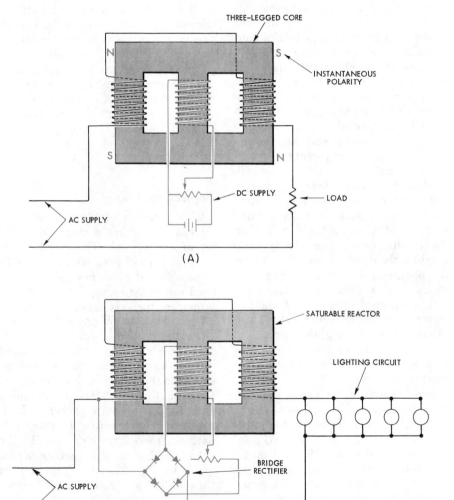

Fig. 30. Saturable reactor and circuit.

direct current it establishes a steady flux in all three legs of the unit, holding many domains in a set position, and thereby reducing the number affecting the alternating current. Suppose the inductance produces a counter-emf of 100 volts. Direct current is now sent through the middle coil, ampere-turns creating a flux density one-half *saturation value*. Thus, direct-current flux takes control of half the total number of domains in outer legs of the core. Lines of force from these domains will not generate counter-emf because they do not cut across turns of wire, but maintain unvarying direction. Domains still under influence of alternating flow will constitute only one-half as many as before, and counter-emf will be reduced fifty percent.

As the inductance of the outer coils is lowered by means of varying the control resistor, more and more current passes to the load, its value limited only by resistance of the circuit when the reactor core is completely saturated by the direct-current flux. Variation of current was less important, with regard to thyratron grid voltage, than was phase displacement. Yet, the two factors are closely related, phase displacement decreasing as current increases, because greater current flow results only from lowering of inductive reactance.

Instantaneous polarities of coils on the outer legs are as indicated in the illustration, the left one having an *N* pole at the top, the right one an *S* pole. With this arrangement, flux set up by the coils circulates only through the two outer legs and the cross members, none of it passing down through the middle leg. Flux from the middle coil, on the other hand, divides at top and bottom, equal amounts passing through

either leg. The middle coil usually has a great number of turns so that only a fraction of an ampere is sufficient to create maximum saturation of the core.

Fig. 30 (B) shows a common application of the saturable reactor. Direct-current supply for the middle coil is obtained here from a bridge rectifier. Outer coils of the reactor are in series with a theatre lighting circuit. If the operator in a remote projection booth wishes to increase lighting intensity, he turns the knob of a small rheostat to increase direct-current excitation. If he wishes to dim the lights, he merely decreases the direct-current excitation until the desired state is obtained.

Magnetic Amplifier

The principle of the saturable reactor is utilized in the *magnetic amplifier* of Fig. 31 (A). The core is made of special alloy steel which is highly permeable and which has a sharply defined *saturation curve*, Fig. 31 (B). A saturation curve is drawn by marking off ampere-turns of magnetizing force along the base line and resulting flux densities above. Flux values are noted on the curve according to values shown on the vertical axis of the figure. Coils on the outer legs are supplied with *high-frequency current* which magnetizes the core to a point near saturation, marked working area in the enlarged view of this portion of the curve in Fig. 31 (C). The coil on the middle leg is excited by the signal which is usually alternating current rather than direct.

The amplification process resembles that studied in connection with the

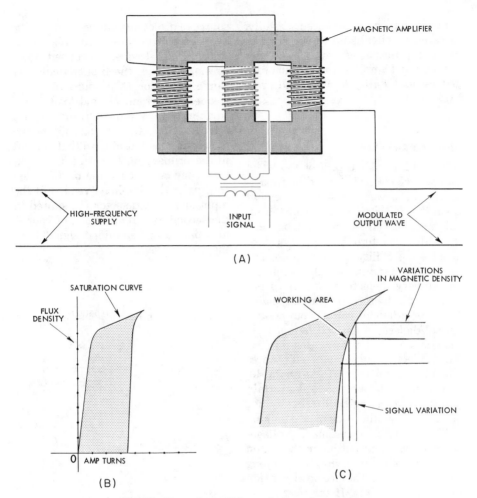

MAGNETIC AMPLIFIER

HIGH-FREQUENCY
SUPPLY

INPUT
SIGNAL

MODULATED
OUTPUT WAVE

(A)

SATURATION CURVE

FLUX
DENSITY

0 AMP TURNS

(B)

VARIATIONS
IN MAGNETIC DENSITY

WORKING AREA

SIGNAL VARIATION

(C)

Fig. 31. Magnetic amplifier and saturation curve.

electron tube. Within the working area indicated on the curve, a slight variation of magnetizing force produces considerable variation in *magnetic density*. Flux created by the middle coil adds to or subtracts from that of the outer coils according to momentary polarity of the signal wave, thus modulating the high frequency wave. Some form of detector is needed in the output circuit to separate the amplified signal from the carrier wave. A diode tube or a solid-state rectifier serves here the same as in a radio receiver.

The magnetic amplifier is not suited to extremely high frequency applications because of high inductive reactance caused by presence of the iron

295

core. It has been used successfully, however, on circuits up to 800 hertz or more. Its principal advantage over other conventional amplifiers is extreme ruggedness and ability to withstand overload.

Dielectric Amplifier

The ability of a capacitor to attract and hold electrons is termed its capacitance. This power rests to a great extent upon the *dielectric constant* of the insulating material, or *dielectric,* between the plates. The term dielectric constant denotes the ability to line up electric dipoles in response to plate voltage. In ordinary dielectrics such as paper or mica, the dielectric constant remains fixed, regardless of voltage, but certain special dielectrics, termed *ferroelectric* materials, do not possess this quality. The dielectric constant of substances like strontium titanate is unstable, reaching the highest degree at a certain voltage, but changing drastically for variations on either side of maximum.

Fig. 32 (A) represents a voltage-dielectric-constant outline in the region of extreme sensitivity, the curve sloping sharply for some distance on either side of its highest value. If the material is subjected to a steady voltage which "polarizes" it to establish a dielectric constant in the vicinity of M, a relatively small voltage may be applied, as indicated on the steep part of the curve, to provide amplification as with the electron tube. It may be seen in the figure that a 2-volt sine wave added to or subtracted from the set voltage produces a large variation in the dielectric constant.

The basic circuit for a *dielectric amplifier* is illustrated in Fig. 32 (B). It consists of an input transformer, T_1, a

source of direct-current polarizing voltage V, a pair of choke coils, X_L, which prevent high-frequency current from passing through the input circuit, ceramic capacitor C, a source of high-frequency current, H, and load transformer, T_2. When an input signal arrives at T_1, this voltage alters the dielectric constant of C so that current in the primary of T_2 varies, the high-frequency current outline modified according to the amplified version of the input signal. A detector is required in the secondary circuit of T_2, of course, to separate the amplified signal from the carrier wave.

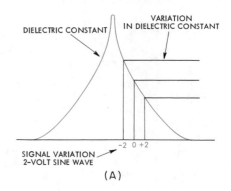

(A)

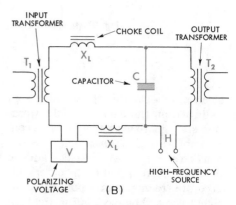

(B)

Fig. 32. Dielectric amplifier and operating curve.

Magnetoresistive Amplifier

When a conductor lies in a magnetic field, Fig. 33 (A) electrons flowing through it tend to deviate in accordance with the familiar motor rule. Here, electrons flowing from left to right in a rectangular conductor, crowd toward the K surface. This development is hardly detectable with ordinary conductors, but a scientific investigator named Hall discovered it to be quite noticeable in certain types of semiconductors.

The *Hall-effect* is employed in the *magnetoresistive amplifier* of Fig. 33 (B). A thin rectangular slab of strongly-resistant *indium* alloy is placed between poles of an electromagnet whose excitation is obtained from a coil supplied by the input signal. As current from a direct-current source passes through the slab from one end to the other, some of the electrons are deflected upward, creating an electrical pressure between top and bottom surfaces. Electron flow in the output circuit provides an amplified version of the minute input control signal.

Transducers

A *transducer* is a device for changing the form of energy. An electric generator, for example, converts mechanical energy of the driving engine into electrical energy. A loudspeaker transforms electrical energy into sound. Solid-state transducers usually provide connecting links between physical motion and recording devices. Quite often, they furnish input signals for analog computers.

Generally, these devices may be divided into tension and compression units. Fig. 34 (A) illustrates the principle of a *tension transducer*. A pattern of hair-fine resistance wire is attached to a flexible cloth or paper surface, the wire in series with a source of direct current. One end of the flexible sheet is fixed, the other attached to the apparatus whose tension is to be measured. Stretching or contracting of the supporting material changes the cross-sectional area of the wire, and consequently its resistance.

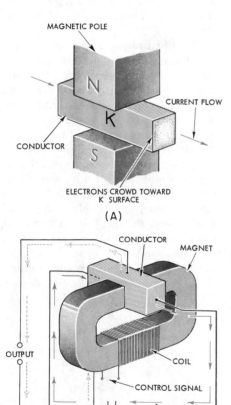

Fig. 33. Hall effect amplifier.

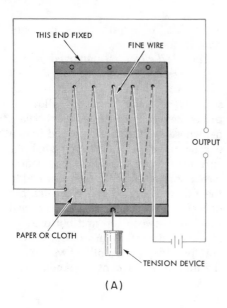

THIS END FIXED

FINE WIRE

OUTPUT

PAPER OR CLOTH

TENSION DEVICE

(A)

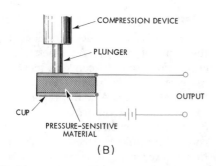

COMPRESSION DEVICE

PLUNGER

OUTPUT

CUP

PRESSURE-SENSITIVE
MATERIAL

(B)

Fig. 34. Tension and compression transducers.

The *compression transducer,* Fig. 34 (B), has a movable element or plunger, and a fixed cup that contains one of the fourteen rare earth metals the resistance of which varies widely under small differences in pressure. The apparatus whose pressure changes are to be noted, is attached to the plunger, and the whole unit is placed in series with output terminals and the direct-current supply. Pressure changes are thus transformed into current changes in the output circuit.

The Thermistor

Fig. 35 shows a simple device that has been found useful in a number of applications. A *thermistor* is a semiconductor which is extremely sensitive to heat and has an atomic vibrational structure such that free-electron paths through the material are severely disturbed. There are two types of thermistors, one having a highly positive temperature coefficient of resistance, the other negative. Resistance of the positive type increases with flow of current; that of the negative type decreases.

The illustration represents a negative thermistor, T, in series with the coil of relay P. When the switch is first closed, resistance of T is so high that very little current passes through the relay coil, and its contacts remain open. After an interval of time, the low current generates enough heat to lower T's resistance and thus permit the relay to operate. This arrangement provides a time-delay feature. If T were a *positive type,* its low initial resistance would allow the relay to close at once, but it would open again as T's resistance mounted. This type of operation would be suitable for many applications, one example being a sign flasher.

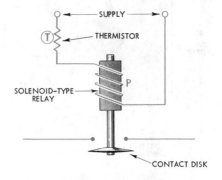

SUPPLY

THERMISTOR

SOLENOID-TYPE
RELAY

P

CONTACT DISK

Fig. 35. Thermistor and relay.

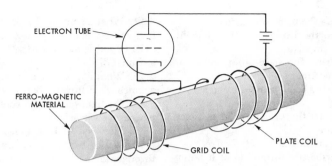

ELECTRON TUBE

FERRO-MAGNETIC
MATERIAL

PLATE COIL

GRID COIL

Fig. 36. Magnetostriction oscillator.

Magnetostriction Oscillator

Operation of the *magnetostriction oscillator* is based on the knowledge that ferromagnetic substances undergo dimensional changes when magnetized. Length increases, while cross-sectional area decreases, volume remaining approximately constant. Two solenoid coils are wound upon a ferromagnetic bar, Fig. 36, one connected between cathode and grid of an electron tube, the other in the cathode-plate circuit.

Plate current magnetizes the bar. As a result, the bar lengthens, its lines of force cutting across turns of the grid coil. This voltage induces a grid current, creating ampere-turns which oppose plate magnetization, and thus cause the material to shrink. Lines of force now cut across the turns of the grid coil in the other direction, reducing the flow of grid current. Oscillation is established, in this manner, its frequency in the high audio range. A transistor may be substituted, of course, for the electron tube. The magnetostriction oscillator is employed in sound navigation and ranging equipment.

REVIEW QUESTIONS

1. How do electrons flow inside a copper-copper-oxide disk?

2. Draw a diagram of a bridge-type rectifier.

3. Define the term semiconductor.

4. What is a co-valent bond?

5. How is germanium doped?

6. What is the essential difference in atomic structure between germanium and arsenic?

7. Distinguish between donors and acceptors.

8. Is *P*-type germanium positively charged?

9. Explain hole-conduction.

10. What is a transducer?

11. How are the batteries connected to provide forward biasing in a *P-N* diode?

12. What are the battery polarities for reverse biasing of an *N-P-N* triode?

13. Define the term transistor.

14. Name the three essential elements of a triode transistor.

15. Describe the base-input, collector-output transistor.

16. State the basic principle of a saturable reactor.

17. Describe the operation of a dielectric amplifier.

18. What is meant by *Hall-effect?*

19. Describe the formation of a potential barrier.

20. For what purpose is a zener diode most useful?

21. What is meant by *avalanche point* in a zener diode?

22. What is meant by the Esaki effect in a tunnel diode?

23. What happens to a varactor diode reactance when forward bias is applied?

24. Do light-emitting diodes have to be placed in sockets?

25. How is a LED installed in a circuit?

26. What happens to current flow through an SCR after you remove gate voltage?

Chapter 15

Solid State Circuits

DIODE CIRCUITS

As was the case with the electron tube diode, we find that the solid state diode does not amplify. It does, however, do a fine job of rectifying. Solid state power supplies are a vast improvement over their electron tube counterparts. The reasons for this are quite numerous. Weight alone is a prime factor in choice of equipment, especially for airborne and space applications. There is no problem of warm-up time. The moment the solid state power supply is turned on, full power is available to the load. Solid state rectifiers operate at a low temperature compared to electron tube rectifiers.

In the next several paragraphs we will deal with the diode as used in various power supply circuits. A comparison should be made to electron tube power supplies in Chapter 13 as the reader delves into this chapter.

Half-Wave Rectifier (See Fig. 1)

The half-wave rectifier is the simplest of all the power supplies. It consists of a transformer and one rectifier diode. The transformer is a step-down type because power for the circuit usually

comes from a 115 VAC or a 220 VAC source. Solid state circuits are typically biased by low voltage d-c (example 15 VDC). Therefore the a-c voltage must be transformed to a low level prior to rectification.

Transformers for power supplies are chosen to have a secondary average voltage output that is near the d-c level required after rectification. Transformers are also chosen to be able to withstand the current demands of the load.

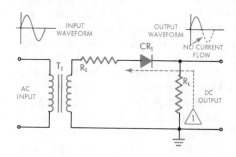

ANALYSIS:

⚠ POSITIVE ALTERNATION of input sine wave forward biases CR1 and electron current will flow through the diode and load.

2 NEGATIVE ALTERNATION of input sine wave reverse biases CR1 and electron current stops flowing.

Fig. 1. Half-wave rectifier.

15

Diodes are chosen for their current capability and peak inverse voltage rating (PIV). Peak inverse voltage is defined as the maximum reverse-bias voltage that may be accepted by the rectifier. This voltage may vary somewhat with temperature. Actual construction of the power supply requires that the diode be installed on a heat sink to help dissipate some of the heat generated by current flow.

The half-wave rectifier is not very efficient. The reason for this, is that the average d-c derived by the circuit is low. In fact the average voltage is only half that of the full-wave rectifier. Without filtering, the circuit is very inefficient to use under any application. Pulsating d-c output ripple is determined by input frequency.

The circuit shown in Fig. 1, is a typical half-wave rectifier. The transformer secondary has one lead tied to reference (ground). The second lead has a resistor R_s in series with the diode. The resistor is a current-limiting resistance. Its purpose is to limit the current flowing through the diode. If the secondary d-c resistance in the transformer is great enough, the resistor is not needed.

On the positive alternation of input voltage the diode is forward biased and current flows through the diode and the load. On the negative alternation of the input voltage, the diode is reverse-biased and current stops flowing except for minor reverse current. This arrangement gives pulsating d-c output in the positive direction. To obtain a negative d-c pulse, the diode is placed in the circuit in the opposite direction.

Full-Wave Rectifier (See Fig. 2)

The full-wave rectifier is similar in operation to the half-wave rectifier. The transformer utilized has a center tap.

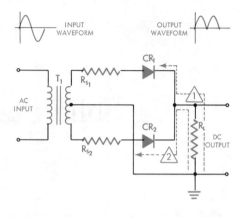

ANALYSIS:

⚠️ 1 POSITIVE ALTERNATION of input sine wave forward biases CR1. Electron current flows through CR1 and the load. CR2 is reverse biased on this alternation.

⚠️ 2 NEGATIVE ALTERNATION of input sine wave forward biases CR2. Electron current flows through CR2 and the load. CR1 is reverse biased on this alternation.

Fig. 2. Full-wave rectifier.

Each leg of the center tap should be capable of an average a-c voltage near the level expected of the d-c level output. The center tap is tied to reference (ground).

Two diodes are used. They are placed in the same direction in series with the transformer secondary winding and series current-limiting resistors. These limiting resistors R_{S1} and R_{S2} are used to limit current through diodes CR_1 and CR_2 if the secondary winding does not have enough resistance.

The full-wave rectifier is much more efficient than the single wave because it utilizes the secondary of the transformer for 360 degrees of the input sine wave. Current flows through one diode for one half cycle and the opposite diode for the second half cycle. Transformer and diodes are chosen in the same manner as in half-wave rectifiers. Diodes in full-wave rectifiers must be mounted on heat sinks as in the single wave configuration.

Output of the full-wave rectifier is usually filtered, then regulated. Pulsating d-c output ripple is twice the input frequency.

The circuit shown in Fig. 2, is a typical full-wave rectifier. On the positive alternation of the input sine wave, the diode CR_1 is forward-biased and the diode CR_2 is reverse-biased. Current flows from the center tap (ground) through the load resistor, the diode CR_1, and the current limiting resistor R_{S1} back to ground. On the negative alternation of the input sine wave, the diode CR_2 is forward-biased and the diode CR_1 is reverse-biased. Current flows through the load resistor, the diode CR_2, and the current limiting resistor R_{S2} back to ground.

The full-wave rectifier arrangement provides pulsating d-c. The d-c power is fed normally through a filter and a regulator. The full-wave rectifier is used in many high-voltage configurations where peak inverse rating is an important factor. In this event a pair of diodes may be placed in series with each winding to prevent rectifier breakdown.

Full-Wave Bridge Rectifier (See Fig. 3)

The bridge rectifier is a full-wave rectifier. It relies on two rectifiers, one on each arm of the transformer, to conduct together on each alternation. The full-wave bridge rectifier does not require a center tap as does the conventional full-wave rectifier. It conducts during the full secondary output of the transformer. There are four diodes used in the bridge. Each diode is connected in series with and operates at the same time as the diode on the opposite transformer winding.

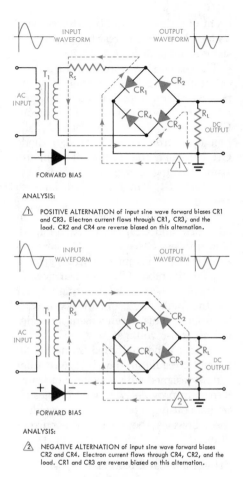

ANALYSIS:

⚠ POSITIVE ALTERNATION of input sine wave forward biases CR1 and CR3. Electron current flows through CR1, CR3, and the load. CR2 and CR4 are reverse biased on this alternation.

ANALYSIS:

⚠ NEGATIVE ALTERNATION of input sine wave forward biases CR2 and CR4. Electron current flows through CR4, CR2, and the load. CR1 and CR3 are reverse biased on this alternation.

Fig. 3. Full-wave bridge rectifier.

In Fig. 3, the reader will see that CR_1 and CR_3 are forward-biased when CR_2 and CR_4 are reverse-biased. This arrangement allows the bridge rectifier to require only one half the peak inverse rating of the conventional full-wave rectifier during conduction. The other advantage is that the bridge rectifier utilizes the entire secondary voltage at all times, making it more efficient. The current-limiting resistor R_S is used to limit current through the diodes in the event of a short. If the transformer secondary has enough resistance, this resistor is not necessary.

The circuit shown in Fig. 3, is a typical full-wave bridge rectifier. On the positive alternation, the diodes CR_1 and CR_3 are forward-biased. The diodes CR_2 and CR_4 are reversed biased. Current flows from ground through diode CR_1, through the current limiting resistor R_S, the transformer secondary, the diode CR_3, and the load, back to ground.

On the negative alternation, the diodes CR_2 and CR_4 are forward-biased. The diodes CR_1 and CR_3 are reverse-biased. Current flows from ground through the diode CR_4, the transformer secondary, the current limiting-resistor R_S, the diode CR_2, and the load, back to ground. This arrangement provides pulsating d-c voltage.

The d-c is normally fed through a filter and a regulator. The bridge rectifier can be designed with a center-tap transformer to provide two load voltages simultaneously. This circuit is seldom used, however, because of the current drain.

Voltage Doubler Circuit (See Fig. 4)

The voltage doubler consists of a full wave rectifier with parallel capacitors and bleeder resistors. The capacitors charge as conduction takes place through the diode at a rate determined by the resistor R_S and the capacitance time constant.

Bleeder resistances R_{B1} and R_{B2} determine the discharge time constant. They are normally very large (two or three megohms). All components are of equal value for equalization. The reason for this is that the capacitor must not discharge too much during the alternate cycle. It must keep its charge because

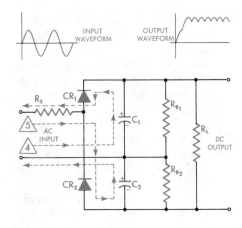

ANALYSIS:

1 CR1 and CR2 act as a full wave rectifier.

2 C1 and C2 charge on alternating half cycles.

3 RB1 and RB2 are stabilizers and act as bleeder resistors when power is turned off.

4 ON POSITIVE ALTERNATION, CR1 is forward biased and C1 is charging at a time constant equal to values of C1 and RS. CR2 is reverse biased on this alternation.

5 ON NEGATIVE ALTERNATION, CR2 is forward biased and C2 is charging at a time constant equal to values of C2 and RS. CR1 is reverse biased on this alternation.

6 RB1 and RB2 are very large values (2 or 3 megohm) therefore, C1 and C2 discharge very little during alternate cycles.

Fig. 4. Voltage doubler circuit.

the output voltage is felt in parallel to both capacitors.

Voltage triplers and voltage quadruplers are also available. These two circuits are used in small current applications and are very inefficient because of power losses and very poor regulation.

The circuit in Fig. 4, requires the following analysis. On the positive alternation of the input sine wave, diode CR_1 is forward-biased and diode CR_2 is reverse-biased. Current flows through the diode CR_1 and the capacitor C_1 is charging at a time constant $TC=RC$ (where TC = time constant; R = resistance; and C = capacitance) equal to the values of C_1 and resistor R_S.

On the negative alternation of the input sine wave diode CR_2 is forward biased and diode CR_1 is reverse-biased. Current flows through the diode CR_2 and the capacitor C_2 is charging at a time constant $(TC=RC)$ equal to the values of C_2 and the Resistor R_s. During the negative alternation the capacitor C_1 tries to discharge but can't because of the large value of R_{B1}. During the positive alternation the capacitor C_2 tries to discharge but can't because of the large value of R_{B2}. The load is in parallel with the bleeder resistances, therefore the doubling action is felt as load voltage. If a meter were placed across the load, the meter would be indicating the charge on the two capacitors.

Filtering (See Fig. 5)

Filtering is a *must* for the solid state d-c power supply. The filter comes in many forms and arrangements and is used to smooth out the ripple voltage caused by rectification. Use of the filter is dependant on space, d-c ripple requirements, and cost.

If a capacitor is used for filtering it is placed in parallel with the load. As you know, the capacitor opposes a change in voltage, so a constant voltage is seen on the load. The larger the capacitor is, the better the filtering capability. However, this causes the capacitor to be heavy and costly.

The choke (inductor) is used in series with the load. As you know, the choke (inductor) opposes a change in current. The capacitor is sometimes used as a filter by itself. (Figs. 5A and 5B.)

Other arrangements are the *L*-filter using one choke in series and one capacitor in parallel, the *T*-filter using two chokes in series and a capacitor between them in parallel, and the *pi* filter using

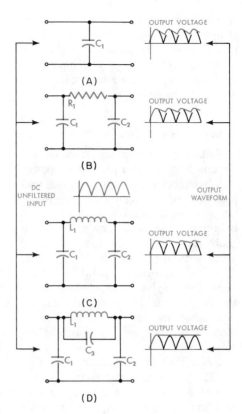

Fig. 5. Filtering.

two capacitors in parallel with a choke in series between them.

Each does a respectable job of opposing current change or voltage change. The choke filter has a disadvantage in that output voltage tends to be lower than with capacitor filters. The capacitor filter draws current. The two together then (Fig. 5C) are the most efficient. The best filter is the tuned filter (Fig. 5D). The capacitor and choke (inductor) have equal impedances and produce practically ripple-free d-c.

Ripple-free d-c is demanded in many solid state circuits. Ripples in power may cause switching transistors or

SCR's in computers to fire accidentally. Ripple d-c can cause noise in audio and radio circuits. In some circuits where exact d-c levels are required, ripple voltage may even cause some solid state devices to be destroyed.

Voltage Regulator Circuits

Voltage regulators are required for circuits that demand a constant power supply output while input voltages and load demands change. There are many types of regulators used with a-c to d-c converters. The series regulator and the shunt regulator are the most common, along with the zener diode regulator. Since the coming of solid state devices, the breakdown diode has taken the limelight in power regulation. For this reason we have utilized the zener diode in our explanation of solid state regulators.

Series Voltage Regulator (See Fig. 6). The series regulator consists basically of a power transistor Q_1, a current-limiting resistor R_1, and a zener (voltage regulator) diode CR_1. The transistor used in the illustration is a PNP power transistor. It is chosen for its current-carrying capabilities. The power transistor must be placed on a heat sink when installed in the power regulator circuit. The heat sink is made of soft metal which absorbs heat. The zener diode on the base of the regulator, along with the load voltage, provides the base-emitter bias to control the transistor's operation. Note that the transistor is placed in the circuit in series with the load.

Bias for the transistor Q_1 is provided by the load voltage and the zener diode CR_1. If the voltage increases across the load, the bias current on the transistor decreases, current through the transistor

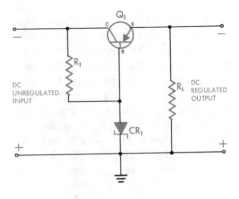

ANALYSIS:

1 CR1 and load voltage provide bias to Q1.

2 R1 limits current for the fixed voltage zener doide.

3 If load voltage increases, bias current on Q1 decreases, current through Q1 decreases, voltage drop across Q1 increases, and load voltage lowers to normal.

4 If load voltage decreases, bias current on Q1 increases, current through Q1 increases, voltage drop across Q1 decreases and load voltage rises to normal.

Fig. 6. Series voltage regulator.

decreases, therefore the voltage drop across the transistor increases, causing the load voltage to lower back to normal.

In the event of a decrease in load voltage, bias current on the transistor Q_1 increases, current through the transistor increases, therefore the voltage drop across the transistor decreases, causing the load voltage to rise back to normal.

It must be noted that all these events take place simultaneously, and the moment a change is felt, regulation takes place. If you were to put a meter across the load it would be difficult to see regulation happening.

Shunt Voltage Regulator (See Fig. 7). The shunt regulator consists of a current-limiting resistor R_1, a zener (voltage regulator) diode CR_1 and a power transistor Q_1. The transistor used in the illustration is a PNP power tran-

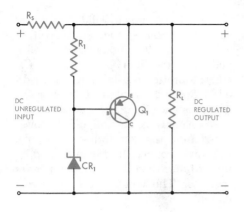

ANALYSIS:

1. R1 and load voltage provide bias to Q1.

2. CR1 is a fixed voltage zener diode.

3. If load voltage increases, voltage drop across R1 increases, bias current increases, current through Q1 increases, voltage drop across Q1 decreases, and load voltage lowers to normal.

4. If load voltage decreases, voltage drop across R1 decreases, bias current decreases, current through Q1 decreases, voltage drop across Q1 increases, and load voltage rises to normal.

5. RS provides short protection for the unregulated input.

Fig. 7. Shunt voltage regulator.

sistor. It is chosen for its current-carrying capabilities. The power transistor here must be placed on a heat sink as in the series regulator circuit. The zener diode is installed in the base circuit, to provide a reference for bias. Bias for base emitter junction is provided by the load voltage and current-limiting resistor R_1. Note that the transistor is placed in the circuit in shunt (parallel) with the load.

If the voltage across the load increases, the bias current increases, current through the transistor increases, and the voltage drop across the transistor decreases, causing the load voltage to lower back to normal.

In the event of a decrease in load voltage, the bias decreases, current through the transistor decreases, therefore the voltage drop across the transistor increases, causing the load voltage to rise to normal.

It must be noted that all these events take place simultaneously, and the moment a change is felt, regulation takes place.

Zener Diode Regulator. (See Fig. 8) The zener diode is sometimes used alone for voltage regulation. When using zener diodes by themselves for voltage regulation, a limiting resistor R_S is always placed in series with the load and the shunted zeners. In this way the unregulated output is protected from overload current.

The power rating of the limiting resistor R_S must be very high because it carries the load current and the current through the zeners. This makes the efficiency of the shunted zener regulator very low. Its simplicity makes up for its inefficiency. The zener diode is chosen for power-carrying capacity and its voltage setting. Large current zeners require heat sinks.

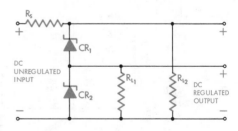

ANALYSIS:

1. CR1 and CR2 are voltage regulator zener diodes in series with each other and in parallel with the load.

2. Any change in load voltage will be felt on RL. Voltage across zener diodes always remains the same.

3. RS provides short protection for the unregulated input.

4. Load no. 1 (RL1) is picked off center tap between zener diodes.

5. Load no. 2 (RL2) is across both zener diodes.

Fig. 8. Zener diode regulator.

The figure shows the limiting resistance R_S in series with two zener diodes CR_1 and CR_2 and two loads. The zeners are regulating two load voltages at set levels. If the unregulated d-c supply is large enough, it is possible to place several zeners in series with each other and provide multiple pickoffs similar to the ones illustrated.

**Complete Power Supply Circuit
(See Fig. 9)**

Fig. 9 is a complete solid-state power supply. It consists, from left to right, of a transformer T_1, diode rectifiers CR_1 and CR_2, the filter $C_1 - L_1 - C_2$, current-limiting resistance R_1, zener diode CR_3, and the power transistor Q_1. The resistance R_L represents the load.

As an analysis, unregulated a-c power is fed from the transformer through a full-wave rectifier. It is then filtered by the *pi* filter, then regulated by a series regulator. The load is therefore d-c regulated.

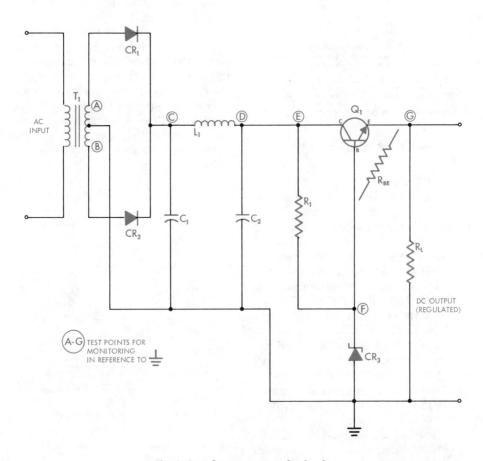

Fig. 9. Complete power supply circuit.

Test points *A* through *G* are shown on this circuit to show the reader where he would monitor the different devices in a trouble-shooting situation.

TRANSISTOR AMPLIFIER CIRCUITS

Transistor Amplifier Configurations (See Fig. 10)

The circuits shown in Fig. 10 are the three basic transistor amplifier configurations. All components except the transistor have been omitted from the diagrams for simplicity.

The configuration in Fig. 10 (A) is the common-emitter (CE) type, so called because the emitter element is common to both the input signal and the output signal. The common emitter is the most utilized configuration because of its high gain capability.

The common-collector (CC) configuration in Fig. 10 (B) is so called because its collector is common to both input signal and output signal. The common collector is best utilized to isolate large input impedance to low output impedance, thereby replacing a step-down transformer.

The common-base (CB) configuration in Fig. 10 (C) is so called because its base is common to both input signal and output signal. The common base configuration has massive voltage gain and the advantage of a large output resistance.

Transistor Biasing Techniques (See Fig. 11)

As stated previously, the forward junction of the transistor configuration must be forward-biased, while the output junction must be reverse-biased.

In Fig. 11, the (A) configuration represents a common-emitter amplifier. The reader will note that +5 volts are applied to its base thereby forward-biasing the input junction. Note also that +9 volts are applied to its collector, thereby reverse-biasing its output junction.

In the (B) common-collector configuraton +5 volts are applied to the base and +9 volts are applied to the collector. This forward-biases the input junction and reverse-biases the collector.

Finally, in the (C) common-base configuration, −5 volts are applied to the emitter and +9 volts are applied to the collector, reverse-biasing the output junction. This forward-biases the input junction and reverse-biases the output junction.

The voltage values shown here are representative and show polarity, but should not be misconstrued as actual value.

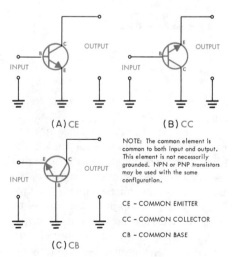

NOTE: The common element is common to both input and output. This element is not necessarily grounded. NPN or PNP transistors may be used with the same configuration.

CE – COMMON EMITTER

CC – COMMON COLLECTOR

CB – COMMON BASE

Fig. 10. Transistor amplifier configurations.

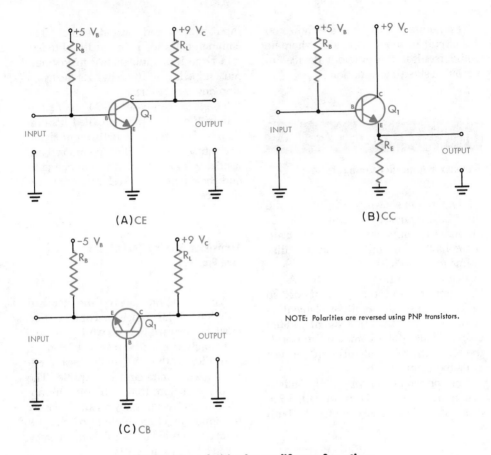

Fig. 11. Bias polarities for amplifier configurations.

Amplier Input and Output Resistance Comparisons (See Fig. 12)

Fig. 12 illustrates input and output resistance comparisons within the transistor configurations.

The CE amplifier has low input resistance and high output resistance. The CC amplifier has very high input resistance and very low output resistance. The CB amplifier has very low input resistance and very high output resistance.

Input and output resistance are important because they must match the input signal which varies with the signal generator such as a phonograph cartridge, a carbon microphone, or a tape head. Output resistances must be matched with second stages as *cascades*, isolated to other low-impedance circuits or fed to speakers or other listening devices.

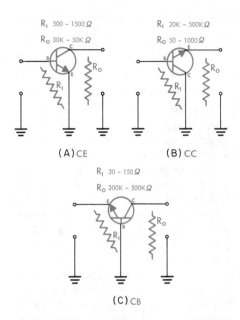

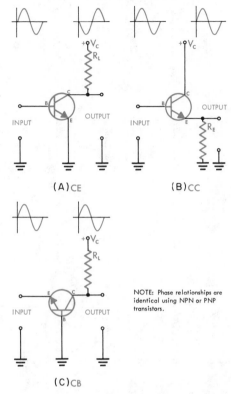

Fig. 12. Comparisons of input and output resistances in different amplifier configurations.

Fig. 13. Phase comparisons of input and output signals in different amplifier configurations.

Amplifier Input and Output Signal Phase Comparisons (See Fig. 13)

As shown in Fig. 13, the CE amplifier is the only configuration that shows a phase reversal of the input signal voltage. This is not necessarily a disadvantage but just the peculiarity of the configuration. If the same polarity is required, two CE amplifiers can be *cascaded* (tied together) so that two phase reversals take place, thereby providing the input-to-output signals in phase.

The other two amplifiers (CC and CB) do not have a phase inversion. You will notice that the signal input in a CE amplifier is applied to the base and the output signal is taken from the collector. The CC amplifier's signal input is applied to the base and the output signal is taken from the emitter.

The CB amplifier signal input is applied to the emitter and the output signal is taken from the collector.

Amplifier Current Flows (See Fig. 14)

Each amplifier configuration has three current flows. The first current flow is the emitter current. This current represents 100% of the current in the circuit. The second current flow is the base-emitter junction, which contains 1% to 5% of the current flow. This value changes with the transistor and the configuration. The third current is the collector current, which represents 95% (plus) of the current flow. Fig. 14, shows

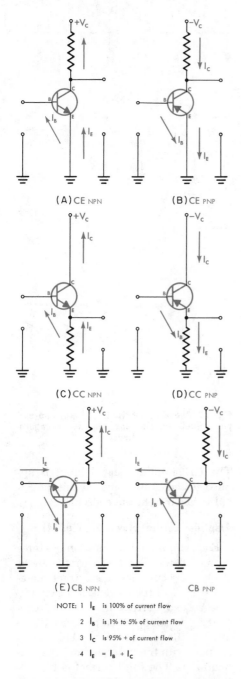

(A) CE NPN   (B) CE PNP

(C) CC NPN   (D) CC PNP

(E) CB NPN   CB PNP

NOTE: 1 I_E is 100% of current flow

2 I_B is 1% to 5% of current flow

3 I_C is 95% + of current flow

4 $I_E = I_B + I_C$

Fig. 14. Current flows in different amplifier configurations.

the paths of current flow in the three configurations CE, CC, and CB using both NPN and PNP transistors.

Thermal Runaway

Most problems that transistors have stems from an action called *thermal runaway*. At the collector, power is dissipated into heat, which causes collector current to increase. An increase in collector current will cause forward junction bias to increase, further increasing collector current. This action is cumulative, thereby causing still more heat and finally thermal runaway and transistor destruction.

The problem is not impossible to handle. There are two methods for preventing thermal runaway. One method is to ensure that the transistor is operated well within the limits suggested by the manufacturer. The other method is by bias stability.

In Fig. 15, three types of bias stability are illustrated. Fig. 15 (A) shows negative feedback. This method returns some of the output signal to the base, reducing gain and thereby reducing thermal runaway. In Fig. 15 (B) a *swamping* resistor is placed into the emitter circuit. This resistance provides the larger part of the base-emitter input resistance, thereby reducing the problem of thermal runaway. Finally the thermistor—another solid state device —is placed in the base circuit of Fig. 15 (C). The thermistor changes resistance with temperature change, thereby holding a constant bias on the base.

The best configuration for thermal stability is the common base, but this configuration may not be able to meet the requirements of the amplifier as to gain, input and output resistance, or phase inversions.

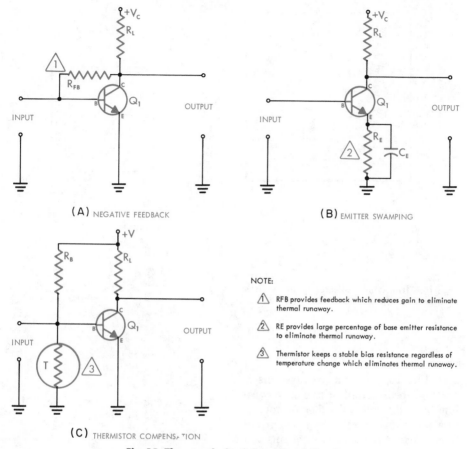

(A) NEGATIVE FEEDBACK

(B) EMITTER SWAMPING

(C) THERMISTOR COMPENSATION

NOTE:

⚠1 RFB provides feedback which reduces gain to eliminate thermal runaway.

⚠2 RE provides large percentage of base emitter resistance to eliminate thermal runaway.

⚠3 Thermistor keeps a stable bias resistance regardless of temperature change which eliminates thermal runaway.

Fig. 15. Three methods of thermal stabilization.

Transistor Collector Characteristic Curves (See Fig. 16)

Each transistor has a numerous amount of curves that may be drawn from its characteristics. These curves are plotted on graphs and give the designer valuable information on the important parameters concerning the transistor.

The curves may be in the form of output current against input current, output voltage against input voltage, voltage gain, current gain, power gain and many more. In fact, there are probably curves for curves. All of them are

useful. We cannot deal with all of them, so we shall deal with the one curve that stands out as the most used and by far the most valuable. This curve is known as the *collector characteristic curve.*

Fig. 16 illustrates a typical curve of this type. You may note at this time that the curve plots the collector voltage against collector current with varying base-bias current. Analysis of this curve will show that selection of a collector voltage and bias current will provide the reader with the value of current in the collector circuit.

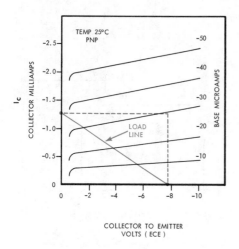

Fig. 16. Transistor collector characteristic curves with load line superimposed.

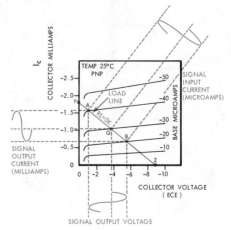

GIVEN:

Supply Voltage	V_{CC}	= 9 VDC
Load Resistance	R_L	= 5KΩ
Input Current	ii	= 20 MICROAMPS P-P
Bias Current	I_B	= 30 MICROAMPS

CONSTRUCTION:

1 Plot point Z = V_{CC}

2 Plot point Y = V_{CC}/R_L

3 Draw load line ---- Y to Z

4 Plot operating point Q = Select I_B (Q)

5 Plot point A and B = Peak to Peak Input Signal current swing

6 Extend dotted line as shown in illustration

7 Draw I_B, I_C, and E_{CE} sine waves

Fig. 17. Load line construction and analysis for Class A amplification.

In Fig. 16, you can see how this is accomplished. Simply draw a vertical line from the selected collector voltage to a point on the selected bias current lines. Then draw a horizontal line left to the collector current scale, where you can read the collector current for that combination. Between these two points draw a solid line. This is the *load line*.

In a transistor, varying either current or voltage will change the entire amplifier analysis. In the next several paragraphs we shall see how the load line determines the class of the amplifier.

Class A Amplification (See Fig. 17)

Fig. 17 shows a typical load line superimposed on a set of collector characteristic curves. To be a class A amplifier, a configuration must allow current to flow during the complete 360 degrees of the input cycle. That is, current must flow in the collector circuit during the entire positive and negative pulses of the input signal.

As shown in the figure, some facts are known prior to construction of the load line. In this case, the supply voltage, load resistance input current, and bias current have been selected. With a collector voltage of 9 volts and a load resistance of 5 kilohms, the collector current, using Ohm's law, is as follows:

$$I_C = \frac{E}{R}$$

$$I_C = \frac{9}{5} = 1.8 \text{ milliamps}$$

This value of 1.8 milliamps represents the extreme case on this graph where collector current would be maximum when collector voltage is zero. This point on the graph is represented by the point Y. Point Z (collector voltage) has been selected as 9 volts. Drawing a line from point Y to point Z completes construction of the load line for a 5 kilohm load resistance.

At this point it is up to the person constructing the amplifier to decide where he should place his operating point Q. This point represents the value of base current that is to be utilized. To get the best fidelity, the Q point should be placed as near the center of the load line as possible. This allows the amplifier to be able to amplify as much in the negative direction as in the positive direction. This is much like an airplane landing. The pilot attempts to stay in the center of the runway so that he will have equal room on either side.

The operating point we have chosen is 30 microamps. This point is found where the load line crosses the base current of 30 microamps. Since the signal current swings 20 microamps peak-to-peak (or 10 microamps in each direction from the Q point), we can establish the points A and B which represent the entire amplification cycle. As shown in the illustration, dotted lines from the A, B, and Q points in three directions will provide us with the collector current, base current and collector voltage swings for this configuration (class A).

As you can see, the sine waves are within the limits of the characteristic curves. Class A amplifiers have very low efficiency because of constant current flow, but have perfect fidelity. Fidelity is the ability of an amplifier to produce an output signal which is an exact likeness of its input signal. Class A amplifiers are used in *preamplifiers* because of their low distortion.

Class B Amplification (See Fig. 18)

Fig. 18 illustrates the load line for Class B amplifier operation. It is constructed in the same manner as the Class A amplifier configuration. In the Class B amplifier you will see that the collector current flows only during the positive cycle of the input current sine wave. That is, the amplifier is biased at

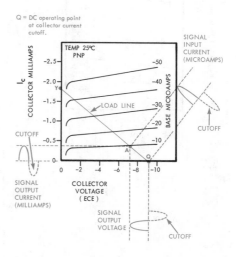

Fig. 18. Load line construction and analysis for Class B amplification.

cutoff. It is also possible for the Class B amplifier to be biased at saturation. This would allow collector current to flow only during the negative cycle of the input current.

The efficiency of the Class B amplifier is very high, but this class lacks fidelity. The Class B amplifiers are utilized primarily for power amplifiers and are set up in push-pull amplifier circuit arrangement. Class B amplifiers are never used for single-stage amplification because of their distortion.

Class AB Amplification (See Fig. 19)

Fig. 19 illustrates the load line for Class AB amplifier operation. It is constructed in much the same manner as the Class A amplifier configuration. In the Class AB amplifier you will see that

collector current flows during all of the positive alternation and about half of the negative alternation of the input current sine wave. That is, the amplifier is biased in between a Class A and Class B amplifier style.

The Class AB amplifier may be used in lieu of Class A or Class B amplifiers. It has fair fidelity and may be used in preamplifier and driver circuits. It is not as efficient as Class B but may be utilized comfortably in push-pull amplifiers.

Class C Amplification (See Fig. 20)

Fig. 20 illustrates the load line for Class C amplifier operation. It is constructed in the same manner as the Class A amplifier configuration. In the Class C amplifier you will see that the col-

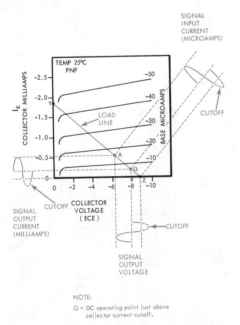

NOTE:
Q = DC operating point just above collector current cutoff.

Fig. 19. Load line construction and analysis for Class AB amplification.

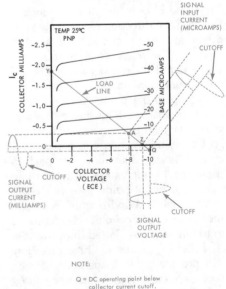

NOTE:
Q = DC operating point below collector current cutoff.

Fig. 20. Load line construction and analysis for Class C amplification.

lector current flows during less than 180 degrees of the input current sine wave. That is, the amplifier is biased somewhat below cutoff. It would be possible to bias the Class C amplifier at saturation for a negative swing of collector current less than 180 degrees; however, this is not the general procedure.

The Class C amplifier is the most efficient of all the amplifier classes, but also has the poorest fidelity. Class C amplifiers are utilized in oscillator circuits. Class C amplifiers are seldom used in audio circuits because of their distortion.

Push-Pull Amplifier Circuit (See Fig. 21)

Fig. 21 illustrates a typical push-pull amplifier. Amplifiers of this type may be operated using any of the classes of amplification, but class B is probably the most efficient and more frequently used. The push-pull amplifier is most useful as the output stage for audio networks.

Base bias for the amplifier is normally provided by a single-stage driver amplifier coupled into the push-pull amplifier by a transformer. The push-pull network utilizes the common emitter con-

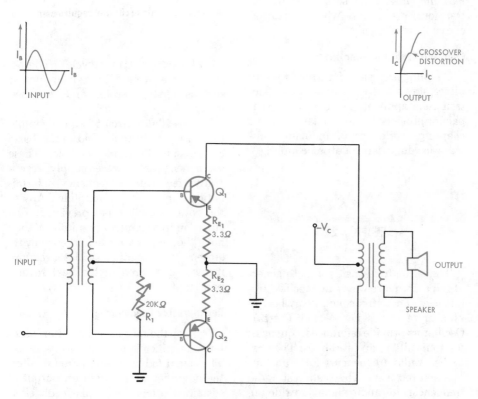

Fig. 21. Push-pull amplifier circuit.

figuration with a small, high-power emitter resistance for temperature stabilization.

The output transformer must match the collector and the output impedance. The variable resistor in the input stage varies the collector operating current to the desired amount.

Each transistor turns on when its base is negative; that is, in every other alternation of the input sine wave.

The two signal outputs of Q_1 and Q_2 are 180 degrees out of phase with each other. One output phase is inverted, so that it adds to the other phase at the transformer primary. The first phase pushes, so to speak, while the second phase pulls. Thereby the name *push-pull amplifier*. The secondary of the transformer sees only what is induced across its windings.

Coupling Stages of Amplifiers

Coupling provisions for transistor amplifier stages are the same as for electron tube applications. Use of the capacitor provides high-impedance output coupling, while use of inductors provides low-impedance output coupling.

OSCILLATOR CIRCUIT BASICS

(SEE FIG. 22.)

The oscillator's basic job is to generate sine or square waves at specified frequencies. These frequencies may be any number of cycles per second (hertz). Oscillators provide reference frequencies for transmitters and receivers. They establish radio frequencies for transmitters and receivers. They establish radio frequencies for audio signals to ride on, hence the different radio bands (such as

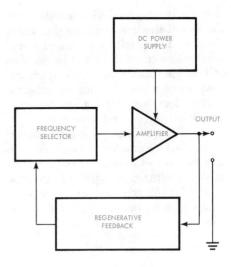

Fig. 22. Basic oscillator requirements.

1000 kilocycles, 1 megacycle). The oscillator is also used in test equipment such as signal generators and frequency meters.

The oscillator circuit is fairly simple to construct, but demands four basic components to fulfill its role. These components are a power supply, regenerative feedback, a frequency determination circuit, and an amplifier. Any of the power supply types previously discussed in this chapter are suitable if they have the proper voltage and current requirements. The other oscillator requirements are covered in the next several paragraphs.

Regenerative Feedback

Part of the output signal of an oscillator amplifier is fed back to its input. The signal fed back is required to allow the amplifier to maintain an output of sustained oscillation. This feedback is called positive or regenerative feedback.

Output signals should be of the same phase as the input; therefore the feedback signal must in some way provide in-phase feedback. This can be done inductively by transformer coupling, as in the Hartley oscillator. Inductive feedbacks are often called *tickler coils*. Feedback can also be accomplished by constructing a two-stage common-emitter amplifier. Finally, feedback can be accomplished with the use of a common-emitter amplifier and a *lag circuit*, as in the Wien-Bridge phase-shift oscillator.

These and other oscillators will be explained in detail later on in this chapter. In any event, regenerative feedback requires an in-phase component large enough to provide the oscillator with self-generating abilities.

Frequency Determination

The oscillator must produce a stable frequency that is undistorted. Therefore some frequency device is required. Basically there are three types of frequency determining networks: the RC network, the tuned circuit, and the crystal.

The RC network (sometimes called a *lag* or *high-pass*) provides a stable frequency input with the use of RC circuits. Each RC combination provides 60 degree phase shift. This frequency determiner is explained in detail with the phase-shift oscillator later in the chapter. RC oscillators are usually used in audio circuits and other low-frequency operations.

The *tank* circuit is merely another name for a capacitor and inductor parallel arrangement, also known as a *tuned* circuit, tuned to a specific frequency. The name *tank* comes from the frequency supply that this circuit produces. It is like having a tank full of

water; in this case it is a tank full of cycles per second.

The base formula for calculating frequency of oscillation is:

$$f = \frac{1}{2\sqrt{LC}}$$

(Where f = frequency in hertz; L = inductance in henrys; and C = capacitance in farads). The feedback signal is taken off the tank circuit and fed back to the input at a frequency determined by the resonant frequency of the tank. The capacitors and inductors in the tank are preselected to resonate at a specific frequency.

Some oscillators have variable capacitors for tuning. Others have variable inductors. The Clapp, Hartley, and Colpitts oscillators have tuned tanks for frequency determination. The larger the capacitors and inductors, the lower the resonant frequency. Tuned circuit (tank) oscillators are used in high-frequency applications.

The final frequency determining device used is the crystal.

Crystals used in electronic oscillators are made of substances like quartz or tourmaline which have a special electrical feature called *piezoelectric effect*. When mechanical pressure is applied to the crystal it will generate a-c voltage. By the same effect, if voltage is applied to the crystal it will vibrate mechanically at a fixed rate.

The crystal, when placed in series, has a low resonant frequency and, when placed in parallel, has a high resonant frequency. Therefore, one crystal could be used in an oscillator circuit to provide one of two separate oscillation frequencies. Crystals have very stable frequency characteristics and are used normally in high-frequency applications.

Oscillator Amplifier Choice

The common-emitter and the common-base amplifier configurations are widely used as oscillator amplifiers. The common-emitter has a phase change from input to output. The common-base does not have this phase change.

Impedance matching is a particular problem in oscillators. For this reason inductive coupling is utilized because of its versatility in matching impedance.

High-power oscillators always use the common-emitter configuration. For high-frequency common collector amplifier configuration is seldom used because it does not produce enough power to allow for feedback voltage and output oscillations.

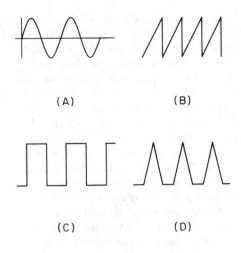

(A)   (B)

(C)   (D)

Fig. 23. Oscillator waveforms: (A) sine wave, (B) saw tooth, (C) square wave, and (D) spike.

Sine Wave Oscillator (See Fig. 23)

A sine wave oscillator produces a-c frequencies (both audio and radio) from d-c power. The oscillator is used in superhetrodyne receivers, and in transmitters, signal generators, and test equipment.

Oscillation can be in the form of perfect sine waves, sawtooth waves, square pulses, or just spikes. See Fig. 23 for some of these waveforms. The waveforms, in order to react normally, should be stable in frequency and amplitude, and also be undistorted. In other words, the oscillator should have an output that is exactly the same at all times. The reader will also note that the output of the oscillator is a sustained output.

Colpitts Oscillator (See Fig. 24)

The Colpitts oscillator is one of the most common and most useful of the oscillators. It is best distinguished because of its method of feedback coupling. Feedback energy from a Colpitts

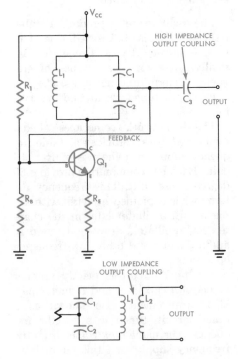

Fig. 24. Colpitts oscillator.

tuned circuit is center-tapped from a pair of split-load capacitors which form the capacitance part of the tuned circuit.

In Fig. 24, a common-emitter amplifier is arranged with a tuned circuit in its collector branch. The resistor R_E is a swamping resistor and the resistors R_1 and R_B form a voltage divider for initial bias.

The tuned circuit, as previously described, provides oscillation for output and also provides regenerative feedback to the base, causing the amplifier to continue to operate. The capacitors in the tuned circuit are usually of the same value. Tuning is accomplished by one of two methods. The inductor L_1 can be variable or the capacitors C_1 and C_2 can be variable. If the capacitors are variable, they must be linked together by a shaft to ensure that both change by the same amount when varied.

The reader will note that the feedback line is sent to the base as in any other common-emitter signal. However, the Colpitts oscillator may be organized in the common base configuration if it is so desired. The basic circuit in the figure makes use of capacitor C_3 for stage coupling.

If the load impedance is high, capacitor coupling (C_3) is the best method of coupling. If the load impedance is low, the oscillation must be inductively coupled (L_2) from the tuned circuit as shown in the figure.

If desired, the Colpitts oscillator may be modified to utilize a crystal in place of the inductor L_1 for oscillation control (reference the Pierce oscillator).

Clapp Oscillator (See Fig. 25)

The Clapp oscillator is a modified Colpitts oscillator. See Fig. 25. Its uniqueness comes from the fact that a capacitor (C_S) is placed in series with

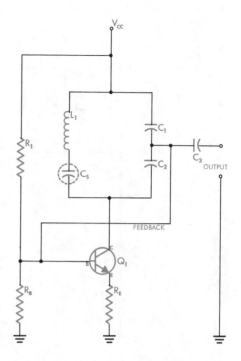

Fig. 25. Clapp oscillator.

the inductor in the tuned circuit. This configuration provides the designer with much greater frequency stability than the basic Colpitts oscillator.

All the other components in the Clapp oscillator are the same as the Colpitts circuit components and operate in the same manner as previously described. Although not shown in the figure, the output of the Clapp oscillator may be inductively coupled from the tuned circuit for low impedance loads.

Hartley Oscillator (See Fig. 26)

The Hartley oscillator is another of the most commonly used and most useful of the oscillator types. Its characteristic feature is the method of feedback coupling. Feedback energy from a Hartley tuned circuit is tapped from the inductor of the tuned circuit.

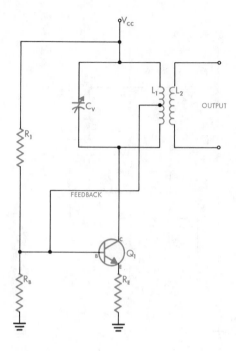

Fig. 26. Hartley oscillator.

In Fig. 26, a common-emitter amplifier is arranged with a tuned circuit in its collector branch. The resistor R_E is a swamping resistor and the resistors R_1 and R_B provide a voltage divider for initial bias. The tuned circuit, as previously described, provides oscillation for output and also provides regenerative feedback to the base of the transistor, causing the amplifier to continue to operate. The capacitor in the tuned circuit is variable and is used for tuning.

The reader will note that the feedback line is sent to the base as in any other common-emitter configuration. The Hartley oscillator is sometimes configured as a common-base amplifier for high-frequency operation. The

circuit in the figure shows the output coupled to a low-impedance load.

Phase-Shift Oscillator (See Fig. 27)

The phase-shift oscillator is utilized when a phase shift is required and when temperature and voltage must remain exceptionally stable. This oscillator is used normally in audio circuits.

In Fig. 27 the transistor Q_1 is configured as a common-emitter amplifier with resistor R_L as the load resistance and resistor R_B as the bias resistance. The high-pass network consisting of the capacitors C_1 C_2, and C_3, and the resistors R_1, R_2, R_V, provide low frequencies, say under 500 hertz. Output voltage is very nearly supply voltage and frequency is quite stable.

The variable resistor R_V provides frequency adjustment, and the circuit provides a phase shift of 360 degrees. The common-emitter amplifier provides 180 degrees of phase shift. The capacitors and their charge and discharge resistors supply 180 degrees (60 degrees for each capacitor) phase shift. Output is taken from the collector which is in parallel with the transistor amplifier and in parallel with the high-pass circuit.

Wien-Bridge Oscillator (See Fig. 28)

The Wien-Bridge oscillator is shown in Fig. 28 is a typical RC phase-shift oscillator that operates much the same as the phase-shift oscillator previously discussed. The Wien-Bridge oscillator is unique in that it utilizes a bridge network to provide frequency determination and feedback.

In the figure, the arrangement of C_3-R_3-R_4, C_1-R_1, and C_2-R_2 provide 60 degree phase shift each, at a predetermined frequency. The capacitive reactance of the circuit changes at all other frequencies, so the 180 degrees

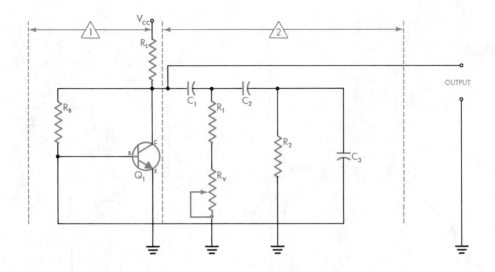

NOTE:

⚠️ **1**   RL, RB, and Q1 are a common emitter amplifier.

⚠️ **2**   C1, C2, C3, R1, R2, and RV are a phase shift circuit.

3   Common emitter amplifier furnishes 180° phase shift and phase shift network furnishes 180°.

4   RV controls and varies frequency.

Fig. 27. Phase-shift oscillator.

phase shift occurs only at resonant frequency. The transistor Q_1 is the input stage and oscillator amplifier. The transistor Q_2 provides 180 degree phase inversion and amplification. Capacitor C_5 couples the output oscillation to the output stage. Capacitor C_4 passes only a-c to the second stage Q_2, R_5, R_6, R_7, R_E, R_{L1}, and R_{L2} provide the necessary bias networks.

It is possible to make the Wien-Bridge oscillator variable frequency by making the two bridge resistors R_1 and R_2 variable. It is also possible to make the capacitors C_1 and C_2 variable to change resonant frequency.

The reader will note that negative feedback is placed on the emitter of Q_1 through resistor R_3. This negative feedback is constant at resonant frequency but tends to change at other frequencies. A change in frequency then is avoided by this negative feedback.

The Wien-Bridge oscillator is used in low-frequency application and is best noted for its frequency stability. As you can see, though, there are more components in this oscillator than others. Therefore cost, weight and space problems are inherent.

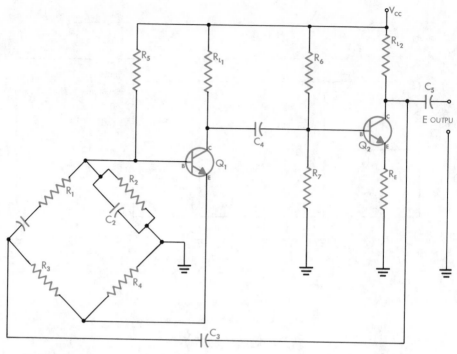

1   Q1 is oscillator amplifier

2   Q2 provides 180° phase shift and amplification.

3   RC NETWORKS C1–R1, C2–R2, and C3–R3–R4 provide 60° phase shift each for regenerative feedback and frequency determination.

4   R3 provides negative feedback to emitter of Q1 for stabilization.

5   C3 provides feedback coupling

6   C5 provides output coupling

7   Bias resistances are obvious

Fig. 28. Wien-Bridge oscillator.

Crystal Oscillators

Crystal oscillators may be in any of the forms discussed in this chapter. Often the crystals are used in place of inductors in tuned circuits. In many cases, the crystal is the basic frequency-controlling device. The crystal is used for frequency stabilization because its frequency is fixed and never changes.

The crystal is generally made of quartz. When electricity is applied to a crystal such as quartz, the crystal changes physical shape which results in mechanical vibrations. These vibrations are almost constant. This mechanical effect is called the piezoelectric effect. The crystal vibrates at separate frequency when placed in series or in parallel. In series, the impedance of crystals is very low. In parallel, their impedance is very high.

The typical crystal oscillator is the Pierce oscillator, which uses the crystal in modification of the Colpitts oscillator.

REVIEW QUESTIONS

1. What is the difference between the transformer used in a conventional full-wave rectifier and a full-wave bridge rectifier?

2. What is meant by peak inverse voltage (PIV) rating?

3. What is the purpose of a current limiting resistor?

4. What characteristics are necessary when choosing a diode for a power supply?

5. What is the advantage of a series regulator over a shunt regulator?

6. What is the advantage of a shunt regulator over a series regulator?

7. What is the advantage of a choke filter?

8. Is the choke filter installed in series or parallel with the load?

9. Is a capacitor filter placed in series or parallel with the load?

10. What effect will a shorted zener diode have on a series or shunt regulator?

11. Why would you select a common collector amplifier configuration? Common emitter? Common base?

12. What can you determine from a set of collector characteristic curves?

13. Which class of amplifier is biased at cutoff?

14. Which class of amplifier is biased below cutoff?

15. Which class of amplifier gives the best fidelity? Poorest?

16. Which class of amplifier gives the best efficiency? Poorest?

17. Which oscillator feedback is most versatile for impedance matching?

18. Which amplifier configuration is best suited for very high frequency oscillation applications?

19. What distinguishes a Colpitts oscillator from other oscillators?

20. What is the basic difference between a Colpitts oscillator and a Hartley oscillator?

21. Which oscillator is the Clapp oscillator a modification of?

22. What are the four basic requirements of an oscillator?

23. Which amplifier configuration is used with the phase shift oscillator?

24. What oscillator application is the crystal best suited for?

25. Why is the common collector amplifier configuration avoided in oscillators?

26. In a tank circuit, how can the resonant frequency be raised? Lowered?

27. What is the purpose of a tickler coil?

28. Where is the feedback connected in a common base Hartley oscillator

29. What general type oscillator is the Wien-Bridge oscillator?

Chapter 16

Automatic Control Systems

INTRODUCTION TO AUTOMATION

The automatic performance of commercial and industrial operations is covered by the term *automation*. Almost every phase of productive labor comes within its scope. Developed in the manufacturing field, automation's range has been extended to include such diverse pursuits as agriculture, shipping, transportation, and education. Through the use of the computer, automation has been applied to clerical and office tasks, involved mathematical calculations associated with scientific investigations and missile control, sorting, and evaluating data, even to the complex region of medical diagnosis.

CONTROL MECHANISMS

Many industrial and military automatic control systems are founded upon the *closed-loop* or *closed-cycle* operation. That is, an operation in which an output is compared continuously or at intervals with an input. Corrective measures are automatically taken when the output deviates from the input. The deviation between the system's output and its input is referred to as an *error signal*. Comparisons are made through application of the feedback principle. That is, system output is fed back to its input for comparison. The whole chain of activity is termed a *servomechanism*.

Suppose, for example, a system output is the voltage on a power line. A voltmeter at the distant end of the power line transmits information to a sensor at a generating station. The sensor is a Wheatstone-Bridge type of instrument in which feedback voltage from the voltmeter is balanced against known voltages. In this case the input is fixed, as is the desired voltage of the transmission line. Any difference between the known voltage and the voltmeter signal voltage results in current flow through a relay which governs generator field excitation, raising or lowering it until the bridge reaches a neutral state. This state is called *null*. It is the point when fixed input and feedback are identical.

In this example, the input is fixed and is the desired voltage of the transmission line. The servomechanism here includes the feedback voltmeter, the bridge sensor, the intervening motors, and the rheostats that cause field current to increase or decrease. The closed loop involves this apparatus plus the generator itself.

Machinery operations in a manufacturing plant follow a similar linkage. Dimensions of a finished shaft, for instance, are fed back to a sensor which compares dimension signals with preset manufacturing tolerance levels. Difference pulses are sent to govern positions of tools being used in the cutting process.

It should be noted that transducer input and output functions may be reversed. They may transform an electrical input into a mechanical output. Transducers are used as inputs or feedback devices. Closed loop systems can be as unique as their designer, and the method will expand as new ideas are generated.

The Transducer

Under feedback procedure, some output indication is checked by a suitable instrument called a *transducer*. This is a device that transforms one form of energy into another form of energy. In control applications, this is characteristically a change from a mechanical input to a proportional electrical output.

An example of one form of transducer is the potentiometer. In a potentiometer (Fig. 1) the mechanical input is the potentiometer shaft rotation. The elec-

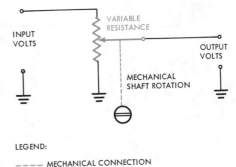

LEGEND:

- - - - MECHANICAL CONNECTION

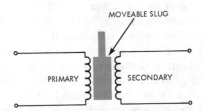

POTENTIOMETER

Fig. 1. The potentiometer.

trical output is the potentiometer wiper voltage, which is proportional to shaft rotation.

Another example of a transducer is the *linear variable differential transformer* (LVDT) (Fig. 2). The LVDT is commonly used in aircraft applications and is basically a transformer with a

MOVEABLE SLUG

PRIMARY   SECONDARY

Fig. 2. The linear variable differential transformer (LVDT).

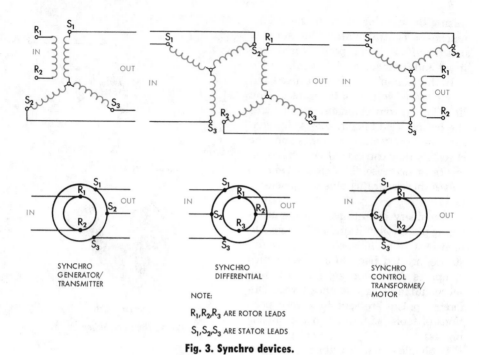

SYNCHRO
GENERATOR/
TRANSMITTER

SYNCHRO
DIFFERENTIAL

SYNCHRO
CONTROL
TRANSFORMER/
MOTOR

NOTE:

R_1, R_2, R_3 ARE ROTOR LEADS

S_1, S_2, S_3 ARE STATOR LEADS

Fig. 3. Synchro devices.

moveable metal slug between its primary and secondary windings. As the slug moves linearly, the secondary voltage varies proportionally. The primary voltage is fixed. This device is usually used as a *position transducer*. That is, with the movable slug attached to a linearly moving piston, the piston position is known by the voltage reading across the secondary winding.

The most common control device is the *synchro* (Fig. 3). As this is the most used device we shall study it in greater detail. From a quick look at Fig. 3 you will notice that the synchro has a three-legged synchro winding and a rotor that may be two- or three-legged. Operation of the synchro is outlined in following paragraphs.

The Synchro

The synchro is probably the most used basic unit in servomechanism construction. A trade name used for the General Electric synchro is the *selsyn*— a word meaning self-synchronous. Bendix synchros may be known as *autosyns*. The following principles apply regardless of the name. However, for our discussion here we shall use the term *synchro*.

Synchros are frequently employed in servomechanisms and control systems. They duplicate, at a remote point, signals derived from motion at a sending station.

A synchro consists of a stator and a rotor resembling those of a small in-

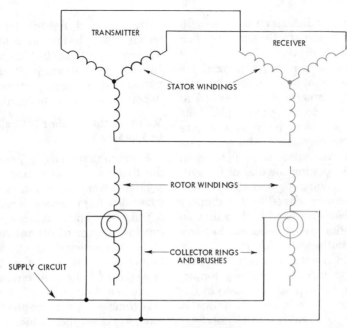

Fig. 4. A basic synchro circuit.

duction motor. The stator winding is three-phase in form, but voltages and currents are actually single-phase. The rotor has a single, two-pole winding that is supplied with a single-phase alternating current.

At least two units are required for synchro remote operation; a transmitter and a receiver. (See Fig. 4.) Transmitters are sometimes referred to as generators, and receivers as motors. In many instances transmitters and receivers are utilized for instrumentation in industrial or aircraft use. In Fig. 4 you will notice that the two rotors may be excited from a single power source. Stator leads of the transmitter are connected to respective leads of the receiver, but not to any external circuit.

When current is supplied to the transmitter rotor, the alternating flux induces in the three stator winding voltages whose relative values depend on position of the rotor with respect to each winding.

A like condition arises at the receiver unit when its rotor is excited. If voltages induced in transmitter and receiver stators are not exactly balanced with each other, stator flux in the receiver applies force to the rotor, causing it to turn until voltages in the two stator windings are identical, at which point current flow ceases. The receiver rotor assumes the identical angular position to that of the transmitter. If the transmitter and receiver rotors drive indicators in a particular application, then the readings indicated on the transmitter indicator are exactly reproduced on the receiver indicator.

Information can be exchanged between one point and another by this means, or power may be exercised at the remote location. For example, if the receiver rotor is attached to a theatre dimmer, an operator in the projection room can manipulate stage lighting as desired by rotating the dial of his synchro. When doing so, mechanical resistance to turning offered by the dimmer contact arm is reflected back to him, so that a distinct effort is needed to move the transmitter dial. More current is required in the stator circuit of the receiver to provide the necessary torque, and since this current flows also in the transmitter stator, a counter-torque is generated there, calling for physical exertion on the part of the operator. The transmitter rotor would be affected identically; however its position is fixed either mechanically or automatically and cannot respond to exerted torques.

In review, synchros are the most widely used devices in control work. This is in part due to their versatility. In most applications, particularly computing devices, the synchro system is an important link in a much greater chain.

Multiple Control Points Controlled by Synchros

Synchro units may be parallel in order that two or more receivers can be regulated from one point, or a number controlled from various locations. Fig. 5 illustrates such an arrangement. A synchro system of several units is located on different floors of a warehouse to govern operation of a conveyor belt system used to load cartons onto trucks at the first-floor level.

If trouble develops on the third-floor level, workmen there can slow down or stop conveyor belts by manually moving the synchro at that point and remotely controlling the other synchros. The synchros control motor speed at each level. If cartons are piling up too

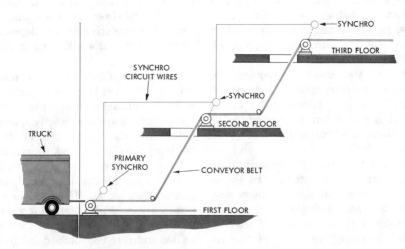

Fig. 5. Synchro control for three separate locations.

fast on the second floor, workmen may slow down the belts by resorting to the nearby unit. Should cartons arrive too fast at the truck location on the first floor, a supervisor slows down belts with the primary synchro, which is designed to be locked with respect to rotation in one direction. The belts cannot be speeded up from one of the other locations, but can be slowed or stopped if it becomes necessary.

Applications discussed thus far have involved the services of a human agency, but such need not be the case. A transmitter synchro whose rotor is turned automatically according to voltage at the end of a transmission line may serve as feedback unit to govern movement of a receiver at the generating plant, providing torque that drives generator field-current apparatus.

Differential Synchro

Units A and C, Fig. 6, are the same as those in Fig. 5, but Unit B differs from them. B is a differential unit with three-phase type windings on both rotor and stator. If rotor and stator windings of B are in an exact alignment, C's rotor will duplicate any movement of A's.

If the rotor of B is mechanically moved so that its windings are displaced with respect to one another, manipulation of A's rotor will cause C's to move through an angle that is either the sum of the other two angles, or their difference, depending upon connections to C's stator. With leads arranged in one way, C's rotation is equal to the sum of the two dial settings. If a pair of leads is interchanged, rotation equals the dif-

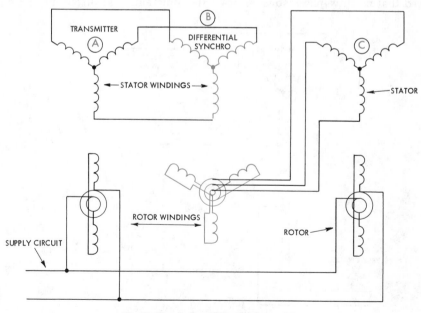

Fig. 6. Circuit for differential synchro.

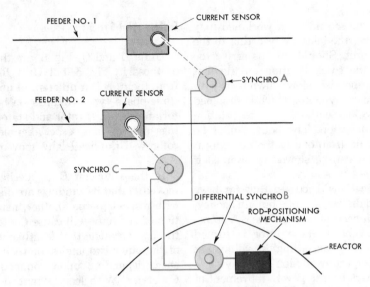

Fig. 7. Simple application of differential principle.

ference between the two values. If *A* and *C* are rotated simultaneously, *B* will turn through an angle which is the sum of the other two. It should be observed that no three-phase voltages appear in the windings of the differential unit, but simply single-phase voltages in a three-part winding.

Fig. 7 offers a practical example of the differential arrangement. Unit *A*

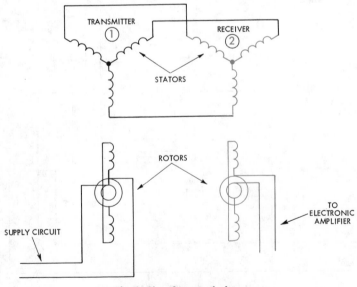

Fig. 8. Signal transmission.

is affected by current in one of the two main feeders from a turbine-driven generator, *C*, by current in the other.

Indications of both devices are totalled in Unit *B*, which governs the position of cadmium rods in an atomic reactor that furnishes steam for the turbine. Raising or lowering of the rods by apparatus under the control of *B*, increases or decreases the amount of steam generated in accordance with load variations, affecting currents in the feeder lines.

Transformer Synchro

A circuit may be designed as in Fig. 8, so that instead of producing motion in the receiver, positioning of the transmitter rotor induces voltages in the receiver rotor. When the rotor of 1 is turned, it creates in the stator winding voltages that are transferred to the re-

ceiver stator. The rotor of unit 2 is not excited from an alternating-current line, and is fixed in the plane of maximum induced voltage. It acts as the secondary of a transformer for stator voltage which is transferred to an electronic amplifier. This application of a synchro is often referred to as a control transformer.

The Power Synchro

As explained earlier, in connection with the dimmer-control arrangement, power furnished to the rotor of a transmitter may be employed by a receiver to drive a load. This principle is utilized in the circuit of Fig. 9. Each of the rotors is attached to a motion picture camera. In this application it is necessary that the two films run in synchronism in order to present a clear picture when projected onto the screen. Motor

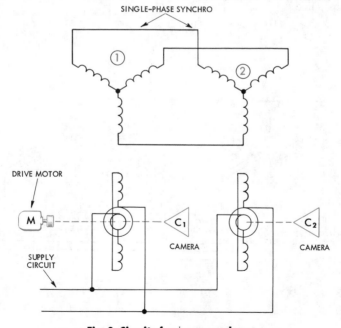

Fig. 9. Circuits for power synchros.

M drives the rotor of unit 1 and also camera C_1, while the rotor of unit 2 drives camera C_2.

If maximum torque is to be obtained from receiver 2, the speed of *M* should not be greater than 1200 rpm where two-pole synchro units are connected to a 60-hertz circuit. This is necessary so that rotor flux of unit 1 cuts turns of its stator winding at a rate sufficient to produce the required voltage. If the transmitter were driven at synchronous speed, 3600 rpm, the receiver would be unable to exert torque, because counter-voltage induced in the stator winding by its rotor would be exactly equal to and opposed to that from stator 1. In this case, load on the rotor of 2 would cause it to fall out of step with 1, and would cause it to operate independently as a single-phase induction motor.

Although single-phase power synchros are used for light duty, industrial three-phase type are also common. Stators and rotors of all units have three-phase windings, the stator windings being supplied with three-phase current. The three lead wires from each rotor winding are connected in parallel, but are not attached to external supply.

Fig. 10 represents a machine-tool set-up that includes a moving belt and ad-

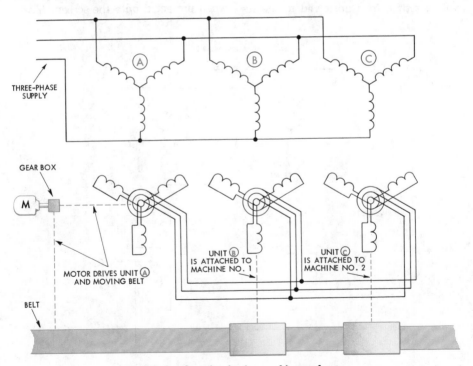

Fig. 10. Synchro circuits for machine tool setup.

jacent machines which perform separate operations upon metal objects that pass along. Motor *M*, besides turning unit *A*, also drives the moving belt which carries pieces through the machines at a steady rate of speed. The sequence of lateral motion and presentation of cutting tools to the work in each of the machines is thus accurately timed so that production goes on without interruption.

In certain instances it is necessary that mechanical devices attached to the rotors must occupy at all times identical angular positions with respect to one another. At starting, therefore, each rotor must be perfectly aligned with the others.

If three-phase current is applied at this time, the desired result is not always obtained, because there is no torque forcing rotors into like positions. For this reason, one stator phase is left open momentarily, so that single-phase flux may provide the required turning effort (as in the elementary synchro circuit), prior to connecting the third wire.

Synchro Null Positioning

In Fig. 11, a synchro null-positioning diagram is shown. Note that as the rotor rotates counterclockwise in relation to the stator, its shaft position in degrees causes the rotor to be electrically in a different position in relation to the stator legs.

As the illustration shows, electrical zero is when the rotor aligns with stator leg *S2*. When the synchro is first installed in equipment, electrical zero is set at this position. There are 360 degrees of rotation between the rotor and stator. The stator legs are 120 degrees apart. As the arrow rotates through

these 360 degrees, the arrow as shown on the rotor and the arrows as shown on the stator legs are in phase when pointing in the same direction and out of phase when pointing in opposite directions.

To check electrical zero in a standard 115-VAC synchro is quite simple. A value of 115 VAC is applied to the rotor of the synchro, 78 VAC between S_2 and S_3, 78 VAC between S_2 and S_1, and 0 VAC between S_3 and S_1. Physically moving the rotor will provide alignment to electrical zero. If the 78 VAC power supply leads are reversed,

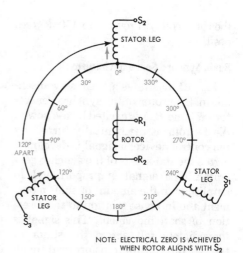

NOTE: ELECTRICAL ZERO IS ACHIEVED WHEN ROTOR ALIGNS WITH S_2

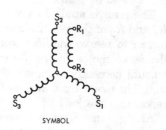

SYMBOL

Fig. 11. Synchro null-positioning diagram.

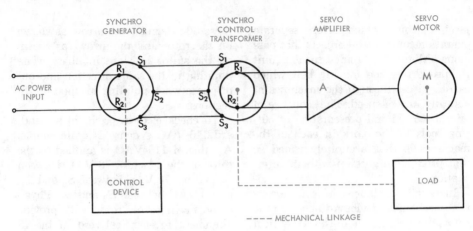

Fig. 12. Basic synchro servomechanism.

the rotor coil will reverse to 180 degrees position.

Basic Synchro Servomechanism

Fig. 12 illustrates a basic servomechanism. The rotor of the synchro generator (R_1 and R_2) is excited by a-c power. As the rotor is mechanically turned by the control device, a signal is developed across the stator legs of the synchro generator. The signal is proportional in amplitude to the amount of shaft movement and in phase relation to the direction of shaft movement. This signal is immediately felt across the stator of the synchro control transformer through S_1, S_2, and S_3 leads.

The error is applied by way of rotor R_1 and R_2 of the synchro control transformer to a servo amplifier which drives the servo motor in the direction and amount necessary to handle the correction. The motor turns and mechanically repositions the load.

At the same time, a feedback shaft attached to the load and the control transformer rotor R_1 and R_2, turns in an amount and direction with which to cancel the error signal and null the synchro control transformer. With error signals removed, the shafts are at a null and the amplifier is in a static condition.

Basic Two-Phase A-C Synchro Servomechanism System

Fig. 13 illustrates a basic two-phase a-c servomechanism system. It consists of a synchro generator, a synchro control transformer, a servo amplifier, and a two-phase shift device for one winding of the motor. Single-phase power is applied to the rotor of the synchro generator while the same voltage is applied to the 90-degree phase-shift device.

With no error, the 2-phase a-c motor does not drive because no rotating magnetic field exists. If the control device sees an error it will either be in phase or out of phase with the shifted voltage to the motor. This error will be imme-

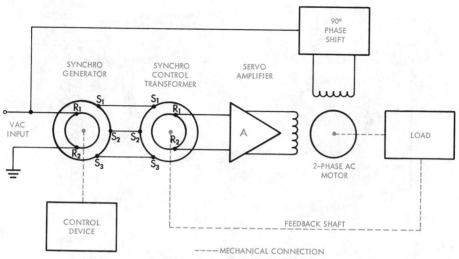

Fig. 13. Basic two-phase a-c synchro servomechanism system.

diately felt on the synchro control transformer, then amplified by the servo amplifier which drives the motor. The motor drives the load because an out-of-null condition exists in the synchro control transformer. The motor will continue to drive the load until the feedback shaft from the load nulls the synchro control transformer.

The speed of the motor is dependent on the magnitude of the error. The direction of rotation of the motor is dependent on shaft rotation direction. As the magnitude of the error decreases the motor will slow down until a null position is located on the control transformer.

MACHINE CONTROL USING CLOSED LOOP TECHNIQUES

The *amplidyne,* employed in numerous closed-cycle operations, is illus-

trated diagrammatically in Fig. 14. View (A) shows a pair of field coils and a direct-current armature whose brushes, midway between the poles, are short-circuited. Field poles and armature conductors are omitted in the illustration to simplify explanation. With field windings polarized as shown, and the armature driven in a clockwise direction, current flowing in the armature conductors sets up magnetic polarities at right angles to those of the field, an S pole on the right, an N pole on the left.

Armature conductors cut across flux created by the short-circuited brushes, generating voltage which appears at another set of brushes in line with field poles, Fig. 14 (B). Current from these brushes flows in the external circuit as indicated by the arrows, at the same time inducing armature polarities N at the top and S at the bottom. Since this flux opposes the original field excitation,

337

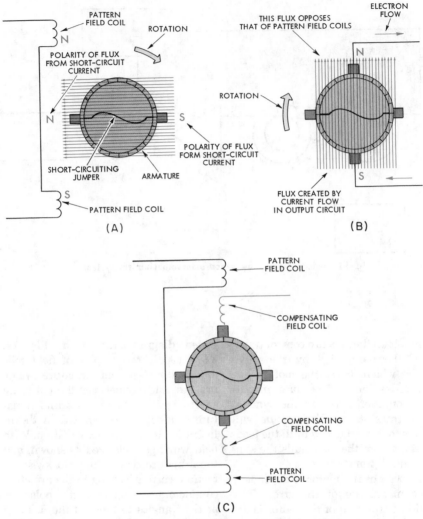

Fig. 14. Amplidyne generator.

it must be nullified by means of a compensating winding, Fig. 14 (C).

Analysis of current flow in armature conductors discloses that opposition exists in a large portion of the winding between that from the short-circuited brushes and that from the load. Laboratory tests show that current in the various conductors has the same total effect in establishing the armature fluxes as though two separate and entirely distinct armature currents were actually present.

A single watt expended in field excitation of the amplidyne may control as much as 10 kw output at the load brushes, and since the energy is supplied by torque of the drive motor, the unit is often termed a dynamic amplifier. The commercial amplidyne has a set of field poles that are provided with central openings, or slots, to aid commutation, and an armature with a standard direct-current winding. Field excitation is supplied usually by the signal from a feedback device, amplified if necessary.

A Closed-Cycle Example With Amplidyne

A closed-cycle system employing an amplidyne is shown in Fig. 15. The amplidyne's speed must be accurately maintained. Power is supplied by current from generator G. Voltage generated by tachometer T (a small permanent-magnet generator) is governed by the speed of M's shaft to which it is attached. This signal is fed to bridge-type sensor, S, which compares it to a standard and transfers any difference voltage to amplifier K. This amplifier is often electronic, but might well be the more rugged magnetic unit. The output of amplifier K is fed to the excitation field of amplidyne A, which acts to increase or decrease field current of generator G. The rise or fall of output voltage compensates for many off-standard variation in M's speed of rotation.

Description here takes much longer than performance, which is practically instantaneous. Speed of correction is one of the essential requirements for any closed-cycle sequence, and the amplidyne is well suited in this respect. Its response to the amplified signal is swift, much faster than that of a relay or other semi-mechanical device.

A single control field was shown in the amplidyne of Fig. 14, but it is common practice to use as many as four if the work pattern so requires. The atomic-reactor example outlined in connection with Fig. 7 will serve again for purpose of illustration. If there are four generator feeders rather than two, a separate field winding can be associated with flow of current in each of them, their total excitation determining the amount and direction in which the cadmium rods will move so the reactor will supply the proper amount of heat to the heat exchanger.

The Rototrol

The *rototrol*, Fig. 16, is closely allied to the amplidyne and performs similar tasks. Self-exciting field B and the armature are in series with the field coils of a generator. Pattern field coils C are in series with an adjustable resistor, across the output terminals of the generator. Pilot field D, the feedback element, is across the output circuit at a remote point where voltage is to be maintained at a given level. Pattern and pilot field excitation is adjusted so that under normal conditions

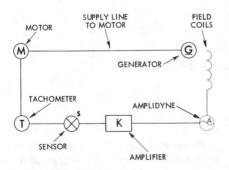

Fig. 15. Diagram of closed cycle operation.

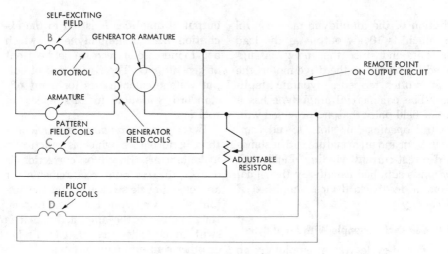

Fig. 16. Rototrol.

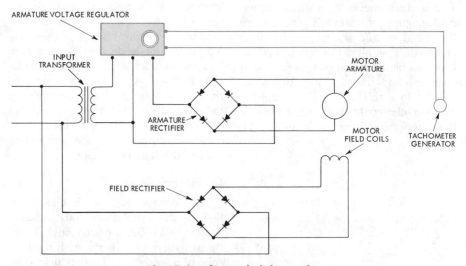

Fig. 17. Speed control of shunt unit.

the armature of the rototrol delivers the proper current to field coils of the main generator.

When voltage at the remote location falls below normal, *D's* opposition is lowered, and voltage of the rototrol armature increases until a new condition of balance is arrived at. If the distant voltage rises above normal, D's opposition increases, and the rototrol voltage is reduced. The feedback signal may be amplified where necessary, as in other closed-cycle arrangements.

Speed Control of Shunt Motors

In many automated processes where maintenance of exact speed is essential, the result is achieved by incorporating certain devices in the motor control circuit to govern armature or field voltage, or both. Fig. 17 illustrates a common circuit of this type, having alternating-current supply, solid-state rectifiers, and a direct-current motor.

The armature rectifier is connected through a voltage regulator to the secondary winding of an input transformer, and also to the armature terminals.

A tachometer attached to the armature shaft feeds back voltage to the regulator which has a dial for presetting speed at any desired value. Field coils, here, are connected to a similar rectifier which is attached to the supply wires. This method ordinarily provides speed variation below normal. When speeds greater than normal are desired, the armature rectifier is connected to the transformer secondary and a regulator is installed in the field circuit.

Fig. 18 illustrates the circuit of a shunt motor whose speed is electronically controlled with the aid of thyratron tubes. The principle is identical to that of Fig. 17, but circuit details are modified. The supply transformer has a primary winding, P, and two center-tapped secondary windings S_1, and S_2. Thyratron tubes V_1 and V_2 which provide full-wave rectification in the arma-

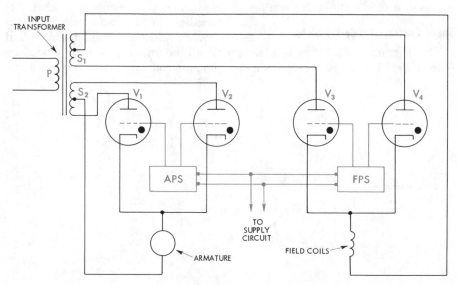

Fig. 18. Thyratron speed control.

ture circuit, are grid-controlled through one of the phase-shifting units, *APS*, discussed in Chapter 12. The field circuit also receives full wave rectification from tubes V_3 and V_4 which are grid-controlled through phase-shifting unit *FPS*.

For speeds below normal, the armature-control arrangement gives a more economical range than the series resistance method studied in Chapter 9, because power is not expended in I^2R loss. There is no comparable gain in efficiency with respect to the field circuit, because field current is so small that I^2R loss in a field resistor is practically negligible.

Thyratron field control for speeds above normal, however, provides a smoother variation than the series-resistance plan. Where only a single-range is desired (that is either below or above normal), phase-shifting procedure need only be applied to one circuit. If the only requirement is for subnormal speeds, field coils are usually excited by means of a solid-state bridge rectifier.

Speed Control of a Series Motor

The diagram of Fig. 19 is similar to that of Fig. 17, except that the shunt motor has been replaced by a series unit. Transformer, armature rectifier, speed regulator, and tachometer are present, but no field rectifier is needed. Use of the series motor in this application is not open to objections because tachometer and regulator combine to hold the speed at preset value even though load on the motor should accidentally vanish.

Further Aspects of Automation

Besides the specialized apparatus discussed above, automation leans heavily upon numerous electronic and solid-state devices. The photoelectric tube, for example, is called upon for process control, counting, and error detection. Electronic time-delay methods are widely used to eliminate necessity for manual control of cycles in heat treatment of metals and plastics. Sensing units are employed to detect, locate, and even anticipate troubles in complex chains of operation.

As industrialization advances, man changes roles from doer to watcher. Automation in the industrial field has moved engineers to build machines that do all the functions of man, guided by programmed tape. The tapes are nu-

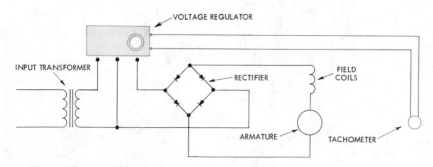

Fig. 19. Circuit for series motor.

merically controlled by computer, in time and action sequences. They provide functions such as cutting, drilling, and milling. The man in the picture becomes the watcher. He fills the machine with raw material or unfinished workpieces and unloads the machine after it has completed its numerically controlled tasks. Also he cares for the machine when any troubles develop.

Numerically controlled machines are of many types, and are used in machine shops, foundries, aircraft and automobile assembly lines, and nearly everywhere that man performs mechanical functions.

The most important contributor to the field of automation is the computer itself, which will be discussed in the next chapter.

REVIEW QUESTIONS

1. What is a closed-loop system?

2. Describe a simple synchro circuit.

3. What is the purpose of a differential synchro?

4. How does a synchro transformer circuit differ from the simple circuit?

5. Should a four-pole, three-phase, power synchro operate at a speed of approximately 1800 RPM?

6. Describe the amplidyne.

7. What is the purpose of the compensating winding in an amplidyne?

8. How does the rototrol differ from the amplidyne?

9. Why is thyratron speed control of a direct-current armature more economical than the series resistor method?

10. Where is automation successfully utilized?

11. What are the three basic control transducers?

12. Describe a closed loop control mechanism.

13. What is the difference between a synchro receiver and a synchro transmitter?

14. How do you check electrical zero in a synchro unit?

15. How many degrees apart are the legs of a synchro stator?

16. Will a synchro operate by application of direct current to its rotor?

17. What is the method of nulling of a synchro rotor and stator?

Chapter 17

Computer Technology

Essentially, a *computer* is a device for making calculations with extreme rapidity. The early lever-operated adding machines performed calculations, mostly addition and subtraction, through the use of mechanical gear chains, executing these tasks at a far greater speed than a man with paper and pencil. But they would hardly be classed as computers because they are far slower relatively, as compared to the modern computer, than the man and the pencil is with respect to them.

The computer is not limited to the solving of mathematical examples. It can store in a small space a vast quantity of information that is instantly available upon demand. It counts, retaining totals and putting them to such use as the operator may require. It can predict the result of a complex assortment of mechanical stresses on the basis of data furnished it from dozens of independent sources.

The speed at which a computer functions almost defies belief. It can multiply 12185 by 585, for example, by adding 12185 to itself 585 times, in the few seconds required to jot the figures on a piece of scratch paper. And it can divide a ten-figure number by a six-figure number, using methods of subtraction, in a like space of time. The computer's services in the field of industrial automation include management of servomechanisms, supervision of processes and methods, checking on costs and efficiency, even the selection among alternatives in case of emergency. These functions are possible, not because the machine is able to think, but because it has been supplied with all data bearing on a particular operation and with set *programs* or methods to be followed in combining it into usable form.

General Nature of Computers

There are two basic types of computers, the *analog* and the *digital*. The analog computer collects information supplied by transducers and other measuring devices. This information varies without interruption. The analog computer assembles and carefully evaluates the information to provide a solution for the kind of problem it has been designed to handle. The digital computer collects data that varies by discrete increments or counts as furnished by punched cards, paper or magnetic tape. The digital computer acts on the data according to pre-arranged instructions to produce answers upon command.

Although both computer types are completely electronic, the digital unit is more generally employed because it is more flexible and accurate. This computer is completely dependent on the data supplied and therefore accuracy of inputs must be totally reliable. Discussion which follows is confined largely to the digital machine.

COMPUTER PRINCIPLES

Strange and simple as it may seem, computer theory rests entirely on the following two types of statements:

(1) Event E will follow if events A *AND* B occur.

(2) Event E will follow if event A *OR* B occurs.

All switching networks, however complex, may be characterized by these two types of statements, or by complex repetitions and combinations of the same two statements.

For instance, the switching circuit of Fig. 1 may be described as a switching network that will light a lamp E if switches A *AND* (B *OR* C) are closed. Because the particles of speech *AND* and *OR* are called logical connectives, switching networks are often called *logical circuits*. The various switching combinations of a switching network are

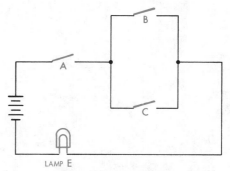

Fig. 1. Switching circuit.

often spoken of as the *switching logic* of the network.

Devices such as lamps and switches that reflect only two possible conditions are very adeptly called *on/off* and *yes/no* devices. Zeros and ones are used to symbolize concisely the two states of these devices. For instance, the table shown in Fig. 2 provides a more com-

	A	B	C	E
0	0	0	0	0
1	0	0	1	0
2	0	1	0	0
3	0	1	1	0
4	1	0	0	0
5	1	0	1	1
6	1	1	0	1
7	1	1	1	1

Fig. 2. Logic table.

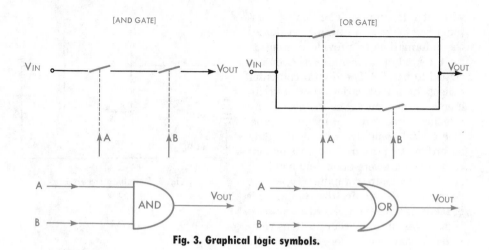

Fig. 3. Graphical logic symbols.

plete description of the switching circuit of Fig. 1 if *O's and 1's* stand, respectively, for off-and-on conditions of the switches and the lamp.

The table in Fig. 2 is referred to as a logic or truth table. This same type of table was originally used by philosophers to analyze the validity or truth of a chain of statements connected by logical connectives. It further explains the philosophical flavor of computer terms.

The English philosopher and mathematician George Boole developed a mathematical technique called Boolean algebra to deal with problems of logic in a systematic manner. The use of truth tables is one of the methods of solving problems in Boolean algebra. Mathematical techniques of Boolean algebra are used extensively in computer technology.

Logic Symbols

Another method of representing switching circuits is by means of the graphical *logic symbols* shown in Fig. 3. These symbols are called **OR** gates and **AND** gates and represent either the purpose or function of a switching circuit or branch. The **AND** gate symbol is read: an output signal is present when both switch-closing inputs *A AND B* are present. The **OR** gate symbol is read: an output signal is present when either switch-closing input *A OR B* is present.

Combinations of logic symbols may be used to represent complex logical circuits as illustrated, for example, in Fig. 4. Here the **AND** gate and the **OR** gate combination is read: An output signal is present when either *A OR B OR C* are present *AND* when *D AND E* are present.

Binary Devices

Two-state devices such as on/off switches and lamps are also referred to as binary devices in analogy to the binary system of numbers. This system of numbers has only two digits, 0 and 1,

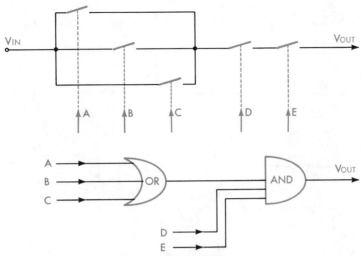

Fig. 4. Complex logic circuit.

and any quantity may be represented by combining 0's and 1's according to rules which will be described at a later time. The rules of arithmetic in this system are somewhat similar to the rules used in the normal decimal system of numbers.

A remarkable fact that is not generally realized is that ordinary relay switching techniques may be used to make computers that work on the same principles as present-day giant-brain computers, and that are able to perform the same variety of tasks. Attempts were made to make such computers, but relay switching operations were excessively sluggish and too power-consuming to be practical. Instead of relay switches, present computers use electronic switches that overcome both of these deficiencies.

As already implied in the above paragraphs, binary switching devices are the basic building units of a modern computer. Therefore, the question arises, "How can simple binary switching operations simulate complex computations and thinking processes?" It has been found that most processes involving practical questions can be broken down into simple chains of plain yes/no type of decisions. Hence, switching operations acting as binary decision elements may be combined to simulate such trains of simple yes/no decisions. Let us see how these statements apply, for example, to computer operations.

COMPUTER OPERATIONS

Fig. 5 illustrates a simple **OR** gate switching circuit. The truth table indicates all the possible combinations of on/off positions of the three switches.

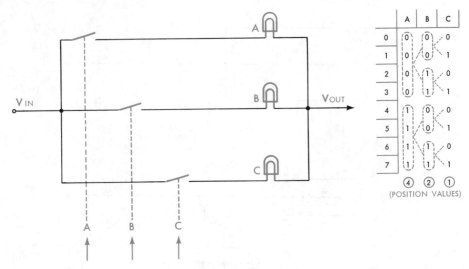

Fig. 5. Binary coding patterns.

The number of possible combinations in this case is eight. That is, each position of B (ON/OFF) may be paired with two possible positions of A (ON/OFF) making, thus, four possible combinations for A and B. These four combinations may be paired with each of two possible positions for C (ON/OFF) making, thus, a total of 2x2x2 or eight possible distinct combinations. Each addition of a binary device doubles the previous number of distinct combinations. This is a useful general rule to remember.

Each combination shown in the truth table of Fig. 5 is distinct from the others. Therefore, these combinations may be used as "binary" codes for the numbers 0 through 7. In fact, if each vertical column from right to left is given position values 1, 2, 4 as shown, the coding pattern may be stated by a simple rule. In any combination, add up the position values for positions containing 1's; the sum is the decimal number on the left. For instance the number values 3 contains 1's in its B and C columns. The position values of these columns are 2 and 1 respectively. The sum of 2 and 1 is 3, thereby proving the procedure.

Fig. 6 shows several examples of arithmetic operations carried in the binary system. The last two examples in Fig. 6 differ from the rules of the decimal system. The reason for this is, that the rule for binary addition of two units is $1 + 1 = 10$ since the binary code for 2 is 10. Hence, there is a 1-carry each time two units are added. For the same reason, $10 - 1 = 1$. This rule also applies when borrowing in binary subtraction as in the last example.

The binary coding patterns may be extended to represent any decimal number by adding more binary elements

ADDITION	SUBTRACTION	MULTIPLICATION
101 = 5	111 = 7	010 = 2
+010 = 2	−011 = 3	×011 = 3
111 = 7	100 = 4	010
		010
		0110 = 6

DIVISION	ADDITION	SUBTRACTION
10 = 2	011 = 3	101 = 5
	+001 = 1	−010 = 2
3 = 011⟌0110 = 6	100 = 4	011 = 3
011 = 3		
0000 = 0		

Fig. 6. Examples of binary arithmetic.

and adding corresponding columns to the logic table. Remember, each time a column is added, the number of possible combinations doubles and the position value of each additional column is double that of the preceding column.

A switching circuit such as shown in Fig. 5 may be used as a *memory* or information storage device. In this case, this switching circuit may store data representing 1's and 0's called *bits* (short for binary digits). It may also store data representing one of the numbers 0 through 7, which in turn may represent codes for data items such as "Don't forget to shop on this date."

If the lights *A, B,* and *C* are located on the front panel of the computer, information *read into* the computer by operating switches *A, B,* and *C* may be *read out* at the panel from the on/off binary patterns of the panel lights. An electric typewriter may be used to operate the relay switches by remote control. Remote control may also be performed by electrical contact made through perforated patterns punched in

cards, or by control signals generated by a magnetic tape.

The binary patterns shown in Fig. 5 work as a complete number system with arithmetic operations similar to the decimal system.

COMPUTER ORGANIZATION

Introduction

As indicated in the block diagram of Fig. 7, a digital computer has five main sections, and one auxiliary section which is common to all electronic devices, the power supply. The input section, designated as *Input* hereafter, takes data which has been recorded on punched cards or paper tape, and transforms it into electrical impulses that pass on to other portions of the machine. An electric typewriter which turns out perforated tape is often found here. The operator, or *programmer,* is trained to convert ordinary information into a sequence of yes/no type of codeable data which can be used by the machine. The tape passes into a slot where vertical contact pins create an electrical signal each time one of them completes a circuit, through a perforation, to the conducting strip underneath.

The storage section, or *Storage,* contains elements for retaining information supplied by Input, these devices being either electronic or magnetic. When a new sequence of operations is being inserted into the unit, all of it usually goes to Storage until fully assembled, after which instructions are furnished to the control section, and portions are removed from Storage as needed. In the office computer, reels of

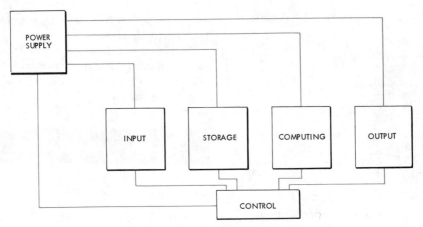

Fig. 7. Block diagram of digital computer.

magnetic tape, in cabinets, hold data indefinitely until called for by the control unit.

In *Computing,* mathematical operations are done on the basis of arithmetic or logical procedures supplied earlier by Input. Problems in higher mathematics can be solved if the necessary steps have been furnished in the shape of simplified routines involving addition, subtraction, multiplication, and division. The principal quality of Computing, however, is not so much the ability to solve intricate mathematical examples, but rather the speed at which it handles simple ones.

Output provides information called for by Input, delivering it in any number of ways, including typed replies, printed tape, punched cards, or even an arrangement of lighted bulbs on a nearby wall panel. In some cases, the answer is then typed out on an electric typewriter.

Control is the heart of the computer, and its guiding influence. It carries out in automatic, clocklike sequence instructions given by Input, utilizing electrical circuits which perform switching and gating operations, combining them to accomplish the purpose. Pulse signals, which carry the information from one part of the machine to another, originate here. Control extracts data from Storage and transfers it to Computing, takes it from Computing and returns it to Storage, or turns it over to Output, as the occasion requires.

This chapter is not concerned with particular sequences of operation, but with electronic circuits which enter into them. Electron tubes and transistors are employed in most cases, but other devices such as *magnetic cores, magnetic drums,* and magnetized wires are sometimes used.

The Bistable or Flip-Flop Circuit

An electronic switch which is usually part of Storage, but which is sometimes found in Control, is illustrated in Fig. 8. Actually there are two identical cir-

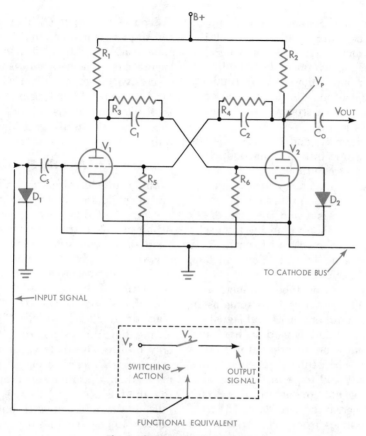

Fig. 8. Flip-flop circuit with tubes.

cuits, with parts of one equal to similarly placed elements of the other. Tubes V_1 and V_2 are matched, plate resistor $R_1 = R_2$, $C_1 = C_2$, $R_3 = R_4$, and $R_5 = R_6$. Cathodes are somewhat positive with respect to grids, being connected to the cathode tap on the voltage divider. The circuit is termed either bistable or flip-flop, and the meaning of these terms will become clear shortly.

Certain facts should be noted. First, when current flows through one of the tubes, say V_1, voltage drop in R_1 causes the voltage at the plate of the tube to be considerably less than before conduction began. When current flow ceases, the plate voltage rises. Also, the plate voltage is fed through a resistor to the grid of the other tube.

Suppose V_2 is conducting at the moment. A negative signal by way of capacitor C_s at the input, has no effect upon V_1 which is inactive, but overcomes the positive voltage on the grid of V_2 so that current flow is stopped. The higher voltage at the grid of V_1, brought on through cessation of current flow in R_2, which raises the potential at its lower end allows V_1 to conduct, which it continues to do until another negative impulse from C_8 interchanges states of the tubes.

Capacitors C_1 and C_2 prevent both tubes falling into a non-conducting state at the same time. When V_2 ceases to conduct, capacitor C_2 reacts quickly to draw electrons up through R_5, making its upper end and the grid of V_1 positive. Any delay resulting from self-inductance in the circuit through R_4 is thus overcome, and plate-current flow in V_1 is initiated.

Since only one tube conducts at a time, and the active tube remains in this state until another input signal arrives, the circuit is said to have two stable states or, in other words, it is *bistable*. When tube V_1 is conducting, the circuit said to be in the *one state*, when it is not conducting, the circuit is said to be in the *zero state*. The term *flip-flop* arises from the fact that the circuit changes so rapidly, or "flips" from one state to the other when a signal appears.

It is now possible to explain action of this circuit in transmitting impulses to the following stage. Start with a negative signal which produces the one state, V_1 is conducting. At the same time, voltage at the plate of V_2 suddenly rises, transmitting a positive impulse to output O, which leads to the next bistable unit. This positive signal is short-circuited by a diode there, similar to D shown at the input here. A second input signal changes the state to zero, V_2 conducting.

When current begins to flow in V_2, its plate voltage drops, sending a nega-

tive pulse to output O. The signal is not short-circuited by the diode, and appears at the grids of tubes in the following stage. The net result is that two successive pulses are required to furnish a signal to the next stage. D_2 is a blocking diode to shield each tube from the steady grid bias of the other.

Transistor Bistable or Flip-Flop Circuit

The transistor circuit of Fig. 9 is equivalent to that of Fig. 8, but there are important differences in polarities. A pair of P-N-P transistors is used instead of vacuum tubes, the emitters biased positively. Collectors are attached to the negative B terminal, the B+ terminal being grounded. Note, also, that diodes D_1 and D_2 which replace the single input diode of the preceding diagram, are polarized to allow only positive signals to go through.

Suppose Q_2 conducting at the moment when a positive signal arrives at the input. Since Q_1 is inactive, the impulse has no effect upon it, but the reduction in negative bias at the base of Q_2 renders it non-conducting, while Q_1 becomes active. Collector Q_2 growing suddenly more negative, sends a negative impulse through C_0, but the next stage is not affected since it accepts only positive signals.

High negative bias at the base of Q_1, supplied through resistor R_2 from the collector terminal of Q_2, has the effect of increasing the emitter bias, and Q_1 conducts to saturation. Flow of current in resistor R_3 lessens the negativity of the collector terminal and the voltage supplied to the base of Q_2 through resistor R_1. This more positive voltage has the effect of decreasing emitter bias of Q_2, and insuring that it does not conduct.

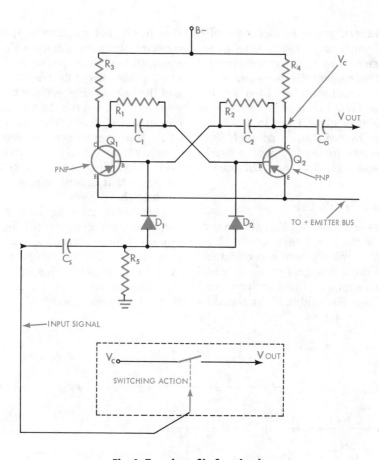

Fig. 9. Transistor flip-flop circuit.

Q_1 remains active until another positive input signal reduces the emitter bias, causing the state to change. When Q_2 now starts to conduct, the drop in resistor R_4 renders the collector less negative, so that a positive signal goes to the output wire, transmitting an impulse to the next transistor stage. P-N-P units were employed here, but N-P-N will serve equally well if biasing rules are observed. Application of the

flip-flop circuit to counting will be explained later, along with another type of counter.

Gating

Control accomplishes its work largely through the use of electronic gates. As explained under *Logic Symbols*, a *gate* is a switch that permits an output signal only when certain conditions have been satisfied. There are two kinds: *AND* gates and *OR* gates. An *AND* impulse is let through only after a series of requirements have been complied with. An *OR* impulse results when any one of multiple conditions has been carried out. The *AND* gate may be represented by a number of switches in series, an *OR* gate by switches in parallel. Control circuits are made up of networks containing both *AND* and *OR* sub-circuits.

Electron Tube *AND* Gate

Plates of V_1 and V_2, Fig. 10, are connected to the B+ terminal through resistor R_5 in which there is considerable voltage drop. The grid of V_1 is biased so that the tube conducts under normal conditions. The grid of V_2 is biased so

that it will not conduct in spite of a constant, low pulse signal to V_2. When a negative signal S_1 arrives at the grid of V_1, plate current flow is interrupted, and the voltage rises sufficiently at the plate of V_2 for the tube to conduct. The output signal, taken from the cathode of V_2, continues to pass so long as the negative input signal is delivered to the grid of V_1. This process is known as *gating-in*, that is, permitting a signal to go through.

The circuit may be used also for *gating-out*, which means cutting off an impulse which, otherwise, would flow continuously. Such a result may be obtained with proper biasing. In this case, V_2 normally conducts, passing signal S_2 to the output circuit. V_1 is biased

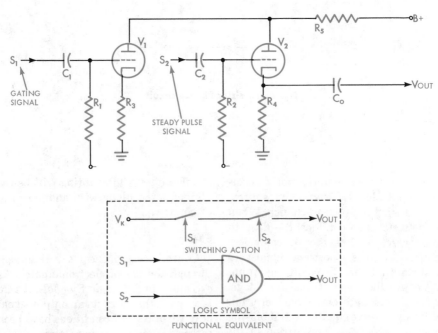

Fig. 10. Electron tube *AND* circuit.

so that it does not conduct, but when positive signal S_1 appears on the grid, V_1 becomes active, current in the plate circuit causing a voltage drop in resistor R_5, lowering the potential at the plate of V_2 and stopping current flow there. In both cases, adjustment of plate voltages and grid biases is rather critical.

Transistor *AND* Gate

A transistorized version of the circuit, using a pair of P-N-P's, is illustrated in Fig. 11. The circuit may be arranged either for gating-in or gating-out as desired, by proper selection of biasing and collector voltages. Voltage drop in resistor R_4 is utilized in much

the same way as in Fig. 10. Similarly placed elements are marked alike in the diagrams to aid in comparison. Note that both input and output here are in the base leads.

Screen-Grid *OR* Circuit

A screen-grid *OR* circuit is shown in Fig. 12. Steady pulse signal S_2 is applied to the cathode, while impulse S_1 may appear at either one of the grids. The two grids are each slightly negative so that their combined charge prevents the tube from conducting. When a positive signal arrives at either one, however, it reduces the total negative potential so that plate current flows.

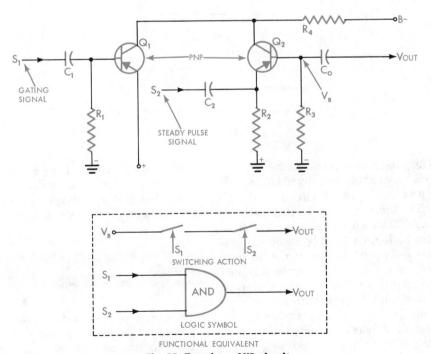

Fig. 11. Transistor *AND* circuit.

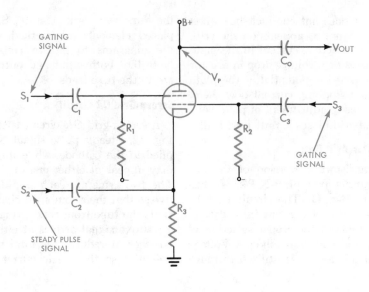

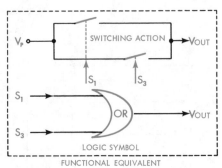

Fig. 12. Screen-grid OR gate.

Thus, the tube becomes active if either grid is excited, thereby fulfilling requirements of an *OR* circuit.

The same tube can be employed for an *AND* circuit if biasing is adjusted so that both grids must be rendered positive before conduction occurs. In both instances, the circuits may be designed for gating-out rather than gating-in. Transistor *OR* gates, with two or more units in parallel, are commonly employed. Diodes, including the solid-state silicon type, are also used for the purpose. The diagrams shown here are merely representative of a great number of circuits devised for switching or gating, but the underlying principles are not essentially different in any case.

MEMORY DEVICES

Memory consists of a delay imposed between reception of an electrical impulse and its transmission to another input device or its return to the sending element. In the course of development, experimenters have applied all possible scientific techniques to the problem of creating delay, including electron-tube and transistor circuits, superconductivity, vibrating crystals, photographic film, magnetostriction, and even chemical cells. Some of these methods provided short-time retardation which is acceptable for many uses, but valueless where long-time retention is required. Cabinets filled with reels of magnetic tape are termed *external memory units,* and are exceedingly important for recording data. Recovery of their information is too slow a process, however, for high-speed computer operations. *Internal memory devices* will now be considered.

Cathode-Ray Memory Tube

The cathode-ray tube, Fig. 13, has vertical and horizontal deflection plates in the neck of the glass tube and cause a beam of electrons to sweep over a fluorescent screen in the expanded end. The same method is applied in the cathode-ray memory unit of Fig. 14. Instead of a light-sensitive coating in

the flared end of the tube, a *capacitor screen,* Fig. 14 (top) is employed. The screen is made up of a metallic back plate to which a sheet of insulating material is attached. The insulator is sprayed with a fine coating of metallic substance that amounts to a microscopic dusting of separate particles, each one forming a small capacitor in conjunction with the insulating material and the back plate.

The face of the screen is divided into a number of definite spaces, each location determined by voltages applied to vertical and horizontal deflecting plates. That is, the electron beam may be directed to any particular space, such as W in the figure, by charging the horizontal plates with voltage H and the vertical plates with voltage V, the proper values being selected by punched tape or cards at computer Input.

Impulses are recorded in the form of *ones* or *zeros.* Suppose it is desired to hold a *one* at position W. The beam will touch there, and electrons deposited at the spot will repel an equal number from the opposite face of the ca-

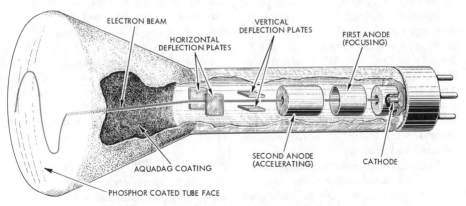

Fig. 13. Cutaway drawing of cathode-ray tube.

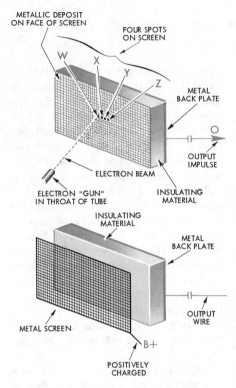

Fig. 14. Cathode-ray memory screens.

change will occur in the output circuit. At *X*, which has no charge, electrons will be deposited, sending a pulse through the output, which is now recorded to show that a *zero* was left there. No pulses will be generated at *Y* or *Z*, indicating that *ones* are present, and a card marked to show information stored here will read: *1011* as it emerges from Output. After all information has been read, the screen is neutralized by a charged plate not shown in the illustration.

Another form of cathode-ray memory screen is shown in Fig. 14 (*bottom*), where the back plate and the insulator are the same as before, but there is no metallic deposit on the latter. A charged grid is placed a short distance in front of the insulating sheet, the electron beam passing between its closely-spaced wires to strike "numbered" spaces on the insulating face. The action is much the same as before except that the electron beam dislodges electrons instead of depositing them, causing secondary emission which is attracted to the positively-charged screen, and leaving the spot positively charged. The grid is deenergized, in order to read the information, and the electron beam focused on desired spots. If a *one* has been deposited, resulting in a positive charge, electrons from the beam will be accepted, giving rise to an impulse in the output circuit. If a *zero* has occurred, no pulse results. Operation of the *iconoscope* TV camera tube is based upon a method like that of Fig. 14 (*top*).

The Magnetic Drum

A metal drum coated with magnetically sensitive material, Fig. 15 (*top*), has proven a useful memory device. Multiple recording heads are poised in

pacitor, which is the back plate. A pulse occurs in the output, but the computer does not take note of it. Suppose further that a *zero* is to be held at spot *X*, directly in line with *W*. When deflection plates are set for this spot, the beam is deenergized, momentarily, so that no charge is deposited. If *ones* are to be left at *Y* and *Z*, the beam remains excited while deflection plates are set for the two locations.

When this information is needed, an Input card punched for these four locations directs the beam to each spot in turn. As it strikes position *W*, which already has an electrical charge, no

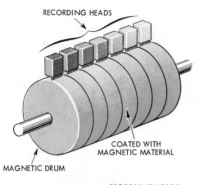

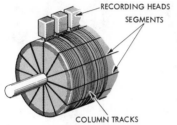

Fig. 15. Magnetic drum.

Ferrite Cores

One of the most important memory devices developed thus far is the *ferrite core*, Fig. 16. Ferrite is a mixture of powdered iron and a ceramic. It is highly magnetic, with a permeability curve more nearly rectangular than that of high-grade steel used in the magnetic amplifier. The core is of *toroidal* shape, Fig. 16 (A), and sometimes has two separate windings, Fig. 16 (B), one for recording or magnetizing, the other for "reading." When current is supplied to a coil, the core is magnetized so that domains line up in, say, a clockwise direction, and will remain in this state indefinitely. When current is passed in the opposite direction, the domains swing quickly to a counterclockwise alignment, lines of force cutting across turns of the coil to generate a voltage there.

Core windings may be dispensed with, as in Fig. 16 (C). Three straight wires pass through the center, X and Y at right angles to one another, S at a 45° angle. The core is magnetized normally, in a certain direction. A pulse current through X or Y alone is insufficient to reverse this polarity, but a pulse current of the same strength through both X and Y produces the required magnetic force. Domains suddenly reverse alignment, the high rate of flux change generating an appreciable voltage in wire S.

Wires and cores are arranged in panels within the computer, Fig. 16 (D). The illustration shows thirty-six cores, six vertical, six horizontal, and one pick-up or sensing wire. Cores are selected, at Input or Control, through the energizing of a certain pair of wires. For example, current through *1-V* and *1-H* at the same time, affects the core at the upper left, and no others, because

line close to the drum which revolves at speeds as high as 4000 rpm. The cylindrical surface is divided into *segments* and *tracks,* a segment marking a portion of the circumference, and a track the complete circle which lies under an individual head as the drum turns. Locations are designated in this manner, which is allied to that noted in connection with the cathode-ray screen.

Electrical impulses in windings of the recording heads induce a certain polarity for a *1,* the opposite polarity for a *0,* so that the track bears a succession of *N* and *S* poles. This information is recovered, when needed, by the same heads which deposited it.

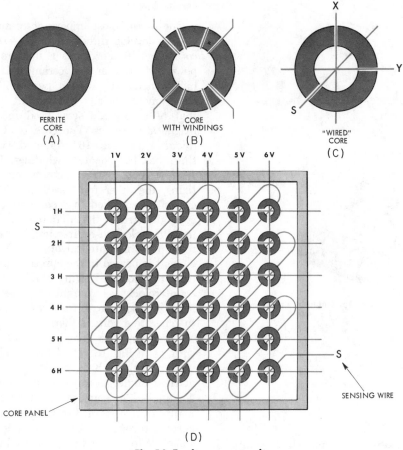

FERRITE
CORE
(A)

CORE
WITH WINDINGS
(B)

"WIRED"
CORE
(C)

Fig. 16. Ferrite memory units.

both magnetizing forces are present only in this one location. If any other pair is chosen at random, for example *3-V* and *4-H*, the only core affected is the third from the left in horizontal row 4. The procedure for depositing or reading *1*'s and *0*'s is the same as in other memory elements discussed above. Experimentation with memory devices continues from day to day. In a recently devised method, seeking to miniaturize and to cheapen core assemblies, the crossed wires and sensor

conductor are placed on a cellophane sheet. Junctions are then sprayed, using a template, so that circles of a quick-drying iron mixture similar to toroids, are deposited there. A deviation from magnetic core storage is the *twistor*, which has a thin piece of special magnetic material taped in helical fashion around a copper wire that carries electrical impulses. A sensing wire around the two of them detects impulses generated by changing magnetic polarities.

Shift Circuit

Note the principle of the decade counter circuit, Fig. 17. Three thyratron tubes are shown in the illustration, all plates attached to B+ supply, all grids to a signal bus through appropriate capacitors. Suppose tube V_1 conducting at the moment. Although a positive voltage which caused V_1 to become active is no longer present, plate current is uninterrupted because current flow in a thyratron is independent of grid voltage, as explained in Chapter 13, once it has started. The upper end of cathode resistor R_2 is attached to the grid of V_2, giving it a positive charge which is not quite large enough to overcome the negative potential grid resistor R_1 delivers there. Voltage drop in resistor R_2, which is in series with plate supply, is not great enough to render V_1 inactive.

A strong positive impulse that arrives on the signal bus passes to grids of all tubes by way of the capacitors. It has no effect on V_1 of course, or on V_3 where it is insufficient to overpower the negative grid charge. At V_2, however, its potential adds to that from the upper end of R_2, starting a flow of plate current. For an instant, both V_1 and V_2 conduct, but the voltage drop in R_7 has doubled, and V_1 is no longer able to carry on. Any tendency for V_2 to drop out is opposed by its positive grid voltage until V_1 ceases to conduct, at which time the voltage drop in R_7 is cut in half, and potential at the V_2 anode rises to the higher level.

Meanwhile, when V_1 becomes inactive, the voltage supplied to the grid of V_2 by cathode resistor R_2 vanishes. At the same time, voltage from the top of R_4 is imposed on the grid of V_3, pre-

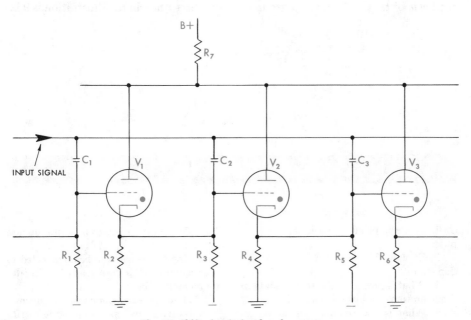

Fig. 17. Shift circuit for decade counter.

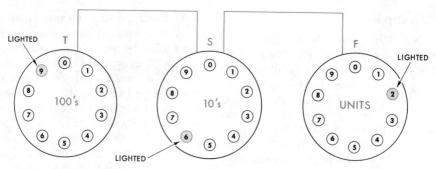

Fig. 18. Shift counter.

paring this tube to become active at the next signal impulse.

A number of tubes can be arranged in a circle, for counting purposes, conduction being shifted to the next tube in line with each input signal. Fig. 18. shows a *decade counter* (counting on the basis of tens) which has a group of three elements: *T*, *S*, and *F*. The one on the right, *F*, shows individual impulses, and delivers a signal to *S* as the circle is completed. Thus, *S* counts in tens, transmitting a signal for each full turn so that *T* counts hundreds. It is apparent that the number recorded by the three units in the illustration is 962.

REVIEW QUESTIONS

1. Describe briefly a basic digital computer.

2. Explain the general nature of a flip-flop circuit.

3. What is meant by the term 1-state in the operation of a bistable circuit?

4. What is the difference between an *AND* gate and an *OR* gate?

5. Differentiate between gating-in and gating out.

6. How is the electron beam directed to a particular spot on the screen of a cathode-ray memory tube?

7. Describe a magnetic memory drum.

8. Explain the operation of a ferrite core.

9. What is a shift circuit?

10. Describe a basic logical circuit.

11. Prepare a truth table for three switches having an on and an off position each.

12. Draw a logic circuit for the statement, "an output signal is present when switch *A AND B* are closed and switch *C OR D* are closed."

13. Add the binary numbers 011 and 101.

14. What is meant by a "bit"?

Chapter 18

Electrical/Electronic Safety

GENERAL

Safety education today has become an important phase of every training program. Under the 1970 Federal *Occupational Safety and Health Act* (OSHA), the employer is required to furnish a place of employment free of known hazards likely to cause death or injury. The *employer* has the specific duty of complying with safety and health standards as set forth under the 1970 act. At the same time, *employees* also have the duty to comply with these standards.

A full treatment of the subject of safety is far beyond the scope of this book, and there is ample justification for a full course in safety procedures, including first aid treatment, for all electrical and electronic technicians. Instructors should be certified by the state and qualified for any special electrical applications. The intent of this chapter is to make students fully aware of the ever-present, invisible, and generally silent hazards in handling electrical apparatus and to point out some fairly common causes of electrical shocks and fires that can easily be overlooked.

Since the time when Benjamin Franklin flew his famous kite it has become more and more apparent that electricity, even in its milder forms, is *dangerous*. A fact not widely mentioned in history is that shortly after Franklin's experiment a Russian experimenter was killed in his attempt to duplicate the kite trick. We can therefore assume that good safety habits are mandatory for all who use, direct the use of, or come in contact with electricity. Electrical equipment is found in every place that the ordinary person may find himself. For this reason it becomes the responsibility of each of us to be knowledgeable of electrical safety and to become his own and his brother's keeper.

The most basic cause of electrical accidents, as other types of accidents, is *carelessness* and the best prevention is *common sense*. However, knowing exactly what to do in an emergency is only achieved through formal education or experience. Unfortunately, the experience could be fatal, so it is more desirable to derive your knowledge through schooling. When an emergency happens it is often accompanied by panic that can cause the mind or muscles to become paralyzed. The antidote for this panic is education.

RESPONSIBILITY

Electrical/electronic safety is especially important to the technician who is exposed to electrical equipment in the raw state. He is likely to be the one who comes into direct contact with electrical and/or electronic components that may do harm to the body. Since this is the case, he finds himself in the precarious position of being continually alert to hazards that may affect him, his associates, and the people who use the equipment he builds or repairs. The supervisor is responsible for enforcing the rules of safety in the area under his direct supervision. Inspectors ensure that equipment is tested before it is released for use, and final testing should include safety precautions. Finally, the user of the equipment should be qualified in its operation so he may know when it is operating in the proper manner. *No one who comes into contact with such equipment is exempt from some degree of responsibility.*

The average layman may not believe he can get a severe shock from an electrical component disconnected from a power source. The electrical or electronic technician should know better. You know from your study of capacitance, for instance, that capacitors hold electrical charges which they later discharge. A fully charged capacitor disconnected from the power source can deliver a severe and possibly fatal shock, so you take the precaution of "shorting" it before removing it from the equipment. Also, as your brother's keeper, you should warn the user of this hazard.

ELECTRICAL SHOCK

There is a common belief that it takes a great volume of electricity to cause a fatal shock and that high voltage is the thing that provides the jolt to do the job. Imaginative stories about people being literally fried or jolted from their shoes are supplied freely by storytellers. Although there are true tales of this type, this is not generally the case. The bulk of electrical shocks come in small packages. Death from electrical shock happens most often from ordinary 60-hertz, 115 VAC house power. The effect, in general, is an instantaneous, violent type of paralysis. The human body contains a great deal of water and is normally somewhat acid and saline. For these reasons it is a fairly good conductor of electric current, which has no regard for human life or feelings.

Our brains are loaded with tiny nerve ends that provide transmitting service to the muscles. Muscles, in turn, provide motion for the body functions such as the heart and lungs. Now, assume that an electrical impulse that was not called for told your heart to speed up its pumping job or to stop its pumping job. Or suppose that a similar impulse of electricity told the lungs to quit taking in air. These situations do occur and with a comparitively small amount of current flow. Depending on a multitude of complex variables, a small current such as 10 milliamps can be very unpleasant. Currents no larger than 20 milliamps can cause muscle tightening or freezing. Currents of 30 milliamps can cause damage to brain tissue and blood vessels. And damage to brain and blood vessels can, of course, be fatal.

As you know, a decrease in resistance (according to Ohm's Law) causes current to increase. Body resistance decreases when perspiring or otherwise wet. A great number of things can cause variation in resistance. The general health of the person is probably the most important variable. A body in good condition has a much better chance of recovering from electrical shock. The muscles, by being in better tone, can recover to normal from paralyzation.

The path of current flow varies the shock. For instance, if the current involves the brain or heart it is naturally more dangerous than another path. The length of exposure can also be a factor, as well as the size or surface area of the electrical contact. Large voltages can cause spastic action, but recovery can be rapid. Also, small currents can cause muscle paralysis.

In the event of paralysis, artificial respiration or massaging must take place as quickly as possible to prevent loss of body functions and damage to the brain because of lack of oxygen. A condition known as ventricular fibrilation (uncoordinated heart beats, both fast and irregular) can occur with high currents, say 50 to 100 milliamps. This action will continue until something is done externally to restore regular heartbeat.

Death normally occurs when currents reach 250 milliamps. This does not have to occur, however, as rapid first aid can save a victim whose heart has actually stopped beating or whose lungs have stopped pumping momentarily. In many cases rapid action by knowledgeable persons can prevent body damage and save lives.

RAPID RESCUE TECHNIQUES FOR ELECTRICAL EXPOSURE

A special course in first aid will ensure the proper methods. A wrong method can be worse than no action at all. The first rule of thumb for an electrical shock victim is to remove the current path. This can be done by turning the power switch off if it is readily accessible. If not accessible, the person can be detached from the current source by using an insulator of some sort such as wood, rubber, cork, or plastic. Sometimes it is more suitable and sensible to remove the power source from the victim. Whichever is the case, isolation from the current source is by far the most important and first move that can be made.

Isolation procedures can cause problems. The rescuer may find himself in the current path. Touching a person who is paralyzed to a current source provides the current source another path through the rescuer's body. Care must be taken to prevent this from happening. After isolation, artificial respiration and/or other first aid should be applied. In all cases speed is vital. Death occurs in direct proportion to time. It is therefore obvious that a person given artificial respiration in the first three minutes has a much better chance of survival than one who is given artificial respiration after five minutes.

SNEAKY ELECTRICAL CONDUCTORS

In every electrical activity there are sneaky conductors of electricity that cause continuous problems of "shorts" and therefore electrical shock. Anywhere that you have electricity it is not just the wire, the source, or the load that provides these paths for current flow. For instance, cement may seem dry and clean but have moisture in it. In this condition the cement could be a sneaky conductor. Metal floors are, of course, good conductors. A sweating body can cause a multitude of problems. Machines of all types in the general area will serve as conductors. Steel building posts can be conductors as well as metal roofs, steel desks, pots and pans, bicycles, automobiles, refrigerators, washing machines, and just about any other metal object you may name. These sneaky conductors can cause current draw.

Make sure, then, that you are properly grounded or that the equipment you are working with or near is grounded. Use insulated tools to prevent current from flowing where it isn't supposed to. Use floor pads and keep water and oil from floors around which you are performing electrical work. Water does not mix with electricity. Keep debris and scraps picked up to avoid similar situations. Cleanliness and alertness will help avoid or eliminate sneak circuits.

BATTERY HANDLING

Three basic safety problems are associated with the handling of batteries. These are: acid burns, fires, and explosions. Acid burns may be prevented by use of battery-handling clothes and equipment. Clothes for this purpose are mostly made of rubber and include such things as aprons, gloves, boots, and special glasses.

Proper tools are essential to perform the correct procedures safely. Proper flooring will prevent falling and spilling acids.

Fire and explosion may be caused by ignition of gases given off from the charging action. These gases, when mixed with air (oxygen), provide a highly flammable and explosive situation. Gases should not be allowed to accumulate. Ventilators should be installed in battery shops to expel the dangerous gases. Smoking in the area should be prohibited. Signs should be installed warning everyone who might enter the battery shop of the dangers that are within.

HOW TO CONTROL AN ELECTRICAL FIRE

An electrical fire is caused by current flow to some circuit that cannot withstand the current level. Also, electrical fires are caused by sneak circuits which accidently draw current, for instance a "short" to the case in a motor-driven furnace. In any event, since the cause of the problem is current flow, disconnecting the current should be the first step in eliminating the problem. Remove power from the circuit preferably by throwing a switch or by isolating the fire, using insulating material such as wood or plastic. Cut wires with wooden handled hatchets or some similar device. Prevent yourself from becoming part of the circuit. After removing the current, call the fire department, then put out the fire.

Electrical fires are best extinguished with the use of carbon dioxide (CO_2) directed toward the base of the fire. Do not use foam, as it conducts electricity.

GOOD SOLDERING HABITS

Soldering irons or soldering guns all have one thing in common: they are hot. Each is hot enough to melt solder joints. The actual temperature varies with the solder type. The speed at which the soldering iron or gun melts the solder joints is dependent on the wattage of the iron or gun and the size or complexity of the joint. In all events, the soldering device must not only be protected from the handler but also from the other circuits or equipment around it.

The soldering iron or gun should be placed in a heat-sink holder between soldering actions. Heat sinks should also be used to protect electrical/electronic circuit components. Danger of work contamination is always present as dripping or stringing of solder may occur during soldering operations. Fire haz-

ards are always present when working with heat.

Electrical fire hazards may be prevented by ensuring that power is removed from equipment being worked on. Soldering operations should take place only after proper preparation of the work area. Clean and dry work areas, the proper wattage iron, and a well laid out soldering plan help prevent soldering accidents.

REVIEW QUESTIONS

1. What type of electrical power is generally the most dangerous?

2. Whose responsibility is good safety practices?

3. What is the basic cause of most electrical accidents?

4. High currents can cause a condition known as ventricular fibrillation. What is another name for this condition?

5. Why is a body in good condition better able to recover from electric shock?

6. Electrical fires are best extinguished by what type of fire extinguisher?

7. Why are ventilators installed in battery shops?

8. What is the first rule of thumb when dealing with an electrical shock condition?

9. What is the basic method of first aid when dealing with an electrical shock victim?

10. Why does a sweating body cause electrical problems?

Appendix A: Letter Symbols, Abbreviations, and Prefixes

Fundamentals of Electricity

TABLE I. ELECTRICAL/ELECTRONIC SYMBOLS*

LETTERS	MEANINGS	BASIC UNITS OR DESCRIPTIONS
A	ampere(s)	coulomb(s) per second, the basic unit of electrical current
AC, ac or a-c	alternating current	changes polarity every half-cycle
AF, af or a-f	audio frequency	16 to 16,000 Hz, within audible range
AM, am or a-m	amplitude modulation	
B	(1) susceptance, (2) base of transistor	susceptance in mho(s)
C	capacitance or capcitor	capacitance in farads
CHC, chc	choke coil	
CR, cr	(1) crystal rectifier, (2) control relay	
CRT, crt	cathode ray tube	
CU	piezoelectric-crystal unit	
DC, dc or d-c	direct current	flows in one direction only
DIEL	dielectric	insulating material
DIO	diode	
DP, dp	double-pole	applies to elec. switches
DPDT, dpdt	double-pole double-throw	applies to elec. switches
DPST, dpst	double-pole single-throw	applies to elec. switches
E, e	electromotive force (emf) voltage, volts, potential	volt(s)
EDP	electronic data processing	
EHF	extremely high frequency	30 GHz to 300 GHz used for weather radar, NASA communications
EMF, emf	electromotive force, voltage, volts, potential	volt(s)
f	frequency	hertz (cycles per second)
F	farad(s)	unit(s) of capacitance

*See Table 2 for Greek Alphabet

TABLE I. ELECTRICAL/ELECTRONIC SYMBOLS*

LETTERS	MEANINGS	BASIC UNITS OR DESCRIPTIONS
FF, ff	flip-flop	type of electronic circuit
FM	frequency modulation	
H	henry(s)	unit(s) of inductance
HF	high frequency	3 MHz to 30 MHz (used for aircraft, amateur radio)
Hz	hertz (singular and plural)	unit of frequency (cycles per second)
I, i	current	ampere(s), coulombs per second
IC	integrated circuit	
L	inductance	henry(s)
LF	low frequency	30 kHz to 300 kHz (used for LF marker beacons)
MF	medium frequency	300 kHz to 3 MHz (used for AM broadcasting)
Q	(1) coulomb, (2) factor merit of induction coil, (3) general symbol for transistor in diagrams	coulomb is electrical charge of approx. 6.25×10^{18} electrons, the basic unit of electrical quantity
R	resistance	ohm(s)
RF, rf or r-f	radio frequency	30 kHz to 30,000 MHz
SP, sp	single-pole	applies to elec. switches
SPDT, spdt	single-pole double-throw	applies to elec. switches
SPST, spst	single-pole single-throw	applies to elec. switches
SW, sw	(electric) switch	
UHF	ultra high frequency	TV channels 14 through 83, GCA, Tacan)
V	volt(s), voltage, emf, potential	unit of electromotive force, electrical pressure, or potential
VHF	very high frequency	30 MHz to 300 MHz (used for TV, FM radio, VHF marker beacons) unit of electrical power

*See Table 2 for Greek Alphabet

TABLE I. ELECTRICAL/ELECTRONIC SYMBOLS*

LETTERS	MEANINGS	BASIC UNITS OR DESCRIPTIONS
VLF	very low frequency	DC to 30 kHz (used for DC, AC power, audio sonar, ultrasonic cleaning)
W	watt(s)	unit(s) of electrical power
X	reactance	ohm(s)
Y	admittance	mho(s)
Z	impedance	ohm(s)

*See Table 2 for Greek Alphabet

TABLE 2. INTERNATIONAL SYSTEM (SI) SYMBOLS
IN GENERAL USE FOR ELECTRONIC WORK

PHYSICAL QUANTITY	UNIT	SYMBOL	OLD NAME & SYMBOL (IF DIFFERENT)
Capacitance	farad	F	f, fd
Current	ampere	A	a, amp
Frequency	hertz	Hz	Cycles c, cps Cycles per second
Inductance	henry	H	h
Power	watt	W	w
Resistance	ohm	Ω	
Time	second	s	sec
Voltage (potential difference, electromotive force)	volt	V	v
Wavelength	meter	m	

TABLE 3. GREEK ALPHABET

NAME	CAPITAL	LOWER CASE	DESIGNATES
ALPHA	A	α	Angles, coefficient of thermal expansion
BETA	B	β	Angles, flux density
GAMMA	Γ	γ	Conductivity
DELTA	Δ	δ	Variation of a quantity, increment
EPSILON	E	ϵ	Base of natural logarithms (2.71828)
ZETA	Z	ζ	Impedance, coefficients, coordinates
ETA	H	η	Hysteresis coefficient, efficiency, magnetizing force
THETA	Θ	θ	Phase angle
IOTA	I	ι	
KAPPA	K	κ	Dielectric constant, coupling coefficient, susceptibility
LAMBDA	Λ	λ	Wavelength
MU	M	μ	Permeability, micro, amplification factor
NU	N	ν	Reluctivity
XI	Ξ	ξ	
OMICRON	O	o	
PI	Π	π	3.1416
RHO	P	ρ	Resistivity
SIGMA	Σ	σ	Summation symbol (cap)
TAU	T	τ	Time constant, time-phase displacement
UPSILON	Υ	υ	
PHI	Φ	ϕ	Angles, magnetic flux
CHI	X	χ	
PSI	Ψ	ψ	Dielectric flux, phase difference
OMEGA	Ω	ω	Ohms (capital), angular velocity ($2\pi f$)

TABLE 4. STANDARD PREFIXES, THEIR SYMBOLS AND MAGNITUDES

FACTOR BY WHICH UNIT IS MULTIPLIED	PREFIX	SYMBOL	OLD PREFIX & SYMBOL (IF DIFFERENT)	
10^{12}	tera–	T		
10^{9}	giga–	G	kilomega–	km
10^{6}	mega–	M		m
10^{3}	kilo–	k		
10^{2}	hecto–	h		
10^{1}	deka–	da		
10^{-1}	deci–	d		
10^{-2}	centi–	c		
10^{-3}	milli–	m		
10^{-6}	micro–	μ		
10^{-9}	nano	n	millimicro–	mμ
10^{-12}	pico–	p	micromicro–	$\mu\mu$
10^{-15}	femto–	f		
10^{-18}	atto–	a	Kl	

EXAMPLES:

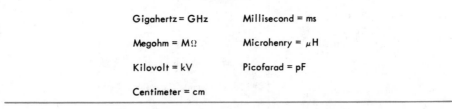

Gigahertz = GHz   Millisecond = ms

Megohm = MΩ Microhenry = μH

Kilovolt = kV   Picofarad = pF

Centimeter = cm

Appendix B: Electrical and Electronic Symbols

ADJUSTABILITY VARIABILITY

These recognition symbols shall be drawn at about 45 degrees across the body of the symbol to which they are applied. Use only if essential to indicate special property.

ADJUSTABILITY (extrinsic adjustability)

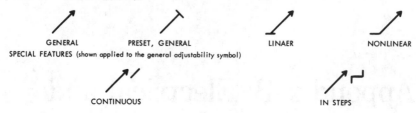

| GENERAL | PRESET, GENERAL | LINAER | NONLINEAR |

SPECIAL FEATURES (shown applied to the general adjustability symbol)

CONTINUOUS   IN STEPS

SHIELD SHIELDING

Normally used for electric or magnetic shielding. When used for other shielding, a note should so indicate.

These are long dashes

GENERAL   OPTICAL

RESISTOR

| GENERAL | TAPPED RESISTOR | ADJUSTABLE CONTACT | ADJUSTABLE; RHEOSTAT |

CAPACITOR

If it is necessary to identify the capacitor electrodes, the curved element shall represent the outside electrode in fixed paper-dielectric and ceramic-dielectric capacitors, the moving element in adjustable and variable capacitors, and the low-potential element in feed-through capacitors.

GENERAL   POLARIZED CAPACITOR   ADJUSTABLE OR VARIABLE CAPACITOR

If it is necessary to identify trimmer capacitors, the letter T should appear adjacent to the symbol.

ANTENNA

Types or functions may be indicated by words or abbreviations adjacent to the symbol. Qualifying symbols may be added to the antenna symbol to indicate polarization, direction of radiation, or special application. If required, the general shape of the main lobes of the antenna polar diagrams may be shown adjacent to the symbol. Notes may be added to show the direction and rate of lobe movement. The stem of the symbol may represent any type of balanced or unbalanced feeder, including a single conductor.

GENERAL   DIPOLE   LOOP

BATTERY

The long line is always positive, but polarity may be indicated in addition.

GENERAL   ONE CELL   MULTICELL   MULTICELL WITH TAPS

GENERALIZED ALTERNATING-CURRENT, DIRECT-CURRENT SOURCE

ALTERNATING-CURRENT GENERATOR   DIRECT-CURRENT GENERATOR

PICKUP HEAD

GENERAL   RECORDING HEAD   PLAYBACK HEAD   MAGNETIC ERASER

RECORDING, PLAYBACK, AND ERASING HEAD   STEREO

PIEZOELECTRIC CRYSTAL UNIT

Including crystal unit, quartz

THERMOCOUPLE

TEMPERATURE-MEASURING   With integral heater internally connected   With integral insulated heater

TRANSMISSION PATH, CONDUCTOR, CABLE, AND WIRING

CROSSING NOT CONNECTED   JUNCTION OF CONNECTED PATHS
The crossing is not necessarily at a   Only if required by
90 degree angle.   layout considerations

TERMINATION

OPEN CIRCUIT   SHORT CIRCUIT

GROUND

A direct conducting connection to the earth or body of water that is a part thereof. A conducting connection
to a structure that serves a function similar to that of an earth ground (that is, a structure such as a frame of
an air, space, or land vehicle that is not conductively connected to earth).

CHASSIS OR FRAME CONNECTION

A conducting connection to a chassis or frame, or equivalent chassis connection of a printed-wiring board. The
chassis or frame (or equivalent chassis connection of a printed-wiring board) may be at substantial potential
with respect to the earth or structure in which this chassis or frame (or printed-wiring board) is mounted.

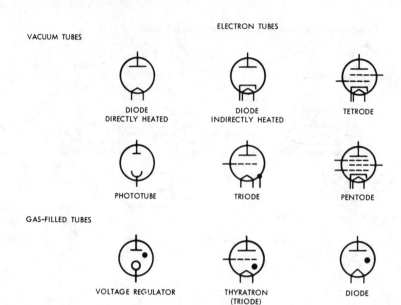

ELECTRON TUBES

VACUUM TUBES

DIODE
DIRECTLY HEATED

DIODE
INDIRECTLY HEATED

TETRODE

PHOTOTUBE

TRIODE

PENTODE

GAS-FILLED TUBES

VOLTAGE REGULATOR

THYRATRON
(TRIODE)

DIODE

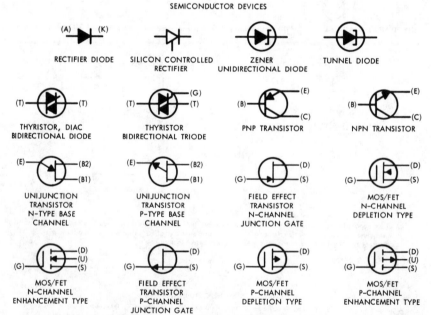

SEMICONDUCTOR DEVICES

(A)   (K)
RECTIFIER DIODE

SILICON CONTROLLED
RECTIFIER

ZENER
UNIDIRECTIONAL DIODE

TUNNEL DIODE

(T)   (T)
THYRISTOR, DIAC
BIDIRECTIONAL DIODE

(T)   (G)
         (T)
THYRISTOR
BIDIRECTIONAL TRIODE

(B)   (E)
         (C)
PNP TRANSISTOR

(B)   (E)
         (C)
NPN TRANSISTOR

(E)   (B2)
         (B1)
UNIJUNCTION
TRANSISTOR
N-TYPE BASE
CHANNEL

(E)   (B2)
         (B1)
UNIJUNCTION
TRANSISTOR
P-TYPE BASE
CHANNEL

(G)   (D)
         (S)
FIELD EFFECT
TRANSISTOR
N-CHANNEL
JUNCTION GATE

(G)   (D)
         (S)
MOS/FET
N-CHANNEL
DEPLETION TYPE

(G)   (D)
         (U)
         (S)
MOS/FET
N-CHANNEL
ENHANCEMENT TYPE

(G)   (D)
         (S)
FIELD EFFECT
TRANSISTOR
P-CHANNEL
JUNCTION GATE

(G)   (D)
         (S)
MOS/FET
P-CHANNEL
DEPLETION TYPE

(G)   (D)
         (U)
         (S)
MOS/FET
P-CHANNEL
ENHANCEMENT TYPE

COMPOSITE ASSEMBLIES

The triangle is pointed in the direction of transmission. The symbol represents any method of amplification
(electron tube, solid-state device, magnetic device, etc.).

AMPLIFIER

Amplifier use may be indicated in the triangle by words, standard abbreviations, or a letter combination from the following list:

*BDG	Bridging	*MON	Monitoring
*BST	Booster	*PGM	Program
*CMP	Compression	*PRE	Preliminary
*DC	Direct-current	*PWR	Power
*EXP	Expansion	*TRQ	Torque
*LIM	Limiting		

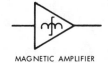

MAGNETIC AMPLIFIER

RECTIFIER

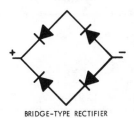

BRIDGE-TYPE RECTIFIER

MECHANICAL FUNCTIONS

MECHANICAL CONNECTION

- - - - - - -

GENERAL

MECHANICAL MOTION

TRANSLATION
ONE DIRECTION

TRANSLATION
BOTH DIRECTIONS

ROTATION
ONE DIRECTION

CIRCUIT PROTECTORS

FUSE

CIRCUIT BREAKER

NETWORK PROTECTOR

SYMBOLS FOR READOUT DEVICES

METER, INSTRUMENT

The asterisk is not part of the symbol. Always replace the asterisk by one of the following letters or letter combinations, depending on the function of the meter or instrument, unless some other identification is provided in the circle and explained on the diagram.

*A	Ammeter		*PF	Power factor meter
*CRO	Oscilloscope, Cathode-ray oscillograph		*PH	Phasemeter
*DB	DB (decibel) meter Audio level/meter		*UA	Microammeter
*F	Frequency meter		*V	Voltmeter
*GD	Ground detector		*VA	Volt–Ammeter
*MA	Milliammeter		*W	Wattmeter
*OHM	Ohmmeter			

LAMPS AND VISUAL-SIGNALING DEVICES

GLOW LAMP, COLD-CATHODE LAMP, NEON LAMP

ALTERNATING–CURRENT TYPE

DIRECT–CURRENT TYPE

INDICATING, PILOT LAMP

GENERAL

ACOUSTIC DEVICES

AUDIBLE-SIGNALING DEVICE

GENERAL

If specific identifications of loudspeaker types is required, the following letter combinations may be added in the symbol.

* HN	Horn, electrical		**EMN	Electromagnetic with moving coil and
* HW	Howler			neutralizing winding (moving–coil
* LS	Loudspeaker			leads should be identified)
* SN	Siren		**MG	Magnetic armature
**EM	Electromagnetic with moving coil (moving–coil leads should be identified)		**PM	Permanent magnet with moving coil

HEADSET

DOUBLE

SINGLE

HAND SET

GENERAL

MICROPHONE

GENERAL

DISCONNECTING DEVICE

The contact symbol is not an arrowhead. It is larger and the
lines are drawn at a 90-degree angle.

FEMALE CONTACT   MALE CONTACT   SEPARABLE CONNECTORS   2-CONDUCTOR JACK
                                          (ENGAGED)

CORE

NO SYMBOL
AIR CORE   MAGNETIC CORE   CORE OF MAGNET

If it is necessary to identify an air core,
a note should appear adjacent to the
symbol of the inductor or transformer.

INDUCTOR

GENERAL   MAGNETIC-CORE   TAPPED   ADJUSTABLE   CONTINUOUSLY
                                                                   ADJUSTABLE

TRANSFORMER

Additional windings may be shown or indicated by a note.

GENERAL   MAGNETIC-CORE   SATURATING TRANSFORMER

SWITCH

Fundamental symbols for contacts, mechanical connections, etc., may be used for switch symbols. The standard
method of showing switches is in a position with no operating force applied. For switches that may be in any of
two or more positions with no operating force applied, and for switches actuated by some mechanical device (as
in air-pressure, liquid-level, rate-of-flow, etc., switches), a clarifying note may be necessary to explain the
point at which the switch functions. When the basic switch symbols are shown in the closed position on a
diagram, terminals must be added for clarity.

SINGLE-THROW   DOUBLE-THROW   MULTIPOSITION SWITCH   ROTARY SWITCH

Any number of transmission
paths may be shown.

TELEGRAPH KEY

SIMPLE   SIMPLE WITH SHORTING SWITCH   OPEN-CIRCUIT

Appendix C: Electrical and Electronic Formulas

TABLE 1. OHM'S LAW FOR DIRECT AND ALTERNATING CURRENTS

Alphabetical List of Symbols Used in Formulas

cos = Cosine (Pure number)
E = Electromotive force or voltage (Volts)
I = Current or amperage (Amps)
P = Power (Watts)
R = Resistance (Ohms)
θ (Theta) = Phase angle (Degrees)
Z = Impedance (Ohms), the vector sum of resistance and reactance(s)

FORMULAS

DIRECT CURRENT

$$E = IR = \frac{P}{I} = \sqrt{PR}$$

$$I = \frac{E}{R} = \frac{P}{E} = \sqrt{\frac{P}{R}}$$

$$P = EI = I^2R = \frac{E^2}{R}$$

$$R = \frac{E}{I} = \frac{P}{I^2} = \frac{E^2}{P}$$

ALTERNATING CURRENT

$$E = IZ = \frac{P}{I\cos\theta} = \sqrt{\frac{PZ}{\cos\theta}}$$

$$I = \frac{E}{Z} = \frac{P}{E\cos\theta} = \sqrt{\frac{P}{Z\cos\theta}}$$

$$P = IE\cos\theta = I^2Z\cos\theta = \frac{E^2\cos\theta}{Z}$$

$$Z = \frac{E}{I} = \frac{P}{I^2\cos\theta} = \frac{E^2\cos\theta}{P}$$

TABLE 2. FORMULAS FOR PURE RESISTIVE CIRCUITS

Alphabetical List of Symbols Used in Formulas

E = Voltage (Volts); E_1, E_2, E_3, etc. = Individual voltages; E_T = Total voltage

I = Current (Amperes); I_1, I_2, I_3, etc. = Individual currents
  I_T = Total current

PF = Power factor (Pure number)

R = Resistance (Ohms); R_1, R_2, R_3, etc. = Individual resistances;
  R_T = Total resistance

FORMULAS

SERIES RESISTIVE CIRCUITS

$$E_T = E_1 + E_2 + E_3 +, \text{etc.}$$
$$I_T = I_1 = I_2 = I_3, \text{etc.}$$
$$R_T = R_1 + R_2 + R_3 +, \text{etc.}$$

PARALLEL RESISTIVE CIRCUITS

$$E_T = E_1 = E_2 = E_3, \text{etc.}$$
$$I_T = I_1 + I_2 + I_3, \text{etc.}$$
$$R_T = \frac{R_1 \times R_2}{R_1 + R_2}$$

$$R_T = \frac{1}{\dfrac{1}{R_1} + \dfrac{1}{R_2} + \dfrac{1}{R_3} +, \text{etc.}}$$

383

$$\frac{1}{R_T} = \frac{1}{R_1} + \frac{1}{R_2} + \frac{1}{R_3} +, \text{ etc.}$$

POWER FACTOR

$$PF = 1$$

TABLE 3. FORMULAS FOR PURE INDUCTIVE CIRCUITS

Alphabetical List of Symbols Used in Formulas

E = Voltage (Volts); E_P = Voltage in primary; E_S = Voltage in secondary

EFF = Efficiency (Percent)

I = Current (Amperes); I_P = Current in primary; I_S = Current in secondary

L = Inductance (Henrys); L_1, L_2, L_3, etc. = Individual inductors;
$\qquad L_M$ = Mutual inductance; L_T = Total inductance

N = Number of turns in coil; N_P = Number of turns in primary;
$\qquad N_S$ = Number of turns in secondary

P = Power (Watts); P_{in} = Power in; P_{out} = Power out

R = Resistance of inductor (Ohms); R_L = Load resistance

TC = Time constant (Seconds)

FORMULAS

INDUCTORS IN SERIES $\qquad L_T = L_1 + L_2 + L_3$, etc.

INDUCTORS IN PARALLEL $\qquad L_T = \dfrac{1}{\dfrac{1}{L_1} + \dfrac{1}{L_2} \cdot \dfrac{1}{L_3}}$, etc.

INDUCTIVE TIME CONSTANT
(CURRENT LAGS VOLTAGE) $\qquad TC = \dfrac{L}{R}$

TRANSFORMER TURNS,
AND EFFICIENCY $\qquad$ Turns ratio $= \dfrac{N_S}{N_P}$

$$\frac{E_S}{E_P} = \frac{N_S}{N_P}$$

$$I_S = \frac{N_S}{R_L}$$

$$\frac{I_P}{I_S} = \frac{E_S}{E_P}$$

$$EFF = \frac{P_{in}}{P_{out}} \times 100$$

TABLE 4. FORMULAS FOR PURE CAPACITIVE CIRCUITS (AC AND PULSATING DC ONLY)

Alphabetical List of Symbols Used in Formulas

C = Capacitance (Farads); C_1, C_2, C_3, etc. = Individual capacitors
C_T = Total Capacitance

E = Voltage (Volts)

Q = Electrical charge (Coulombs)

R = Resistance (Ohms)

TC = Time constant (Seconds)

FORMULAS

CAPACITANCE CHARGE

$$Q = CE$$

$$E = \frac{Q}{C}$$

$$C = \frac{Q}{E}$$

SERIES CAPACITIVE CIRCUITS

$$C_T = \frac{C_1 C_2}{C_1 + C_2}$$

$$\frac{1}{C_T} = \frac{1}{C_1} + \frac{1}{C_2} + \frac{1}{C_3} \text{, etc.}$$

$$C_T = \frac{1}{\frac{1}{C_1} + \frac{1}{C_2} + \frac{1}{C_3} + \text{, etc.}}$$

PARALLEL CAPACITIVE CIRCUITS

$$C_T = C_1 + C_2 + C_3 \text{, etc.}$$

CAPACITIVE TIME CONSTANT (CURRENT LEADS VOLTAGE)

$$TC = R \times C$$

TABLE 5. FORMULAS FOR RL CIRCUITS
(COMBINED RESISTANCE AND INDUCTANCE)

Alphabetical List of Symbols Used in Formulas

cos = Cosine (Pure number)

E = Voltage (Volts); E_A = Applied voltage; E_1, E_2, etc. = Individual voltages; E_{X_L} = Inductor voltage; E_R = Voltage across resistance; E_T = Total voltage

f = Frequency (Hertz)

I = Current (Amperes); I_1, I_2, etc. = Individual currents; I_R = Current through resistor; I_T = Total current; I_{X_L} = Inductive current

L = Inductance (Henrys); L_1, L_2, etc. = Individual inductors; L_T = Total inductance

R = Resistance (Ohms); R_1, R_2, etc. = Individual resistors; R_T = Total resistance

TC = Time constant (Seconds)

tan = Tangent (Pure number)

θ (Theta) = Phase angle between applied voltage and current (Degrees)

X = Reactance (Ohms); X_L = Inductive reactance (Ohms); X_{L_1}, X_{L_2}, etc. = Individual inductive reactances; X_{L_T} = Total inductive reactance

Z = Impedance (Ohms)

FORMULAS

INDUCTIVE REACTANCE
(Single inductor)

$$X_L = 2\pi fL$$

$$L = \frac{X_L}{2\pi f}$$

$$E = I \times X_L$$

$$X_L = \frac{E}{I}$$

$$I = \frac{E}{X_L}$$

NOTE: $2\pi = 6.28$ (approx.)

SERIES INDUCTIVE REACTANCE

$$X_{L_T} = X_{L_1} + X_{L_2} + X_{L_3} +, \text{etc.}$$

PARALLEL INDUCTIVE REACTANCE

$$X_{L_T} = \frac{X_{L_1} \times X_{L_2}}{X_{L_1} + X_{L_2}}$$

$$\frac{1}{X_{L_T}} = \frac{1}{X_{L_1}} + \frac{1}{X_{L_2}} + \frac{1}{X_{L_3}} +, \text{ etc.}$$

$$X_{L_T} = \frac{1}{\dfrac{1}{X_{L_1}} + \dfrac{1}{X_{L_2}} + \dfrac{1}{X_{L_3}}}, \text{ etc.}$$

X_L AND RESISTANCE IN SERIES

$$Z = \sqrt{R^2 + (X_L)^2}$$

$$E_A = \sqrt{(E_R)^2 + (E_{X_L})^2}$$

$$\text{TAN } \theta = \frac{X_L}{R}$$

X_L AND RESISTANCE IN PARALLEL

$$I_T = \sqrt{(I_R)^2 + (I_{X_L})^2}$$

$$\text{TAN } \theta = \frac{-I_{X_L}}{I_R}$$

INDUCTIVE TIME CONSTANT

$$TC = \frac{L}{R}$$

TABLE 6. FORMULAS FOR RC CIRCUITS
(COMBINED RESISTANCE AND CAPACITANCE)

Alphabetical List of Symbols Used in Formulas

C = Capacitance (Farads); C_1, C_2, etc. = Individual capacitances; C_T = Total capacitance

cos = Cosine (Pure number)

E = Voltage (Volts); E_A = Applied voltage; E_1, E_2, etc. = Individual voltages; E_R = Voltage across resistance; E_T = Total voltage; E_{X_C} = Capacitor voltage

f = Frequency (hertz)

I = Current (Amperes); I_1, I_2, etc. = Individual currents; I_R = Current through resistor; I_T = Total current; I_{X_C} = Capacitive current

R = Resistance (Ohms); R_1, R_2, etc. = Individual resistances; R_T = Total resistance

TC = Time constant (Seconds)

tan = Tangent (Pure number)

θ (Theta) = Phase angle between applied voltage and current (Degrees)

X = Reactance (Ohms); X_C = Capacitive reactance (Ohms);

X_{C_1}, X_{C_2}, etc. = Individual capacitive reactances;
C_T = Total capacitive reactance

Z = Impedance (Ohms)

FORMULAS

CAPACITIVE REACTANCE
(SINGLE CAPACITOR)

$$X_C = \frac{1}{2\pi fC} = \frac{E}{I}$$

$$C = \frac{1}{2\pi fX_C}$$

$$E = I \times X_C$$

NOTE: $2\pi = 6.28$ (APPROX.)

$$I = \frac{E}{X_C}$$

SERIES CAPACITIVE
REACTANCE

$$X_{C_T} = X_{C_1} + X_{C_2} + X_{C_3} + \text{etc.}$$

PARALLEL CAPACITIVE
REACTANCE

$$\frac{1}{X_{C_T}} = \frac{1}{X_{C_1}} + \frac{1}{X_{C_2}} + \text{etc.}$$

$$X_{C_T} = \frac{X_{C_1} \times X_{C_2}}{X_{C_1} + X_{C_2}}$$

$$X_{C_T} = \frac{1}{\dfrac{1}{X_{C_1}} + \dfrac{1}{X_{C_2}} + \dfrac{1}{X_{C_3}}} \text{ . etc.}$$

X_C AND RESISTANCE
IN SERIES

$$Z = \sqrt{R^2 + (X_C)^2}$$

$$E_A = \sqrt{(E_R)^2 + (E_{X_C})^2}$$

$$\cos\theta = \frac{R}{Z}$$

$$\tan\theta = -\frac{X_C}{R}$$

X_C AND RESISTANCE
IN PARALLEL

$$I_T = \sqrt{(I_R)^2 + (I_{X_C})^2}$$

$$Z = \frac{E}{I_T}$$

$$\tan\theta = \frac{I_{X_C}}{I_R}$$

CAPACITIVE
TIME CONSTANT

$$TC = R \times C$$

TABLE 7. FORMULAS FOR RCL CIRCUITS
(COMBINED RESISTANCE, CAPACITANCE, AND INDUCTANCE)

Alphabetical List of Symbols Used in Formulas

C = Capacitance (Farads); C_1, C_2, etc. = Individual Capacitances;
  C = Total Capacitance

cos = Cosine (Pure number)

E = Voltage (Volts); E_A = Applied voltage; E_1, E_2, etc. = Individual voltages;
  E_{X_L} = Inductor voltage; E_R = Voltage across resistance;
  E_T = Total voltage; E_{X_C} = Capacitor voltage; E_{X_L} = Inductor voltage

f = Frequency (Hertz); f_r = resonant frequency

I = Current (Amperes); I_1, I_2, etc. = Individual currents;
  I_R = Current through resistor; I_T = Total current;
  I_{X_C} = Capacitive current; I_{X_L} = Inductive current

L = Inductance (Henrys); L_1, L_2, etc. = Individual inductors;
  L_T = Total inductance

R = Resistance (Ohms); R_1, R_2, etc. = Individual resistors;
  R_T = Total resistance

TC = Time constant (Seconds)

tan = Tangent (Pure number)

θ (Theta) = Phase angle between applied voltage and current (Degrees)

X = Reactance (Ohms); X_C = Capacitive reactance (Ohms);
  X_{C_1}, X_{C_2}, etc. = Individual capacitive reactances;
  X_{C_T} = Total capacitive reactance; X_L = Inductive reactance (Ohms);
  X_{L_1}, X_{L_2}, etc. = Individual inductive reactances;
  X_{L_T} = Total inductive reactance

Z = Impedance (Ohms)

FORMULAS

SERIES RCL CIRCUITS

(Angle is minus ($-$)
if X_C is larger and plus
($+$) if X_L is larger.)

$$Z = \sqrt{R^2 + (X_L - X_C)^2}$$

$$I = \frac{E}{Z}$$

$$\text{Tan } \theta = \frac{X_L - X_C}{R}$$

PARALLEL RCL CIRCUITS

(Angle is minus ($-$)
if I_{X_L} is larger and plus
($+$) if I_{X_C} is larger)

$$I_T = \sqrt{(I_R)^2 + (I_{X_L} - I_{X_C})^2}$$

$$Z = \frac{E}{I_T}$$

$$\text{Tan } \theta = \frac{I_{X_L} - I_{X_C}}{I_R}$$

389

RESONANCE

$$f_r = \frac{1}{2\pi \sqrt{LC}}$$

$$L = \frac{1}{(2\pi)^2 (f_R)^2 C}$$

NOTE: $2\pi = 6.28$ (approx.)
$\quad\quad (2\pi)^2 = 39.5$ (approx.)

$$C = \frac{1}{(2\pi)^2 (f_r)^2 L}$$

TABLE 8. ELECTRICAL POWER CALCULATIONS

Alphabetical List of Symbols Used in Formulas

AC = Alternating current

C = Capacitance (Farads)

cos = Cosine (Pure number)

DC = Direct Current

E = Voltage (Volts)

EFF* = Efficiency (as a decimal fraction); E% = Percentage efficiency

G = Conductance (Mhos)

HP = Horsepower (550 ft-lb per second, or 746 Watts)

I = Current (Amperes)

L = Inductance (Henrys)

P = Power (Watts); P_{in} = Power in; P_{out} = Power out

PF = Power factor (a decimal fraction)

R = Resistance (Ohms)

θ (Theta) = Phase angle (Degrees) between voltage and current waves

X = Reactance (Ohms); X_C = Capacitive reactance; X_L = Inductive reactance

Z = Impedance (Ohms)

*(NOTE: The EFF factor applies to *motors* only. It is usually between 0.8 and 0.9 and is determined by tests. Other formulas in this section are used in other electrical power applications.)

FORMULAS

POWER FACTOR (AC ONLY)
$$PF = Cos\,\theta = \frac{R}{Z} = \frac{P}{EI}$$

$$PF = \frac{R}{\sqrt{R^2 + X^2}}$$

EFFICIENCY (FOR MOTORS
ONLY DETERMINED
BY TESTS)
$$EFF = \frac{P_{out}}{P_{in}}$$

$$EFF\% = \frac{P_{out}}{P_{in}} \times 100$$

CONDUCTANCE

$$G = \frac{I}{R}$$

VOLTAGE

$$E\,(DC\ Only) = \frac{P}{I \times EFF}$$

$$E\,(DC\ Only) = \frac{HP \times 746}{I \times EFF}$$

$$E\,(AC,\ 1\text{-Phase}) = \frac{P}{I \times PF}$$

$$E\,(AC,\ 1\text{-Phase}) = \frac{HP \times 746}{I \times PF \times EFF}$$

$$E\,(AC,\ 2\text{-Phase}) = \frac{P}{I \times PF \times 2}$$

$$E\,(AC,\ 2\text{-Phase}) = \frac{HP \times 746}{I \times PF \times EFF \times 2}$$

$$E\,(AC,\ 3\text{-Phase}) = \frac{P}{I \times PF \times 1.73}$$

$$E\,(AC,\ 3\text{-Phase}) = \frac{HP \times 746}{I \times PF \times EFF \times 1.73}$$

CURRENT

$$I\,(AC\ and\ DC) = \sqrt{\frac{P}{R}}$$

$$I\,(DC\ Only) = \frac{E}{R} = \frac{P}{E}$$

$$I\,(AC\ Only) = \frac{P}{E \times Cos\ \theta}$$

$$I\,(AC\ Only) = \sqrt{\frac{P}{Z \times Cos\ \theta}}$$

$$I\,(AC,\ 1\text{-Phase}) = \frac{P}{E \times PF}$$

$$I\,(AC,\ 1\text{-Phase}) = \frac{HP \times 746}{E \times PF \times EFF}$$

$$I\,(AC,\ 2\text{-Phase}) = \frac{P}{E \times PF \times 2}$$

$$I\,(AC,\ 2\text{-Phase}) = \frac{HP \times 746}{E \times PF \times EFF \times 2}$$

$$I \ (AC, 3\text{-Phase}) = \frac{P}{E \times PF \times 1.73}$$

$$I \ (AC, 3\text{-Phase}) = \frac{HP \times 746}{E \times PF \times EFF \times 1.73}$$

POWER
(WATTS AND HP)

$$P \ (AC \text{ and } DC) = I^2R$$

$$P \ (DC \text{ Only}) = EI = \frac{E^2}{R}$$

$$P \ (AC \text{ Only}) = EI \times \cos\theta$$

$$P \ (AC \text{ Only}) = \frac{E^2 \times \cos\theta}{Z} = I^2Z \times \cos\theta$$

$$P \ (AC, 1\text{-Phase}) = EI \times PF$$

$$HP \ (AC, 1\text{-Phase}) = \frac{EI \times PF \times EFF}{746}$$

$$P \ (AC, 2\text{-Phase}) = EI \times PF \times 2$$

$$HP \ (AC, 2\text{-Phase}) = \frac{EI \times PF \times EFF \times 2}{746}$$

$$P \ (AC, 3\text{-Phase}) = EI \times PF \times 1.73$$

$$HP \ (AC, 3\text{-Phase}) = \frac{EI \times PF \times EFF \times 2}{746}$$

RESISTANCE

$$R \ (AC \text{ and } DC) = \frac{E}{I} = \frac{1}{G}$$

$$R \ (DC \text{ Only}) = \frac{E^2}{P} = \frac{P}{I^2}$$

$$R \ (AC \text{ Only}) = \frac{E^2 \times \cos\theta}{P} = \frac{P}{I^2 \times \cos\theta}$$

$$R \ (AC \text{ Only}) = Z \cos\theta$$

REACTANCES
(USUALLY AC)

$$(\text{Inductive}) \ X_L = 2\pi fL$$

$$(\text{Capacitive}) \ X_L = \frac{1}{2\pi fC}$$

IMPEDANCE
(USUALLY AC)

$$Z = \sqrt{R^2 + X^2} = \sqrt{R^2 + (X_L - X_C)^2}$$

REAL POWER

REAL POWER $= I^2R$ (Non-reactive circuit)
REAL POWER $= EI \times \cos\theta$ (Circuit with reactance)

TABLE 9. CALCULATIONS OF CYCLES, TIME, VELOCITY AND SINE WAVE VALUES

Alphabetical List of Symbols Used in Formulas

AVE = Average sine wave value

C = Capacitance (Farads)

f = Frequency (hertz)

L = Inductance (Henrys)

λ (Lambda) = Wavelength; λ_{cm} = Wavelength in centimeters;
$\qquad \lambda_{ft}$ = Wavelength in feet; λ_m = Wavelength in meters

π (Pi) = 3.1416

PK = Peak sine wave value; PK to PK = Peak-to-Peak sine wave value

R = Resistance (Ohms)

RMS = Root-Mean-Square sine wave value

T = Time (Seconds)

V = Velocity; V_{cm} = Velocity in centimeters per second;
$\qquad V_{ft}$ = Velocity in feet per second; V_m = Velocity in meters per second

FORMULAS

$$\text{NUMBER OF CYCLES} = f \times T$$

CYCLES AND FREQUENCY $\qquad f = \dfrac{1}{T}$

TIME $\qquad T = \dfrac{1}{f}$

VELOCITY $\qquad V_m = \lambda_m f$

$\qquad V_{cm} = \lambda_{cm} f$

$\qquad V_{ft} = \lambda_{ft} f$

WAVELENGTH $\qquad \lambda = \dfrac{V}{f}$

$$\lambda_m = \frac{3 \times 10^8 \text{ m/sec}}{f}$$

$$\lambda_{cm} = \frac{3 \times 10^{10} \text{ cm/sec}}{f}$$

$$\lambda_{ft} = \frac{1130 \text{ ft/sec}}{f}$$

SINE WAVE VALUES $\qquad \text{AVE} = 0.637 \times \text{PK}$

$\qquad \text{RMS} = 0.707 \times \text{PK}$

$\qquad \text{PK} = 1.414 \times \text{RMS}$

$\qquad \text{PK TO PK} = 2.828 \times \text{RMS}$

TABLE 10. MAGNETIC AND ELECTROMAGNETIC
CALCULATIONS

Alphabetical List of Symbols Used in Formulas

A $\quad=$ Area (Square centimeters)
B $\quad=$ Flux density (Gausses)
d $\quad=$ Distance (Centimeters)
F $\quad=$ Force (Dynes)
ϕ $\quad=$ Magnetic flux (Maxwells)
H $\quad=$ Magnetic intensity (Oersteds)
I $\quad=$ Current (Amperes)
J $\quad=$ Unit poles per centimeter
l $\quad=$ Length (centimeters)
m $\quad=$ Unit pole
mmf $=$ Magnetomotive force (Gilberts)
μ $\quad=$ Permeability of the medium
N $\quad=$ Number of turns per unit length
π $\quad=$ 3.1416
R $\quad=$ Reluctance (Rels)
r $\quad=$ Radius (Centimeters)
W $\quad=$ Field energy (Ergs)

FORMULAS

FORCE BETWEEN TWO
MAGNETIC POLES
$$F = \frac{m_1 m_2}{\mu d^2}$$

MAGNETIC FIELD INTENSITY
$$H = \frac{F}{m}$$

MAGNETIC INTENSITY AT
A POINT FROM A POLE
$$H = \frac{m}{d^2}$$

MAGNETIC INTENSITY
ABOUT A CONDUCTOR
$$H = \frac{2I}{10r}$$

MAGNETIC INTENSITY
FOR LONG COILS
$$H = \frac{4\pi NI}{10\,l}$$

MAGNETIC INTENSITY
FOR SHORT COILS
$$H = \frac{2\pi NI}{10\,r}$$

MAGNETOMOTIVE FORCE
$$mmf = \frac{4\pi NI}{10}$$

FLUX

$$\phi = \frac{mmf}{R}$$

$$\phi = 4\pi m$$

RELUCTANCE

$$R = \frac{l}{\mu A}$$

FLUX DENSITY

$$B = \frac{\phi}{A}$$

$$B = \mu H$$

$$B = H + 4\pi J$$

FORCE EXERTED ON
A CONDUCTOR

$$F = \frac{BIl}{10}$$

FORCE BETWEEN
TWO POLES

$$F = \frac{B^2 A}{8\pi\mu}$$

PERMEABILITY

$$\mu = \frac{B}{H}$$

FORCE ACTING ON A
MAGNETIC FIELD IN
UNIFORM FIELD DENSITY

$$F = mH$$

ENERGY OF MAGNETIC
FIELD PER CUBIC
CENTIMETER

$$W = \frac{B^2}{8\pi\mu}$$

$$W = \frac{\mu H^2}{8\pi}$$

Appendix D: Solving Electrical Problems with Trigonometry Tables or the Slide Rule

The following tables of natural trigonometric functions (sines, cosines, and tangents) are needed to find phase angles and power factors, but they also can be used for solving vector problems by simple trigonometry. The "trig" method is much simpler than calculations using root-mean-squares. Also the "trig" method is readily adaptable to machine calculation and slide rule.

The slide rule best suited for this work is of the decimal trig type, with the S, ST, and T or T_1 and T_2 scales of angles decimally divided. Rules with the T_2 scale are particularly recommended because the tangents of angles larger than $45°$ can be found on the T_2 scale reading from left to right in the normal manner. Space does not allow a full discussion of plane trigonometry or slide rule operation. However, the best procedures are outlined in the following paragraphs.

It is a good practice to make a small, hand-drawn diagram of the vector triangle to be solved, labeling the parts of the triangle with their electrical symbols. Before doing this, however, examine the four triangles in Fig. 1. These are properly labeled with standard mathematical symbols. Whatever the position of the triangle, small c always stands for the hypotenuse opposite the $90°$ angle of a right triangle, which is labeled as large C. Small a stands for altitude (whether up or down) opposite the angle of elevation or depression, which is labeled as A. Small b always stands for base, and the angle opposite the base is labeled as B. With this system the functions are:

$$\sin A = \frac{a}{c} \qquad \csc A = \frac{c}{a}$$

$$\cos A = \frac{b}{c} \qquad \sec A = \frac{c}{b}$$

$$\tan A = \frac{a}{b} \qquad \cot A = \frac{b}{a}$$

You will notice that the functions in the righthand column are simply the reciprocals of those in the lefthand column. If it is necessary to know the secant of angle A for any reason, no special tables are needed, because

$$\sec A = \frac{1}{\cos A}.$$

Similarly,

$$\csc A = \frac{1}{\sin A}.$$

Also,

$$\cot A = \frac{\cdot\,1}{\tan A}.$$

In Fig. 1 you will notice that angle A shows a minus sign before it when it is in the 4th quadrant. Also you will notice that side a also has a minus side

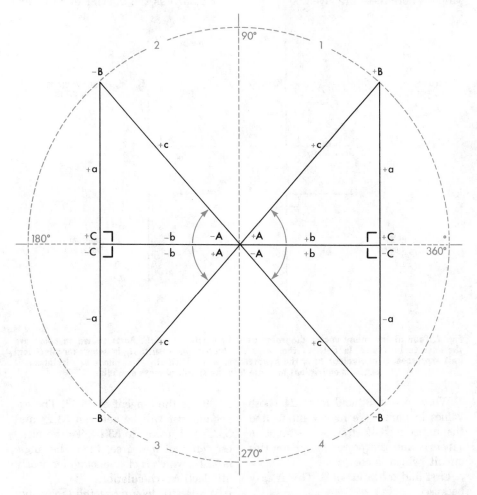

Fig. 1. Right triangles, labeled with standard mathematical symbols, in the four quadrants. Vector triangles for electrical calculations are generally in the 1st or 4th quadrant, and parts are labeled with the appropriate electrical symbols.

when the triangle is in the 4th quadrant. The minus signs do not affect the numerical values; they only indicate the direction.

Fig. 2 shows diagrams of some typical electrical problems, re-labeled with the appropriate electrical symbols. Let us consider one illustrating impedance of an RL circuit and show not only the electrical symbols but their numerical values where these are given.

This is exactly the same as tan A $= \dfrac{a}{b}$ in the standard math symbols. Using numbers, tan $\theta = \dfrac{32}{24} = 1.3333$.

Because this tangent is larger than 1, the angle is larger than 45°.

From the tables you will not find this number, but you will find the number 1.3319 as the tangent of 53.1° and

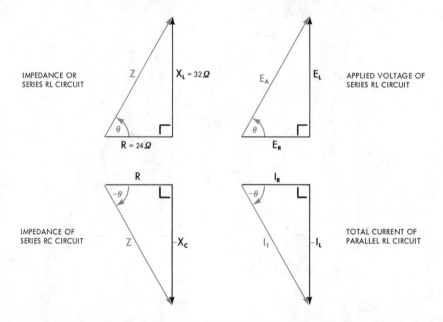

Fig. 2. Four of the many vector diagrams used in electrical work. Parts shown in color are the ones to be found. In each of the cases illustrated the resultant, in whatever electrical unit expressed, is equivalent to c, the hypotenuse, while Θ (theta), the phase or displacement angle, is equivalent to angle A in the standard math symbols.

Where $X_L = 32$ and $R = 24$ (both values in ohms) we may want to find the impedance Z, the phase angle θ (theta), and the power factor for this circuit, which is cos θ.

First find the value of θ. This is easy because

$$\tan \theta = \frac{X_L}{R}$$

1.3367 as the tangent of 53.2°. The exact number will be between 53.1° and 53.2° but nearer to 53.1°. To the nearest tenth of a degree, then, the angle is 53.1°, which is close enough for nearly all electrical calculations.

While still looking at the tables, notice that cos 53.1° = 0.6004, which is almost exactly 0.600. This is the power factor.

Knowing the phase angle θ is 53.1°, it is now very easy to find the impedance Z, using the Law of Sines, that is:

$$\frac{c}{\sin C} = \frac{a}{\sin A}$$, or, in electrical sym-

bols, $\frac{Z}{\sin 90°} = \frac{X_L}{\sin \theta}$. But the sine of

90° = 1. This simplifies the equation

to $Z = \frac{X_L}{\sin \theta}$. Using numbers, $Z =$

$\frac{32}{\sin 53.1°}$. From the tables, sin 53.1°

= 0.7997, which is almost exactly 0.8.

$$Z = \frac{32}{.8} = 40 \text{ ohms.}$$

These various operations can be performed on the slide rule in a small fraction of the time required for looking into tables, etc.

Depending on the construction of the slide rule used, there are some differences in the method of finding tangents. Suppose in this example we use a slide rule with a T_2 as well as a T_1 and ST scale.

Divide 32 by 24 using the D and C scales. This gives 1.333 as the tangent (read on the D scale below the left index of the C scale), but it is not necessary to make any note of this number. At a glance we see that it is larger than 1, know that the angle is therefore larger than 45° and will be read on the T_2 scale.

We move the hairline so it is directly over the left index of the C scale, indicating 1.333. The T_2 scale is on the other side of the slide rule, so we flip it over, leaving the hairline in the same position. Still leaving the hairline sta-

tionary, we move the slide so that the left index is aligned with the left index of the D scale just below the S scale. The hairline, which has not been moved, is directly over 53.1 on the T_2 scale, indicating that the angle is 53.1°. We make a note of this for future reference.

If your slide rule has only one T scale, an extra step is needed to find tangents larger than 45°. After finding the tangent (in this case 1.333) on the D scale you must reset this number on the CI scale (an inverse scale with numbers reading from right to left) by moving the hairline. Then read the angle on the T scale directly below the hairline on the inverse set of numbers (also reading from right to left) on the T scale.

Next we find Z, the impedance, in this manner. Move the hairline to 32 on the D scale. Then move the slide until the angle 53.1° on the S scale is directly beneath the hairline. Read the answer as 40 (meaning in this case 40 ohms) on the D scale beneath the right index of the S scale.

To find the power factor, cos θ, we simply align the slide with the left index of the D scale, move the hairline along the set of numbers reading *from right to left* on the S scale until it is directly above the cosine number 53.1. At this point the hairline is also directly over .600 on the D scale. The power factor cos θ is .600.

Although the slide rule does not indicate decimal location, this is no problem in solving triangles, because the magnitude of the quantities is very obvious.

In calculating impedance for RCL circuits, where inductive reactance X_L operates opposite to capacitive reactance X_C, simplify to an equivalent circuit as follows:

1. When values of X_L and X_C are equal they effectively leave a pure resistive circuit at the specified frequency. Voltage and current in this case are in phase, there is no phase angle to consider, no reactance, and no impedance.
2. When X_L is greater than X_C, subtract the value of X_C from X_L. (Subtraction and addition are two operations not performed on the slide rule.) The remainder will be X_L, used for calculating the equivalent RL circuit.
3. When X_C is greater than X_L, subtract X_L from X_C (but not with the slide rule, of course). The remainder will be X_C, used for calculating the equivalent RC circuit.

NATURAL TRIGONOMETRIC FUNCTIONS FOR DECIMAL FRACTIONS OF A DEGREE

Deg.	Sin	Cos	Tan	Cot	Deg.	Deg.	Sin	Cos	Tan	Cot	Deg.
0.0	.00000	1.0000	.00000	∞	**90.0**	**6.0**	.10453	0.9945	.10510	9.514	**84.0**
.1	.00175	1.0000	.00175	573.0	.9	.1	.10626	.9943	.10687	9.357	.9
.2	.00349	1.0000	.00349	286.5	.8	.2	.10800	.9942	.10863	9.205	.8
.3	.00524	1.0000	.00524	191.0	.7	.3	.10973	.9940	.11040	9.058	.7
.4	.00698	1.0000	.00698	143.24	.6	.4	.11147	.9938	.11217	8.915	.6
.5	.00873	1.0000	.00873	114.59	.5	.5	.11320	.9936	.11394	8.777	.5
.6	.01047	0.9999	.01047	95.49	.4	.6	.11494	.9934	.11570	8.643	.4
.7	.01222	.9999	.01222	81.85	.3	.7	.11667	.9932	.11747	8.513	.3
.8	.01396	.9999	.01396	71.62	.2	.8	.11840	.9930	.11924	8.386	.2
.9	.01571	.9999	.01571	63.66	.1	.9	.12014	.9928	.12101	8.264	.1
1.0	.01745	0.9998	.01746	57.29	**89.0**	**7.0**	.12187	0.9925	.12278	8.144	**83.0**
.1	.01920	.9998	.01920	52.08	.9	.1	.12360	.9923	.12456	8.028	.9
.2	.02094	.9998	.02095	47.74	.8	.2	.12533	.9921	.12633	7.916	.8
.3	.02269	.9997	.02269	44.07	.7	.3	.12706	.9919	.12810	7.806	.7
.4	.02443	.9997	.02444	40.92	.6	.4	.12880	.9917	.12988	7.700	.6
.5	.02618	.9997	.02619	38.19	.5	.5	.13053	.9914	.13165	7.596	.5
.6	.02792	.9996	.02793	35.80	.4	.6	.13226	.9912	.13343	7.495	.4
.7	.02967	.9996	.02968	33.69	.3	.7	.13399	.9910	.13521	7.396	.3
.8	.03141	.9995	.03143	31.82	.2	.8	.13572	.9907	.13698	7.300	.2
.9	.03316	.9995	.03317	30.14	.1	.9	.13744	.9905	.13876	7.207	.1
2.0	.03490	0.9994	.03492	28.64	**88.0**	**8.0**	.13917	0.9903	.14054	7.115	**82.0**
.1	.03664	.9993	.03667	27.27	.9	.1	.14090	.9900	.14232	7.026	.9
.2	.03839	.9993	.03842	26.03	.8	.2	.14263	.9898	.14410	6.940	.8
.3	.04013	.9992	.04016	24.90	.7	.3	.14436	.9895	.14588	6.855	.7
.4	.04188	.9991	.04191	23.86	.6	.4	.14608	.9893	.14767	6.772	.6
.5	.04362	.9990	.04366	22.90	.5	.5	.14781	.9890	.14945	6.691	.5
.6	.04536	.9990	.04541	22.02	.4	.6	.14954	.9888	.15124	6.612	.4
.7	.04711	.9989	.04716	21.20	.3	.7	.15126	.9885	.15302	6.535	.3
.8	.04885	.9988	.04891	20.45	.2	.8	.15299	.9882	.15481	6.460	.2
.9	.05059	.9987	.05066	19.74	.1	.9	.15471	.9880	.15660	6.386	.1
3.0	.05234	0.9986	.05241	19.081	**87.0**	**9.0**	.15643	0.9877	.15838	6.314	**81.0**
.1	.05408	.9985	.05416	18.464	.9	.1	.15816	.9874	.16017	6.243	.9
.2	.05582	.9984	.05591	17.886	.8	.2	.15988	.9871	.16196	6.174	.8
.3	.05756	.9983	.05766	17.343	.7	.3	.16160	.9869	.16376	6.107	.7
.4	.05931	.9982	.05941	16.832	.6	.4	.16333	.9866	.16555	6.041	.6
.5	.06105	.9981	.06116	16.350	.5	.5	.16505	.9863	.16734	5.976	.5
.6	.06279	.9980	.06291	15.895	.4	.6	.16677	.9860	.16914	5.912	.4
.7	.06453	.9979	.06467	15.464	.3	.7	.16849	.9857	.17093	5.850	.3
.8	.06627	.9978	.06642	15.056	.2	.8	.17021	.9854	.17273	5.789	.2
.9	.06802	.9977	.06817	14.669	.1	.9	.17193	.9851	.17453	5.730	.1
4.0	.06976	0.9976	.06993	14.301	**86.0**	**10.0**	.1736	0.9848	.1763	5.671	**80.0**
.1	.07150	.9974	.07168	13.951	.9	.1	.1754	.9845	.1781	5.614	.9
.2	.07324	.9973	.07344	13.617	.8	.2	.1771	.9842	.1799	5.558	.8
.3	.07498	.9972	.07519	13.300	.7	.3	.1788	.9839	.1817	5.503	.7
.4	.07672	.9971	.07695	12.996	.6	.4	.1805	.9836	.1835	5.449	.6
.5	.07846	.9969	.07870	12.706	.5	.5	.1822	.9833	.1853	5.396	.5
.6	.08020	.9968	.08046	12.429	.4	.6	.1840	.9829	.1871	5.343	.4
.7	.08194	.9966	.08221	12.163	.3	.7	.1857	.9826	.1890	5.292	.3
.8	.08368	.9965	.08397	11.909	.2	.8	.1874	.9823	.1908	5.242	.2
.9	.08542	.9963	.08573	11.664	.1	.9	.1891	.9820	.1926	5.193	.1
5.0	.08716	0.9962	.08749	11.430	**85.0**	**11.0**	.1908	0.9816	.1944	5.145	**79.0**
.1	.08889	.9960	.08925	11.205	.9	.1	.1925	.9813	.1962	5.097	.9
.2	.09063	.9959	.09101	10.988	.8	.2	.1942	.9810	.1980	5.050	.8
.3	.09237	.9957	.09277	10.780	.7	.3	.1959	.9806	.1998	5.005	.7
.4	.09411	.9956	.09453	10.579	.6	.4	.1977	.9803	.2016	4.959	.6
.5	.09585	.9954	.09629	10.385	.5	.5	.1994	.9799	.2035	4.915	.5
.6	.09758	.9952	.09805	10.199	.4	.6	.2011	.9796	.2053	4.872	.4
.7	.09932	.9951	.09981	10.019	.3	.7	.2028	.9792	.2071	4.829	.3
.8	.10106	.9949	.10158	9.845	.2	.8	.2045	.9789	.2089	4.787	.2
.9	.10279	.9947	.10334	9.677	.1	.9	.2062	.9785	.2107	4.745	.1
6.0	.10453	0.9945	.10510	9.514	**84.0**	**12.0**	.2079	0.9781	.2126	4.705	**78.0**
Deg.	Cos	Sin	Cot	Tan	Deg.	Deg.	Cos	Sin	Cot	Tan	Deg.

NATURAL TRIGONOMETRIC FUNCTIONS FOR DECIMAL FRACTIONS OF A DEGREE

Deg.	Sin	Cos	Tan	Cot	Deg.	Deg.	Sin	Cos	Tan	Cot	Deg.
12.0	0.2079	0.9781	0.2126	4.705	**78.0**	**18.0**	0.3090	0.9511	0.3249	3.078	**72.0**
.1	.2096	.9778	.2144	4.665	.9	.1	.3107	.9505	.3269	3.060	.9
.2	.2113	.9774	.2162	4.625	.8	.2	.3123	.9500	.3288	3.042	.8
.3	.2130	.9770	.2180	4.586	.7	.3	.3140	.9494	.3307	3.024	.7
.4	.2147	.9767	.2199	4.548	.6	.4	.3156	.9489	.3327	3.006	.6
.5	.2164	.9763	.2217	4.511	.5	.5	.3173	.9483	.3346	2.989	.5
.6	.2181	.9759	.2235	4.474	.4	.6	.3190	.9478	.3365	2.971	.4
.7	.2198	.9755	.2254	4.437	.3	.7	.3206	.9472	.3385	2.954	.3
.8	.2215	.9751	.2272	4.402	.2	.8	.3223	.9466	.3404	2.937	.2
.9	.2233	.9748	.2290	4.366	.1	.9	.3239	.9461	.3424	2.921	.1
13.0	0.2250	0.9744	0.2309	4.331	**77.0**	**19.0**	0.3256	0.9455	0.3443	2.904	**71.0**
.1	.2267	.9740	.2327	4.297	.9	.1	.3272	.9449	.3463	2.888	.9
.2	.2284	.9736	.2345	4.264	.8	.2	.3289	.9444	.3482	2.872	.8
.3	.2300	.9732	.2364	4.230	.7	.3	.3305	.9438	.3502	2.856	.7
.4	.2317	.9728	.2382	4.198	.6	.4	.3322	.9432	.3522	2.840	.6
.5	.2334	.9724	.2401	4.165	.5	.5	.3338	.9426	.3541	2.824	.5
.6	.2351	.9720	.2419	4.134	.4	.6	.3355	.9421	.3561	2.808	.4
.7	.2368	.9715	.2438	4.102	.3	.7	.3371	.9415	.3581	2.793	.3
.8	.2385	.9711	.2456	4.071	.2	.8	.3387	.9409	.3600	2.778	.2
.9	.2402	.9707	.2475	4.041	.1	.9	.3404	.9403	.3620	2.762	.1
14.0	0.2419	0.9703	0.2493	4.011	**76.0**	**20.0**	0.3420	0.9397	0.3640	2.747	**70.0**
.1	.2436	.9699	.2512	3.981	.9	.1	.3437	.9391	.3659	2.733	.9
.2	.2453	.9694	.2530	3.952	.8	.2	.3453	.9385	.3679	2.718	.8
.3	.2470	.9690	.2549	3.923	.7	.3	.3469	.9379	.3699	2.703	.7
.4	.2487	.9686	.2568	3.895	.6	.4	.3486	.9373	.3719	2.689	.6
.5	.2504	.9681	.2586	3.867	.5	.5	.3502	.9367	.3739	2.675	.5
.6	.2521	.9677	.2605	3.839	.4	.6	.3518	.9361	.3759	2.660	.4
.7	.2538	.9673	.2623	3.812	.3	.7	.3535	.9354	.3779	2.646	.3
.8	.2554	.9668	.2642	3.785	.2	.8	.3551	.9348	.3799	2.633	.2
.9	.2571	.9664	.2661	3.758	.1	.9	.3567	.9342	.3819	2.619	.1
15.0	0.2588	0.9659	0.2679	3.732	**75.0**	**21.0**	0.3584	0.9336	0.3839	2.605	**69.0**
.1	.2605	.9655	.2698	3.706	.9	.1	.3600	.9330	.3859	2.592	.9
.2	.2622	.9650	.2717	3.681	.8	.2	.3616	.9323	.3879	2.578	.8
.3	.2639	.9646	.2736	3.655	.7	.3	.3633	.9317	.3899	2.565	.7
.4	.2656	.9641	.2754	3.630	.6	.4	.3649	.9311	.3919	2.552	.6
.5	.2672	.9636	.2773	3.606	.5	.5	.3665	.9304	.3939	2.539	.5
.6	.2689	.9632	.2792	3.582	.4	.6	.3681	.9298	.3959	2.526	.4
.7	.2706	.9627	.2811	3.558	.3	.7	.3697	.9291	.3979	2.513	.3
.8	.2723	.9622	.2830	3.534	.2	.8	.3714	.9285	.4000	2.500	.2
.9	.2740	.9617	.2849	3.511	.1	.9	.3730	.9278	.4020	2.488	.1
16.0	0.2756	0.9613	0.2867	3.487	**74.0**	**22.0**	0.3746	0.9272	0.4040	2.475	**68.0**
.1	.2773	.9608	.2886	3.465	.9	.1	.3762	.9265	.4061	2.463	.9
.2	.2790	.9603	.2905	3.442	.8	.2	.3778	.9259	.4081	2.450	.8
.3	.2807	.9598	.2924	3.420	.7	.3	.3795	.9252	.4101	2.438	.7
.4	.2823	.9593	.2943	3.398	.6	.4	.3811	.9245	.4122	2.426	.6
.5	.2840	.9588	.2962	3.376	.5	.5	.3827	.9239	.4142	2.414	.5
.6	.2857	.9583	.2981	3.354	.4	.6	.3843	.9232	.4163	2.402	.4
.7	.2874	.9578	.3000	3.333	.3	.7	.3859	.9225	.4183	2.391	.3
.8	.2890	.9573	.3019	3.312	.2	.8	.3875	.9219	.4204	2.379	.2
.9	.2907	.9568	.3038	3.291	.1	.9	.3891	.9212	.4224	2.367	.1
17.0	0.2924	0.9563	0.3057	3.271	**73.0**	**23.0**	0.3907	0.9205	0.4245	2.356	**67.0**
.1	.2940	.9558	.3076	3.251	.9	.1	.3923	.9198	.4265	2.344	.9
.2	.2957	.9553	.3096	3.230	.8	.2	.3939	.9191	.4286	2.333	.8
.3	.2974	.9548	.3115	3.211	.7	.3	.3955	.9184	.4307	2.322	.7
.4	.2990	.9542	.3134	3.191	.6	.4	.3971	.9178	.4327	2.311	.6
.5	.3007	.9537	.3153	3.172	.5	.5	.3987	.9171	.4348	2.300	.5
.6	.3024	.9532	.3172	3.152	.4	.6	.4003	.9164	.4369	2.289	.4
.7	.3040	.9527	.3191	3.133	.3	.7	.4019	.9157	.4390	2.278	.3
.8	.3057	.9521	.3211	3.115	.2	.8	.4035	.9150	.4411	2.267	.2
.9	.3074	.9516	.3230	3.096	.1	.9	.4051	.9143	.4431	2.257	.1
18.0	0.3090	0.9511	0.3249	3.078	**72.0**	**24.0**	0.4067	0.9135	0.4452	2.246	**66.0**
Deg.	Cos	Sin	Cot	Tan	Deg.	Deg.	Cos	Sin	Cot	Tan	Deg.

NATURAL TRIGONOMETRIC FUNCTIONS FOR DECIMAL FRACTIONS OF A DEGREE

Deg.	Sin	Cos	Tan	Cot	Deg.	Deg.	Sin	Cos	Tan	Cot	Deg.
24.0	0.4067	0.9135	0.4452	2.246	**66.0**	**30.0**	0.5000	0.8660	0.5774	1.7321	**60.0**
.1	.4083	.9128	.4473	2.236	.9	.1	.5015	.8652	.5797	1.7251	.9
.2	.4099	.9121	.4494	2.225	.8	.2	.5030	.8643	.5820	1.7182	.8
.3	.4115	.9114	.4515	2.215	.7	.3	.5045	.8634	.5844	1.7113	.7
.4	.4131	.9107	.4536	2.204	.6	.4	.5060	.8625	.5867	1.7045	.6
.5	.4147	.9100	.4557	2.194	.5	.5	.5075	.8616	.5890	1.6977	.5
.6	.4163	.9092	.4578	2.184	.4	.6	.5090	.8607	.5914	1.6909	.4
.7	.4179	.9085	.4599	2.174	.3	.7	.5105	.8599	.5938	1.6842	.3
.8	.4195	.9078	.4621	2.164	.2	.8	.5120	.8590	.5961	1.6775	.2
.9	.4210	.9070	.4642	2.154	.1	.9	.5135	.8581	.5985	1.6709	.1
25.0	0.4226	0.9063	0.4663	2.145	**65.0**	**31.0**	0.5150	0.8572	0.6009	1.6643	**59.0**
.1	.4242	.9056	.4684	2.135	.9	.1	.5165	.8563	.6032	1.6577	.9
.2	.4258	.9048	.4706	2.125	.8	.2	.5180	.8554	.6056	1.6512	.8
.3	.4274	.9041	.4727	2.116	.7	.3	.5195	.8545	.6080	1.6447	.7
.4	.4289	.9033	.4748	2.106	.6	.4	.5210	.8536	.6104	1.6383	.6
.5	.4305	.9026	.4770	2.097	.5	.5	.5225	.8526	.6128	1.6319	.5
.6	.4321	.9018	.4791	2.087	.4	.6	.5240	.8517	.6152	1.6255	.4
.7	.4337	.9011	.4813	2.078	.3	.7	.5255	.8508	.6176	1.6191	.3
.8	.4352	.9003	.4834	2.069	.2	.8	.5270	.8499	.6200	1.6128	.2
.9	.4368	.8996	.4856	2.059	.1	.9	.5284	.8490	.6224	1.6066	.1
26.0	0.4384	0.8988	0.4877	2.050	**64.0**	**32.0**	0.5299	0.8480	0.6249	1.6003	**58.0**
.1	.4399	.8980	.4899	2.041	.9	.1	.5314	.8471	.6273	1.5941	.9
.2	.4415	.8973	.4921	2.032	.8	.2	.5329	.8462	.6297	1.5880	.8
.3	.4431	.8965	.4942	2.023	.7	.3	.5344	.8453	.6322	1.5818	.7
.4	.4446	.8957	.4964	2.014	.6	.4	.5358	.8443	.6346	1.5757	.6
.5	.4462	.8949	.4986	2.006	.5	.5	.5373	.8434	.6371	1.5697	.5
.6	.4478	.8942	.5008	1.997	.4	.6	.5388	.8425	.6395	1.5637	.4
.7	.4493	.8934	.5029	1.988	.3	.7	.5402	.8415	.6420	1.5577	.3
.8	.4509	.8926	.5051	1.980	.2	.8	.5417	.8406	.6445	1.5517	.2
.9	.4524	.8918	.5073	1.971	.1	.9	.5432	.8396	.6469	1.5458	.1
27.0	0.4540	0.8910	0.5095	1.963	**63.0**	**33.0**	0.5446	0.8387	0.6494	1.5399	**57.0**
.1	.4555	.8902	.5117	1.954	.9	.1	.5461	.8377	.6519	1.5340	.9
.2	.4571	.8894	.5139	1.946	.8	.2	.5476	.8368	.6544	1.5282	.8
.3	.4586	.8886	.5161	1.937	.7	.3	.5490	.8358	.6569	1.5224	.7
.4	.4602	.8878	.5184	1.929	.6	.4	.5505	.8348	.6594	1.5166	.6
.5	.4617	.8870	.5206	1.921	.5	.5	.5519	.8339	.6619	1.5108	.5
.6	.4633	.8862	.5228	1.913	.4	.6	.5534	.8329	.6644	1.5051	.4
.7	.4648	.8854	.5250	1.905	.3	.7	.5548	.8320	.6669	1.4994	.3
.8	.4664	.8846	.5272	1.897	.2	.8	.5563	.8310	.6694	1.4938	.2
.9	.4679	.8838	.5295	1.889	.1	.9	.5577	.8300	.6720	1.4882	.1
28.0	0.4695	0.8829	0.5317	1.881	**62.0**	**34.0**	0.5592	0.8290	0.6745	1.4826	**56.0**
.1	.4710	.8821	.5340	1.873	.9	.1	.5606	.8281	.6771	1.4770	.9
.2	.4726	.8813	.5362	1.865	.8	.2	.5621	.8271	.6796	1.4715	.8
.3	.4741	.8805	.5384	1.857	.7	.3	.5635	.8261	.6822	1.4659	.7
.4	.4756	.8796	.5407	1.849	.6	.4	.5650	.8251	.6847	1.4605	.6
.5	.4772	.8788	.5430	1.842	.5	.5	.5664	.8241	.6873	1.4550	.5
.6	.4787	.8780	.5452	1.834	.4	.6	.5678	.8231	.6899	1.4496	.4
.7	.4802	.8771	.5475	1.827	.3	.7	.5693	.8221	.6924	1.4442	.3
.8	.4818	.8763	.5498	1.819	.2	.8	.5707	.8211	.6950	1.4388	.2
.9	.4833	.8755	.5520	1.811	.1	.9	.5721	.8202	.6976	1.4335	.1
29.0	0.4848	0.8746	0.5543	1.804	**61.0**	**35.0**	0.5736	0.8192	0.7002	1.4281	**55.0**
.1	.4863	.8738	.5566	1.797	.9	.1	.5750	.8181	.7028	1.4229	.9
.2	.4879	.8729	.5589	1.789	.8	.2	.5764	.8171	.7054	1.4176	.8
.3	.4894	.8721	.5612	1.782	.7	.3	.5779	.8161	.7080	1.4124	.7
.4	.4909	.8712	.5635	1.775	.6	.4	.5793	.8151	.7107	1.4071	.6
.5	.4924	.8704	.5658	1.767	.5	.5	.5807	.8141	.7133	1.4019	.5
.6	.4939	.8695	.5681	1.760	.4	.6	.5821	.8131	.7159	1.3968	.4
.7	.4955	.8686	.5704	1.753	.3	.7	.5835	.8121	.7186	1.3916	.3
.8	.4970	.8678	.5727	1.746	.2	.8	.5850	.8111	.7212	1.3865	.2
.9	.4985	.8669	.5750	1.739	.1	.9	.5864	.8100	.7239	1.3814	.1
30.0	0.5000	0.8660	0.5774	1.732	**60.0**	**36.0**	0.5878	0.8090	0.7265	1.3764	**54.0**
Deg.	Cos	Sin	Cot	Tan	Deg.	Deg.	Cos	Sin	Cot	Tan	Deg.

NATURAL TRIGONOMETRIC FUNCTIONS FOR DECIMAL FRACTIONS OF A DEGREE

Deg.	Sin	Cos	Tan	Cot	Deg.	Deg.	Sin	Cos	Tan	Cot	Deg.
36.0	0.5878	0.8090	0.7265	1.3764	**54.0**	**40.5**	0.6494	0.7604	0.8541	1.1708	**49.5**
.1	.5892	.8080	.7292	1.3713	.9	.6	.6508	.7593	.8571	1.1667	.4
.2	.5906	.8070	.7319	1.3663	.8	.7	.6521	.7581	.8601	1.1626	.3
.3	.5920	.8059	.7346	1.3613	.7	.8	.6534	.7570	.8632	1.1585	.2
.4	.5934	.8049	.7373	1.3564	.6	.9	.6547	.7559	.8662	1.1544	.1
.5	.5948	.8039	.7400	1.3514	.5	**41.0**	0.6561	0.7547	0.8693	1.1504	**49.0**
.6	.5962	.8028	.7427	1.3465	.4	.1	.6574	.7536	.8724	1.1463	.9
.7	.5976	.8018	.7454	1.3416	.3	.2	.6587	.7524	.8754	1.1423	.8
.8	.5990	.8007	.7481	1.3367	.2	.3	.6600	.7513	.8785	1.1383	.7
.9	.6004	.7997	.7508	1.3319	.1	.4	.6613	.7501	.8816	1.1343	.6
37.0	0.6018	0.7986	0.7536	1.3270	**53.0**	.5	.6626	.7490	.8847	1.1303	.5
.1	.6032	.7976	.7563	1.3222	.9	.6	.6639	.7478	.8878	1.1263	.4
.2	.6046	.7965	.7590	1.3175	.8	.7	.6652	.7466	.8910	1.1224	.3
.3	.6060	.7955	.7618	1.3127	.7	.8	.6665	.7455	.8941	1.1184	.2
.4	.6074	.7944	.7646	1.3079	.6	.9	.6678	.7443	.8972	1.1145	.1
.5	.6088	.7934	.7673	1.3032	.5	**42.0**	0.6691	0.7431	0.9004	1.1106	**48.0**
.6	.6101	.7923	.7701	1.2985	.4	.1	.6704	.7420	.9036	1.1067	.9
.7	.6115	.7912	.7729	1.2938	.3	.2	.6717	.7408	.9067	1.1028	.8
.8	.6129	.7902	.7757	1.2892	.2	.3	.6730	.7396	.9099	1.0990	.7
.9	.6143	.7891	.7785	1.2846	.1	.4	.6743	.7385	.9131	1.0951	.6
38.0	0.6157	0.7880	0.7813	1.2799	**52.0**	.5	.6756	.7373	.9163	1.0913	.5
.1	.6170	.7869	.7841	1.2753	.9	.6	.6769	.7361	.9195	1.0875	.4
.2	.6184	.7859	.7869	1.2708	.8	.7	.6782	.7349	.9228	1.0837	.3
.3	.6198	.7848	.7898	1.2662	.7	.8	.6794	.7337	.9260	1.0799	.2
.4	.6211	.7837	.7926	1.2617	.6	.9	.6807	.7325	.9293	1.0761	.1
.5	.6225	.7826	.7954	1.2572	.5	**43.0**	0.6820	0.7314	0.9325	1.0724	**47.0**
.6	.6239	.7815	.7983	1.2527	.4	.1	.6833	.7302	.9358	1.0686	.9
.7	.6252	.7804	.8012	1.2482	.3	.2	.6845	.7290	.9391	1.0649	.8
.8	.6266	.7793	.8040	1.2437	.2	.3	.6858	.7278	.9424	1.0612	.7
.9	.6280	.7782	.8069	1.2393	.1	.4	.6871	.7266	.9457	1.0575	.6
39.0	0.6293	0.7771	0.8098	1.2349	**51.0**	.5	.6884	.7254	.9490	1.0538	.5
.1	.6307	.7760	.8127	1.2305	.9	.6	.6896	.7242	.9523	1.0501	.4
.2	.6320	.7749	.8156	1.2261	.8	.7	.6909	.7230	.9556	1.0464	.3
.3	.6334	.7738	.8185	1.2218	.7	.8	.6921	.7218	.9590	1.0428	.2
.4	.6347	.7727	.8214	1.2174	.6	.9	.6934	.7206	.9623	1.0392	.1
.5	.6361	.7716	.8243	1.2131	.5	**44.0**	0.6947	0.7193	0.9657	1.0355	**46.0**
.6	.6374	.7705	.8273	1.2088	.4	.1	.6959	.7181	.9691	1.0319	.9
.7	.6388	.7694	.8302	1.2045	.3	.2	.6972	.7169	.9725	1.0283	.8
.8	.6401	.7683	.8332	1.2002	.2	.3	.6984	.7157	.9759	1.0247	.7
.9	.6414	.7672	.8361	1.1960	.1	.4	.6997	.7145	.9793	1.0212	.6
40.0	0.6428	0.7660	0.8391	1.1918	**50.0**	.5	.7009	.7133	.9827	1.0176	.5
.1	.6441	.7649	.8421	1.1875	.9	.6	.7022	.7120	.9861	1.0141	.4
.2	.6455	.7638	.8451	1.1833	.8	.7	.7034	.7108	.9896	1.0105	.3
3	.6468	.7627	.8481	1.1792	.7	.8	.7046	.7096	.9930	1.0070	.2
.4	.6481	.7615	.8511	1.1750	.6	.9	.7059	.7083	.9965	1.0035	.1
40.5	0.6494	0.7604	0.8541	1.1708	**49.5**	**45.0**	0.7071	0.7071	1.0000	1.0000	**45.0**
Deg.	Cos	Sin	Cot	Tan	Deg.	Deg.	Cos	Sin	Cot	Tan	Deg.

Appendix E: Squares, Cubes, Roots and Reciprocals

SQUARES, CUBES, SQUARE ROOTS, CUBE ROOTS, AND RECIPROCALS
OF NUMBERS FROM 1 THROUGH 50

No.	Square	Cube	Sq. Root	Cube Root	Reciprocal
1	1	1	1.00000	1.00000	1.0000000
2	4	8	1.41421	1.25992	0.5000000
3	9	27	1.73205	1.44225	0.3333333
4	16	64	2.00000	1.58740	0.2500000
5	25	125	2.23607	1.70998	0.2000000
6	36	216	2.44949	1.81712	0.1666667
7	49	343	2.64575	1.91293	0.1428571
8	64	512	2.82843	2.00000	0.1250000
9	81	729	3.00000	2.08008	0.1111111
10	100	1,000	3.16228	2.15443	0.1000000
11	121	1,331	3.31662	2.22398	0.0909091
12	144	1,728	3.46410	2.28943	0.0833333
13	169	2,197	3.60555	2.35133	0.0769231
14	196	2,744	3.74166	2.41014	0.0714286
15	225	3,375	3.87298	2.46621	0.0666667
16	256	4,096	4.00000	2.51984	0.0625000
17	289	4,913	4.12311	2.57128	0.0588235
18	324	5,832	4.24264	2.62074	0.0555556
19	361	6,859	4.35890	2.66840	0.0526316
20	400	8,000	4.47214	2.71442	0.0500000
21	441	9,261	4.58258	2.75892	0.0476190
22	484	10,648	4.69042	2.80204	0.0454545
23	529	12,167	4.79583	2.84387	0.0434783
24	576	13,824	4.89898	2.88450	0.0416667
25	625	15,625	5.00000	2.92402	0.0400000
26	676	17,576	5.09902	2.96250	0.0384615
27	729	19,683	5.19615	3.00000	0.0370370
28	784	21,952	5.29150	3.03659	0.0357143
29	841	24,389	5.38516	3.07232	0.0344828
30	900	27,000	5.47723	3.10723	0.0333333
31	961	29,791	5.56776	3.14138	0.0322581
32	1,024	32,768	5.65685	3.17480	0.0312500
33	1,089	35,937	5.74456	3.20753	0.0303030
34	1,156	39,304	5.83095	3.23961	0.0294118
35	1,225	42,875	5.91608	3.27107	0.0285714
36	1,296	46,656	6.00000	3.30193	0.0277778
37	1,369	50,653	6.08276	3.33222	0.0270270
38	1,444	54,872	6.16441	3.36198	0.0263158
39	1,521	59,319	6.24500	3.39121	0.0256410
40	1,600	64,000	6.32456	3.41995	0.0250000
41	1,681	68,921	6.40312	3.44822	0.0243902
42	1,764	74,088	6.48074	3.47603	0.0238095
43	1,849	79,507	6.55744	3.50340	0.0232558
44	1,936	85,184	6.63325	3.53035	0.0227273
45	2,025	91,125	6.70820	3.55689	0.0222222
46	2,116	97,336	6.78233	3.58305	0.0217391
47	2,209	103,823	6.85565	3.60883	0.0212766
48	2,304	110,592	6.92820	3.63424	0.0208333
49	2,401	117,649	7.00000	3.65931	0.0204082
50	2,500	125,000	7.07107	3.68403	0.0200000
No.	Square	Cube	Sq. Root	Cube Root	Reciprocal

SQUARES, CUBES, SQUARE ROOTS, CUBE ROOTS, AND RECIPROCALS
OF NUMBERS 51 THROUGH 100

No.	Square	Cube	Sq. Root	Cube Root	Reciprocal
51	2,601	132,651	7.14143	3.70843	0.0196078
52	2,704	140,608	7.21110	3.73251	0.0192308
53	2,809	148,877	7.28011	3.75629	0.0188679
54	2,916	157,464	7.34847	3.77976	0.0185185
55	3,025	166,375	7.41620	3.80295	0.0181818
56	3,136	175,616	7.48331	3.82586	0.0178571
57	3,249	185,193	7.54983	3.84850	0.0175439
58	3,364	195,112	7.61577	3.87088	0.0172414
59	3,481	205,379	7.68115	3.89300	0.0169492
60	3,600	216,000	7.74597	3.91487	0.0166667
61	3,721	226,981	7.81025	3.93650	0.0163934
62	3,844	238,328	7.87401	3.95789	0.0161290
63	3,969	250,047	7.93725	3.97906	0.0158730
64	4,096	262,144	8.00000	4.00000	0.0156250
65	4,225	274,625	8.06226	4.02073	0.0153846
66	4,356	287,496	8.12404	4.04124	0.0151515
67	4,489	300,763	8.18535	4.06155	0.0149254
68	4,624	314,432	8.24621	4.08166	0.0147059
69	4,761	328,509	8.30662	4.10157	0.0144928
70	4,900	343,000	8.36660	4.12129	0.0142857
71	5,041	357,911	8.42615	4.14082	0.0140845
72	5,184	373,248	8.48528	4.16017	0.0138889
73	5,329	389,017	8.54400	4.17934	0.0136986
74	5,476	405,224	8.60233	4.19834	0.0135135
75	5,625	421,875	8.66025	4.21716	0.0133333
76	5,776	438,976	8.71780	4.23582	0.0131579
77	5,929	456,533	8.77496	4.25432	0.0129870
78	6,084	474,552	8.83176	4.27266	0.0128205
79	6,241	493,039	8.88819	4.29084	0.0126582
80	6,400	512,000	8.94427	4.30887	0.0125000
81	6,561	531,441	9.00000	4.32675	0.0123457
82	6,724	551,368	9.05539	4.34448	0.0121951
83	6,889	571,787	9.11043	4.36207	0.0120482
84	7,056	592,704	9.16515	4.37952	0.0119048
85	7,225	614,125	9.21954	4.39683	0.0117647
86	7,396	636,056	9.27362	4.41400	0.0116279
87	7,569	658,503	9.32738	4.43105	0.0114943
88	7,744	681,472	9.38083	4.44797	0.0113636
89	7,921	704,969	9.43398	4.46475	0.0112360
90	8,100	729,000	9.48683	4.48140	0.0111111
91	8,281	753,571	9.53939	4.49794	0.0109890
92	8,464	778,688	9.59166	4.51436	0.0108696
93	8,649	804,357	9.64365	4.53065	0.0107527
94	8,836	830,584	9.69536	4.54684	0.0106383
95	9,025	857,375	9.74679	4.56290	0.0105263
96	9,216	884,736	9.79796	4.57886	0.0104167
97	9,409	912,673	9.84886	4.59470	0.0103093
98	9,604	941,192	9.89949	4.61044	0.0102041
99	9,801	970,299	9.94987	4.62607	0 0101010
100	10,000	1,000,000	10.00000	4.64159	0.0100000
No.	Square	Cube	Sq. Root	Cube Root	Reciprocal

INDEX

A

Nickel-cadmium cell, 83
Non-magnetic materials, 58-60
NPN transistors, 285
N-type germanium, 262, 263, 275-276
Nucleus of atom, 2-6, 15
Null positioning synchro, 335-336
Numerically controlled machines, 342-343

O

Occupational Safety and Health Act
  (OSHA), 364
Ohm, definition of, 20
Ohm's law in direct current applications,
  31-42
Ohmmeter, 208-210
Orbits, atomic, 2-6, 57-58
Operating voltages, electron tube, 245-246
OR gates, 355-356
Oscillator, Clapp, 321
  Colpitts, 320-321
  crystal, 272, 324
  electron-tube, 269-272
  Hartley, 270-272; 321-322
  magnetostrictive, 290
  phase-shift, 322
  sine-wave, 320
  Wien-bridge, 322-323
Oscillator circuit basics, 318-324
Oscilloscope, 225-228
  operation, 227-228
  use of, 228
Output, computer, 350

P

Parallel circuits, 43, 45-47
Paramagnetic materials, 58
Pentode electron tube, 238, 247-248
Permanent magnets, 61
Permeability, 58-59, 68-70
Phase angle, 125-128; 135-137
Phase-shift oscillator, 322
Photoelectric semiconductors, 292

Photoelectricity, 86
Phototube, 239
Pi filters, 261
Piezoelectricity, 85-86
Pins, electron tube, 237-238
Plate characteristics curves, 249-251
Plate, electron tube, 235-237
PNP transistors, 285
Poles, commutating, 103
Poles, magnetic, 59-62, 67-68
Potential transformers, 222-223
Potential barrier, 276-277
Potentiometer, 327
Power,
  atomic, 17
  electrical, 40, 41
  synchros, 333-335
  transformers, 170-188
  three-phase, 184-185
Power supply circuit, 262-263; 308-309
Precipitators, dust, 12, 13
Prefixes, metric, 374
Primary cells, 73-79
Primary winding, transformer, 172
Programmer, computer, 349
Protons, 2-4, 15
P-type germanium, 277-278
Pulsating d-c, 257
Push-pull amplifiers, 269; 317-318

Q

Quartz crystals, 85-86

R

Ratio arms, 216-218
RC circuit electrical summaries, 137-138
Reactance,
  capacitive, 131-132
  inductive, 121-122
Reciprocals, 406-407
Rectifier bridge circuit, 192-194
Rectifier, full-wave, 257-258; 302-303
  full-wave bridge, 303-304
  half-wave, 301-302, 256-257; 301-302